W9-BAG-069

Flowering and its Manipulation

Annual Plant Reviews

A series for researchers and postgraduates in the plant sciences. Each volume in this series focuses on a theme of topical importance and emphasis is placed on rapid publication.

Editorial Board:

Titles in the series:

1. **Arabidopsis** Edited by M. Anderson and J.A. Roberts
2. **Biochemistry of Plant Secondary Metabolism** Edited by M. Wink
3. **Functions of Plant Secondary Metabolites and their Exploitation in Biotechnology** Edited by M. Wink
4. **Molecular Plant Pathology** Edited by M. Dickinson and J. Beynon
5. **Vacuolar Compartments** Edited by D.G. Robinson and J.C. Rogers
6. **Plant Reproduction** Edited by S.D. O'Neill and J.A. Roberts
7. **Protein–Protein Interactions in Plant Biology** Edited by M.T. McManus, W.A. Laing and A.C. Allan
8. **The Plant Cell Wall** Edited by J.K.C. Rose
9. **The Golgi Apparatus and the Plant Secretory Pathway** Edited by D.G. Robinson
10. **The Plant Cytoskeleton in Cell Differentiation and Development** Edited by P.J. Hussey
11. **Plant–Pathogen Interactions** Edited by N.J. Talbot
12. **Polarity in Plants** Edited by K. Lindsey
13. **Plastids** Edited by S.G. Moller
14. **Plant Pigments and their Manipulation** Edited by K.M. Davies
15. **Membrane Transport in Plants** Edited by M.R. Blatt
16. **Intercellular Communication in Plants** Edited by A.J. Fleming
17. **Plant Architecture and its Manipulation** Edited by C. Turnbull
18. **Plasmodesmata** Edited by K.J. Oparka
19. **Plant Epigenetics** Edited by P. Meyer
20. **Flowering and its Manipulation** Edited by C. Ainsworth
21. **Endogenous Plant Rhythms** Edited by A. Hall and H. McWatters
22. **Control of Primary Metabolism in Plants** Edited by W.C. Plaxton and M.T. McManus
23. **Biology of the Plant Cuticle** Edited by M. Riederer

Flowering and its Manipulation

Edited by

CHARLES AINSWORTH
Department of Agricultural Sciences
Imperial College
Wye Campus
Ashford
Kent
UK

Blackwell
Publishing

Editorial Offices:
Blackwell Publishing Ltd, 9600 Garsington Road, Oxford OX4 2DQ, UK
Tel: +44 (0)1865 776868
Blackwell Publishing Professional, 2121 State Avenue, Ames, Iowa 50014-8300, USA
Tel: +1 515 292 0140
Blackwell Publishing Asia, 550 Swanston Street, Carlton, Victoria 3053, Australia
Tel: +61 (0)3 8359 1011

First published 2006 by Blackwell Publishing Ltd

ISBN-10: 1-4051-2808-9
ISBN-13: 978-14051-2808-7

Library of Congress Cataloging-in-Publication Data

Flowering and its manipulation/edited by Charles Anisworth.
p. cm.
Includes bibliographical references.
ISBN-13: 978-1-4051-2808-7 (hardback : alk. paper)
ISBN-10: 1-4051-2808-9 (hardback : alk. paper)
1. Plants, Flowering of. I. Ainsworth, C. C. (Charles Colin), 1954-
SB126.8.F58 2006
635.9′15233–dc22

2005022778

A catalogue record for this title is available from the British Library

Set in 10/12 pt Times
by Newgen Imaging Systems (P) Ltd. Chennai, India
Printed and bound in India
by Replika Press Pvt, Ltd, Kundli

The publisher's policy is to use permanent paper from mills that operate a sustainable forestry policy, and which has been manufactured from pulp processed using acid-free and elementary chlorine-free practices. Furthermore, the publisher ensures that the text paper and cover board used have met acceptable environmental accreditation standards.

For further information on Blackwell Publishing, visit our website:
www.blackwellpublishing.com

Contents

Contributors

Dr Maria Albani Max Planck Institute for Plant Breeding Research, Cologne, Germany

Dr Martine Batoux Entwicklungsbiologie der Pflanzen, Wissenschaftszentrum Weihenstephan, Technische Universität München, Freising, Germany

Dr Nick Battey School of Plant Sciences, University of Reading, Whiteknights, UK

Professor David A. Baum Department of Botany, University of Wisconsin, Madison, Wisconsin, USA

Dr Filippa Brugliera Florigene Ltd, Collingwood, Victoria, Australia

Dr Françoise Budar Plant Breeding and Genetics Department, Institut Jean-Pierre Bourgin – INRA, Versailles, France

Dr Joseph Colasanti Department of Molecular and Cell Biology, University of Guelph, Ontario, Canada

Dr George Coupland Max Planck Institute for Plant Breeding Research, Cologne, Germany

Dr Professor Natalia Dudareva Purdue University, Department of Horticulture and Landscape Architecture, West Lafayette, Indiana, USA

Dr Lynette Fulton Entwicklungsbiologie der Pflanzen, Wissenschaftszentrum Weihenstephan, Technische Universität München, Freising, Germany

Dr Lena C. Hileman Department of Molecular, Cellular and Developmental Biology, Yale University, New Haven, Connecticut, USA

Dr M.L. Jones Floriculture/ Molecular Biology, Horticulture and Crop Science Department, The Ohio State University, Wooster, Ohio, USA

Dr Ram Kishor Yadav Entwicklungsbiologie der Pflanzen, Wissenschaftszentrum Weihenstephan, Technische Universität München, Freising, Germany

Dr Elena M. Kramer Department of Organismic and Evolutionary Biology, Harvard University, Cambridge, Massachusetts, USA

Dr Georges Pelletier Plant Breeding and Genetics Department, Institute Jean-Pierre Bourgin – INRA, Versailles, France

Dr Rafael Perl-Treves Faculty of Life Sciences, Bar-Ilan University, Ramat Gan, Israel

Dr Prem Anand Rajagopalan Faculty of Life Sciences, Bar-Ilan University, Ramat Gan, Israel

Professor Kay Schneitz Entwicklungsbiologie der Pflanzen, Wissenschaftszentrum Weihenstephan, Technische Universität München, Freising, Germany

Dr Jennifer Schnepp Purdue University, Department of Horticulture and Landscape Architecture, West Lafayette, Indiana, USA

Professor Susan R. Singer Department of Biology, Carleton College, Northfield, Minnesota, USA

Dr Anthony D. Stead Biological Sciences, Royal Holloway (University of London), Egham, Surrey, UK

Professor Yoshikazu Tanaka Institute for Advanced Technology, Suntory Ltd, Osaka, Japan

Dr Pascal Touzet Laboratoire de Génétique et Evolution des Populations Végétales, Université de Lille, Villeneuve d'Ascq, France

Dr Theresa Townsend School of Plant Sciences, University of Reading, Whiteknights, UK

Dr Reynald Tremblay Department of Molecular and Cell Biology, Office 330, University of Guelph, Ontario, Canada

Dr Wouter G. van Doorn Wageningen University and Research Centre, Wageningen, The Netherlands

Dr C. Wagstaff School of Biological Sciences, University of Southampton, Southampton, UK

Dr Mike Wilkinson School of Plant Sciences, University of Reading, Whiteknights, UK

Preface

Flowering plants now dominate the terrestrial ecosystems of the planet, and there are good reasons for supposing that the flower itself has been a major contributing factor to the spread of the Angiosperms. Although the Angiosperm flower is based on four whorls of organs – petals, sepal, stamens and carpels – the 300 000 odd species exhibit a huge array of different floral morphologies. Consider, e.g., the humble periwinkle flower and contrast it with the complex passion flower. Since the early 1990s, the powerful tools of the molecular biologist have been applied to the study of the control of Angiosperm flower development. Two plant species, *Antirrhinum majus* (the snapdragon) and *Arabidopsis thaliana* (thale cress) – the former having a remarkable array of floral mutants collected by plant breeders over the years and the latter being easily mutagenised and genetically transformed – were initially studied in respect of the control of floral organ identity. The result was the now-historic ABC model. These two plant species have been responsible for huge advances in our understanding of the control of flower development that go far beyond mere organ identity.

The 11 chapters of this book begin with a chapter describing a model for the evolution of the Angiosperm flower. Chapters 2–5 describe the core development of the flower and include floral induction, floral pattering and organ initiation, floral shape and size, and inflorescence architecture. Chapters 6–8 focus on more specialised aspects of floral development: monoecy, cytoplasmic male sterility and flowering in perennials. Chapters 9 and 10 address more functional aspects: flower colour and scent. The book concludes, appropriately, with flower senescence.

The flowers of higher plants not only contain the organs of plant reproduction but are of fundamental importance in giving rise to fruits and seeds which constitute a major component of the human diet. Thus, wherever possible, this book seeks to explore those aspects of flowers and their development that are important to man.

Charles Ainsworth

Part I Core development and genetics

1 A developmental genetic model for the origin of the flower

David A. Baum and Lena C. Hileman

The angiosperms arose in the early Cretaceous and rapidly came to dominate almost all terrestrial ecosystems. Among the many features contributing to the angiosperm's success is the flower: a compressed, bisexual reproductive axis composed of stamens, carpels and a sterile perianth. With advances in our understanding of seed plant phylogeny over the last decade, some important new fossil finds and great improvements in our knowledge of floral developmental genetics, a fuller understanding of floral origins is now possible. In this chapter we review this new knowledge and offer a developmental genetic model for the origin of the flower. We propose that the first step was the evolution of a bisexual axis via a gynomonoecious intermediate. We speculate that homeotic conversion of distal microsporophylls (stamens) into megasporophylls (carpels) within a pollen cone arose due to differences in maximal expression levels of B- and C-class floral organ identity genes combined with competition among their gene products for partners in multimeric complexes. The second step was the evolution of floral axis compression and determinacy. We hypothesize that this was achieved when C-class genes became negative regulators of the meristem maintenance gene *WUSCHEL*. The third step was the evolution of a petaloid perianth by sterilization of the outer stamens. We suggest that this could have arisen when *WUSCHEL* was co-opted as a coregulator of C-class genes. Last, we speculate that the origin of the dimorphic perianth of core eudicots (such as *Arabidopsis*) coincided with B-class function becoming dependent on *UFO* expression. Although our model is speculative, it makes testable predictions and serves to identify promising avenues for further research.

1.1 Introduction

The angiosperms or flowering plants are the dominant group of land plants both in terms of species numbers (ca. 300 000) and ecological importance. The reasons for their great success remain controversial. Angiosperms show many innovations relative to other seed plants, which means that it is difficult to determine which features were responsible for their dominance. Xylem vessels (e.g. Feild *et al.*, 2002), reticulate venation (e.g. Givnish, 1979), endosperm development (e.g. Floyd and Friedman, 2000), and seed enclosure in fruit (e.g. Regal, 1977) have all been highlighted as possible key innovations. Nonetheless, there is good reason to suppose that the flower itself contributed to the increased rate of diversification seen in the angiosperms. First, by increasing the potential for reproductive isolation (e.g. via changes in flowering time, pollinator identity, location of pollen deposition on pollinators and pollen–stigma recognition), flowers allow lineages to become immune to secondary fusion with their close relatives more quickly (Grant, 1971). Second, flowers promote efficient animal pollination, which allows for accurate pollen targeting and, hence, permits populations to persist at lower densities without extinction (Regal, 1977). Consequently, understanding the origin of the flower and its subsequent diversification is central to explaining why almost all modern terrestrial environments are dominated by angiosperms.

The origin of the flowering plants has been difficult to resolve because the phylogeny of the seed plants was poorly understood, the fossil progenitors of angiosperms were unknown and the homology of floral organs with the reproductive organs found in other seed plants was unclear. Nonetheless, advances in paleontology, phylogeny and developmental genetics have generated renewed interest in the origin of the flower. In this chapter we review some of these advances in the understanding of floral origins. We highlight some attempts to provide a genetic scenario for the evolutionary origin of the flower and offer a synthetic hypothesis of our own. In the latter regard we are unashamedly speculative. To be sure, the foundation of science is hard empirical data and its objective interpretation. Nonetheless, a field can be advanced by risky hypothesis building, provided the resulting hypotheses are consistent with current data and testable with future evidence. It is in the latter spirit that we take this opportunity to propose a detailed genetic model for the origin of the flower.

1.2 What is a flower?

Flowers of extant angiosperms are generally composed of four distinct floral organ types, although there have been extensive evolutionary modifications to this basic floral plan. The outer (1st whorl) is usually comprised of bract-like, protective sepals, constituting the calyx. The 2nd whorl is usually comprised of showy, attractive petals, constituting the corolla. Together the calyx and corolla constitute the perianth. Except in taxa that have secondarily derived unisexual flowers, the 3rd whorl is comprised of pollen-bearing structures, stamens, and the inner, 4th whorl is comprised of ovule-bearing structures, carpels.

Although it is quite easy to describe the morphology of a typical flower, the flower is not just a single characteristic, but a complex of innovations. In order to describe this complex in such a way that its origin can be elucidated, it is necessary to make commitments as to the homologies of floral organs with the reproductive structures of non-flowering seed plants. A central problem is whether a flower should be considered a simple shoot bearing sporophylls (modified leaves bearing sporangia), the uniaxial model, or a complex shoot system with the floral reproductive organs (stamens and carpels) being homologous to shoots rather than sporophylls, the polyaxial model. The paleontological and phylogenetic data are equivocal regarding these two interpretations (Doyle, 1994). In contrast, the developmental genetic data seem most consistent with the uniaxial model, because simple genetic mutations can result in the homeotic conversion of floral organs into one another and into leaf-like sepals. For example, in the *Arabidopsis agamous* mutant background, stamens and carpels are transformed into petals and sepals, respectively (Yanofsky *et al.*, 1990), and in the *Arabidopsis sepalata 1,2,3* triple mutant background, the three inner whorls of floral organs are transformed into sepals (Pelaz *et al.*, 2000). Thus, although it is difficult to formally reject the polyaxial model, here we will assume the uniaxial model under which a flower is a simple shoot bearing microsporophylls (stamens), megasporophylls (carpels) and modified leaves (the perianth). This view follows Arber and Parkin's (1907) definition of a flower as 'an amphisporangiate strobilus of determinate growth and with an involucrum of modified bracts'.

1.3 Phylogenetic and paleontological context

Despite a history of great confusion, the advent of molecular data has allowed confident resolution of the relationships among the extant seed plant groups. It has become clear that the angiosperms are monophyletic and that its earliest branches demarcate a paraphyletic group that includes *Amborella*, Nymphaeales, Illiciaceae, Trimeniaceae and Austrobaileyaceae – the 'ANITA' grade (Mathews and Donoghue, 1999; Qiu *et al.*, 1999, 2000; Soltis *et al.*, 1999; Doyle and Endress, 2000; Graham and Olmstead, 2000; Zanis *et al.*, 2002). Angiosperms are sister to a clade comprising all four lineages of extant gymnosperms, with Cycads most likely sister to the remaining gymnosperms (Winter *et al.*, 1999; Bowe *et al.*, 2000; Chaw *et al.*, 2000; Frohlich, 2002). Given this topology and the fossil evidence of Cycads in the Pennsylvanian, it seems that the lineages leading to the angiosperms and the extant gymnosperms diverged at least 300 million years ago (Savard *et al.*, 1994; Goremykin *et al.*, 1997). The earliest undisputed angiosperm fossils are not found, however, until the early Cretaceous, about 120 million years ago (Walker *et al.*, 1983; Friis *et al.*, 1986; Doyle *et al.*, 1990; Doyle, 2000). Evolutionary events in the angiosperm stem lineage are unknown due to the lack of clear transitional fossils and difficulties of interpretation of those fossil plants that might attach to the angiosperm lineage (e.g. Caytoniales, Bennetitales).

Given the rapid radiation of the angiosperms in the Cretaceous and the great morphological gap between angiosperms and known extant and fossil seed plants, the origin of the angiosperms is hard to understand. One possibility is that the flower and other unique angiosperm features evolved gradually along the angiosperm stem lineage and that the intermediate fossils have yet to be found. This explanation has become more plausible with evidence that the most-recent common ancestors of the extant angiosperms were probably small-seeded, shade-tolerant, under-storey shrubs of moist tropical forests (Feild *et al.*, 2004) and would, thus, have low fossilization potential. That being said, a complete lack of fossils is difficult to imagine given the dense record after angiosperm fossils first appeared. Alternatively, the flower could have evolved rapidly, just before the first definitive angiosperm fossils appeared. But under this scenario we would need to understand how such a superficially complex innovation could arise so quickly. Thus, the sudden appearance and rise to dominance of the angiosperms remain enigmatic, justifying Darwin's moniker for the event: 'an abominable mystery' (in a letter to J.D. Hooker in 1879, Darwin and Seward, 1903).

A possible breakthrough was the discovery of an early Cretaceous aquatic plant *Archaefructus*, which has been interpreted as bearing carpels and stamens (but no perianth) on an elongated floral axis (Sun *et al.*, 1998, 2002). Based on the original interpretation of its morphology, *Archaefructus* was inferred to fall outside the crown-group angiosperms, rendering it a candidate for a true transitional form (Sun *et al.*, 2002). However, *Archaefructus* can also be interpreted as comprising an inflorescence of reduced, ebracteate, unisexual flowers, in which case it is more likely to represent a member of the crown-group angiosperms, probably closely related to water lilies (Friis *et al.*, 2003). The jury is still out as to whether *Archaefructus* represents a transitional step in the evolution of flowers or a derived angiosperm. In the former case, *Archaefructus* would provide direct evidence of how the flower evolved. In the latter it would not provide direct evidence of floral origins, but might, nonetheless, provide a useful analog for the evolution of flowers.

1.4 Evolutionary novelties of the flower

Viewing the flower as a compressed reproductive shoot that bears (or ancestrally bore, in the case of species with secondarily reduced flowers) highly modified megasporophylls (carpels) and microsporophylls (stamens), three evolutionary innovations can be highlighted: (i) bisexuality, (ii) compression of the shoot and (iii) the addition of a perianth. Here, we clarify these three innovations, suggest an order in which they mostly likely accreted and propose a scenario for the developmental genetic transitions entailed.

1.4.1 Bisexuality

Most non-angiospermous seed plants have unisexual primary reproductive axes. Bisexual reproductive units are found in Gnetales in which ovule-bearing structures (usually sterile) may be borne in close association with pollen-bearing structures. However, fossil and morphological evidence show that these bisexual units are polyaxial, being composed of multiple, reduced unisexual axes (Hufford, 1996).

The only seed plants besides angiosperms that have been interpreted as having simple bisexual axes are Bennetitales, and, under this interpretation, they have been referred to as the 'hemiangiosperms' (Arber and Parkin, 1907). If Bennetitales are ultimately shown to be sister to angiosperms, then bisexuality would be viewed as a synapomorphy of Bennetitales + angiosperms, rather than a uniquely angiosperm innovation.

1.4.2 Determinate/compressed axes

A distinctive feature of angiosperm flowers relative to other reproductive shoots in seed plants is that they are telescoped into the shoot apical meristem (SAM). The sporophylls of other living seed plants initiate apically but as they develop they are left behind by the still-active shoot meristem. In contrast, the apical meristematic zone of most flowers ceases to proliferate extensively beyond flower initiation and, consequently, floral organs develop at the SAM and then retain a more or less constant position on the apical dome.

Compression at the floral apex may be separated into two distinct aspects of development: (i) termination of the SAM through its consumption by reproductive structures (e.g. carpels) and (ii) reduced internode elongation near the shoot apex. Although the determinate nature of the flower may be a shared trait among all angiosperms, the underlying developmental patterning leading to internode compression may vary. In early branching angiosperm lineages (e.g. Nymphaeales), floral organs tend to initiate centripetally and the flower meristems may expand as new floral organs are initiated. But, even in such cases, it is unusual to see extensive elongation of the floral axis except in a few, probably derived, groups (e.g. in Magnoliales and Ranunculales). In many eudicots there is no expansion of the flower meristem between the initiation of successive whorls, which often emerge more or less simultaneously. Recent developmental genetic work on one eudicot species, *Arabidopsis thaliana* (Brassicaceae), has begun to uncover the developmental genetic basis for this pattern of floral organ initiation, and these data have been used to argue that the fixed position of floral organ primordia with respect to the shoot apex is critical in the regulation of floral organ identity (Lee *et al.*, 1997; Lohmann *et al.*, 2001; Lohmann and Weigel, 2002).

1.4.3 Perianth

There is a history of terminological confusion in discussing perianth evolution. Here we will use the term 'petal' for any sterile floral organ that is flattened and leaf-like, yet has a coloration, reflectance, texture and vasculature resembling a staminal filament. A sepal, in contrast, is a perianth organ that is green and bract-like but that develops within the floral meristem. These definitions emphasize morphology and, implicitly, genetic identity, without reference to position within the flower. This reflects our view that there is no value in distinguishing petaloid tepals (as in *Tulipa*) from petals in the traditional sense, which must be surrounded by a whorl of sepals (as in *Trillium*). Although it is important to distinguish monomorphic from dimorphic perianths, the identity of the organs in the inner perianth whorl does not change when the outer whorl shifts between a sepaloid and petaloid form. We would like to stress, however, that the classification of perianth organs into the categories sepal versus petal is graded and not absolute.

Recent phylogenetic studies of early branching angiosperms suggest a paraphyletic grade composed of 'ANITA' taxa, most of which have numerous petals and no sepals (Doyle and Endress, 2000; Endress, 2001). This result implies that the plesiomorphic condition in angiosperms is a monomorphic perianth of petals (Albert *et al.*, 1998; Doyle and Endress, 2000; Kuzoff and Gasser, 2000; Soltis *et al.*, 2000; Endress, 2001; Zanis *et al.*, 2003). From this ancestral condition, phylogenetic inference suggests several independent losses of petaloidy (especially in wind-pollinated taxa), several independent origins of a dimorphic perianth (including one near the base of the core eudicots) and reinvention of petals in a few clades that had previously lost petals (e.g. Caryophyllales, Rosaceae, De Craene, 2003; De Craene *et al.*, 1998).

1.5 Ordering the key steps in floral evolution

In order to explore the origin of the flower as a genetic and developmental phenomenon, we need to consider the order in which the three aforementioned innovations evolved. It is natural to assume that the perianth was the last to evolve. As argued by Theissen *et al.* (2002), the evolution of a bisexual axis probably facilitated insect pollination because the pollination droplets of ovules borne on megasporophylls could attract insects to the flower. Therefore, because a perianth ancestrally played an attractive role, it is most likely that it evolved in a lineage that was already insect-pollinated and, thus, that it evolved after a bisexual axis (Theissen *et al.*, 2002). Furthermore, a showy perianth would not be very effective in a plant with an elongated axis. Thus, we believe that a perianth was a late addition to the flower, occurring after the acquisition of bisexuality and at a time when substantial axis compression had already occurred.

If we assume that the perianth was the last of the three innovations, which was the first? If we accept the hypothesis that *Archaefructus* is sister to the extant angiosperms and that it had elongated bisexual flowers (Sun *et al.*, 2002), then it is most parsimonious to assume that bisexuality predates the compressed floral axis. If *Archaefructus* turns out to be a derived angiosperm (Friis *et al.*, 2003) it would become irrelevant to the ordering of floral innovations. In that case one could still argue for bisexuality being the earliest innovation by proposing that Bennetitales had simple axes and are sister-group to the angiosperms (Arber and Parkin, 1907). Therefore, while acknowledging some uncertainty, we will assume here that bisexuality predated both the origin of the perianth and substantial axis compression.

For the purposes of visualization, Figure 1.1 shows a hypothetical reconstruction of some of the intermediate forms that this progression might have involved.

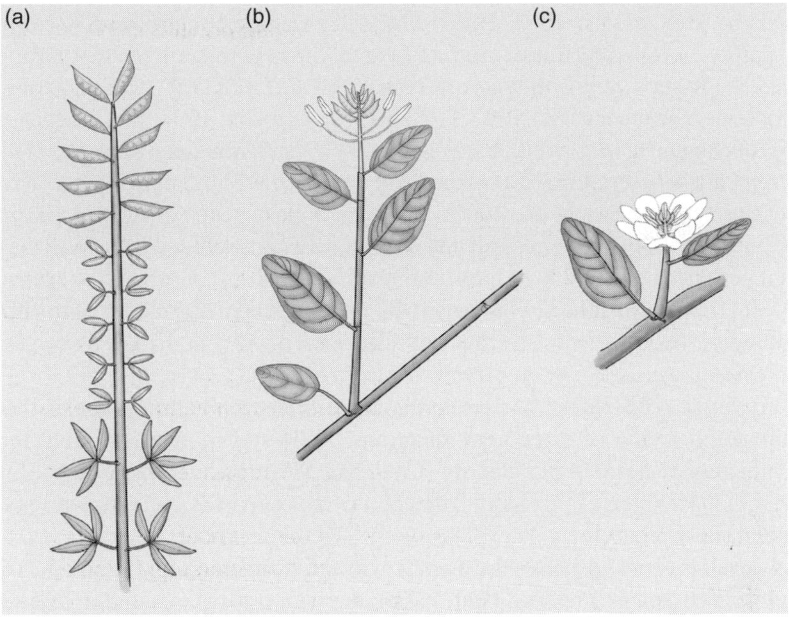

Figure 1.1 Artist reconstruction of intermediate steps in floral evolution. (a) Evolution of a bisexual axis. (b) Evolution of meristem determinacy and axis compression. (c) Perianth evolution.

1.6 Developmental genetic background

1.6.1 Position and identity of the reproductive organs

The transition from vegetative to reproductive development (flowering) in model angiosperms results from the expression of flower meristem identity genes, which respond to, and integrate, environmental and endogenous signals in order to convert meristems that would otherwise have produced vegetative shoots into floral meristems. A key integrator of exogenous and endogenous signaling is the *FLORICAULA* (*FLO*) gene of *Antirrhinum majus* (Coen *et al.*, 1990) and its well-studied ortholog in *Arabidopsis*, LEAFY (*LFY*) (Weigel *et al.*, 1992). Over-expression of *FLO/LFY* in diverse eudicot species results in some degree of conversion of vegetative meristems into flower meristems. In contrast, *flo/lfy* mutants tend to show conversion of flower meristems into shoot meristems. Circumstantial evidence suggests that *FLO/LFY* plays a role in reproductive development throughout the seed plants (Frohlich and Meyerowitz, 1997; Frohlich and Parker, 2000).

FLO/LFY acts to promote floral organ identity by turning on a number of genes involved in the development of stamens and carpels, specifically MADS-box genes in the B- and C-classes (Coen and Meyerowitz, 1991). FLO/LFY protein is needed in *Arabidopsis* to activate expression of the B-class gene *APETALA* (*AP3*;

Weigel and Meyerowitz, 1993; Hill *et al*., 1998) and the C-class gene *AGAMOUS* (*AG*; Parcy *et al*., 1998; Busch *et al*., 1999). In those core eudicots that have been studied, the B-class gene orthologs of *AP3* and *PISTILLATA* (*PI*) function as obligate heterodimers (Trobner *et al*., 1992; Goto and Meyerowitz, 1994; McGonigle *et al*., 1996; Riechmann *et al*., 1996a; Yang *et al*., 2003; Vandenbussche *et al*., 2004). This requirement for heterodimerization probably evolved within angiosperms (Kramer and Irish, 2000; Winter *et al*., 2002). The two B-class genes of *Arabidopsis*, *AP3* and *PI*, are generally co-expressed only in a torus, being excluded from both the center and periphery of the flower meristem (Jack *et al*., 1992; Goto and Meyerowitz, 1994). In *Arabidopsis,* the C-class gene, *AG*, is expressed in the center of the flower, extending outward so as to overlap with the zone of *AP3* and *PI* activity (Drews *et al*., 1991).

The classic ABC model of floral organ identity determination suggests that the combinatorial action of three gene functions, A, B and C, is responsible for the determination of floral organ identity (Coen and Meyerowitz, 1991). According to the traditional version of this model, the joint expression of B + C promotes stamen development whereas the sole expression of C in the center of the flower promotes carpel development and causes the meristem to cease proliferating (Figure 1.2). This model is discussed and revised later, but is adequate for this discussion and serves to show that the C-class proteins specify reproductive organ identity (stamens + carpels), while the B-class proteins act as a switch between stamen and carpel identity. In effect, given a uniaxial interpretation of the flower, the B-genes can be seen as being heterospory genes in that they toggle organs between microsporophyll and megasporophyll fates (Baum, 1998). However, two questions concerning the establishment of stamens and carpels during flower development have remained enigmatic, being solved in *Arabidopsis* only in the last five years.

The first problem is that B- and C-class functions are necessary, but not sufficient to specify reproductive identity. The ectopic expression of C-class genes (*AG*) and/or B-class genes (*AP3* and *PI*) in *Arabidopsis* does not convert vegetative leaves to floral organs, suggesting that an additional factor(s) is required to specify floral organ identity. A solution to this problem came with the characterization of the *SEPALLATA* (*SEP*) family of MADS-box transcription factors. *SEP* genes are expressed in the flower (Flanagan and Ma, 1994; Savidge *et al*., 1995; Mandel and Yanofsky, 1998), and in *SEP* mutant backgrounds (*sep1*, *sep2* and *sep3* triple mutant background in *Arabidopsis*), or in *SEP* co-supression lines (petunia), the inner three whorls of the flower are transformed into sepal-like organs (Angenent *et al*., 1993; Pelaz *et al*., 2000, 2001; Honma and Goto, 2001; Ferrario *et al*., 2003). Therefore, *SEP* proteins are necessary, in addition to B- and C-class proteins, to properly establish reproductive organ identity. *Arabidopsis* plants over-expressing *SEP3* in conjunction with B-class genes (*AP3* and *PI*) show a conversion of vegetative leaves into petals (Honma and Goto, 2001; Pelaz *et al*., 2001), and plants over-expressing *SEP3* in conjunction with B- and C-class genes (*AP3*, *PI* and *AG*) show a conversion of leaves into stamen-like organs (Honma and Goto, 2001). These data show that *SEP* function plus B- and C-class functions are sufficient to specify floral organ identity.

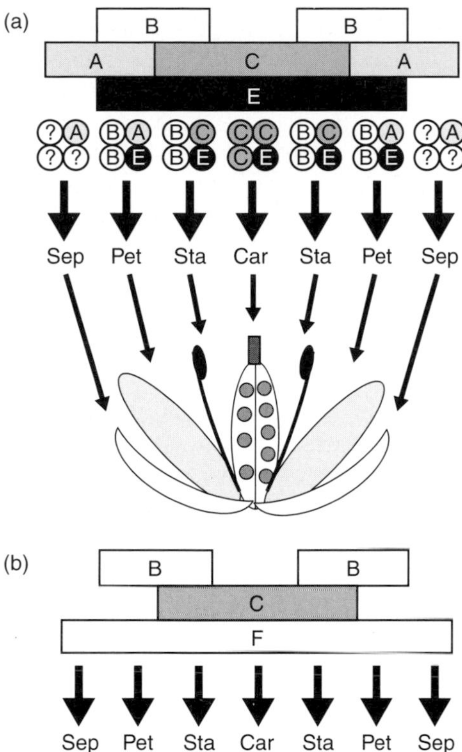

Figure 1.2 Comparison between the ABC model and the alternative FBC model. (a) The classical ABC model for the establishment of floral organ identity (Coen and Meyerowitz, 1991) combined with information of the distribution of E-class (*SEP*) gene expression (Honma and Goto, 2001; Pelaz *et al*., 2001). The boxed areas depict the distribution of A-, B- and C-class gene expressions across a floral meristem. The putative protein quartets arising in each of the overlapping domains are shown below with arrows to indicate the floral organ identity promoted (Theissen, 2001; Theissen and Saedler, 2001). (b) In the FBC model, the ground state of flower development, F, is encoded by floral meristem identity genes (e.g. *FLO/LFY*, *AP1*), which collectively are active throughout the floral meristem. The lack of C-gene expression in the periphery of the flower, rather than the presence of A-gene expression, is seen as conferring organ identity to the sterile floral organs (petals and/or sepals). The FBC and ABC models are identical with regard to the regulation of the reproductive organs by B-, C- and E-class genes. Sep-, sepals/calyx; Pet-, petals/corolla; Sta-, stamens/androecium; Car-, carpels/gynoecium.

These findings have resulted in the addition of an 'E'-class to the ABC model of floral development (Figure 1.2, Goto *et al*., 2001; Theissen, 2001; Theissen and Saedler, 2001).

But how do E-class gene products function with B- and C-class gene products to specify floral organ identity? It has been hypothesized that E-class proteins form multiprotein complexes with the A-, B- and C-class proteins (Theissen, 2001; Theissen and Saedler, 2001) in order to specify petal, stamen and carpel development (Figure 1.2). For instance, quartets consisting of B-, C- and E-class proteins may

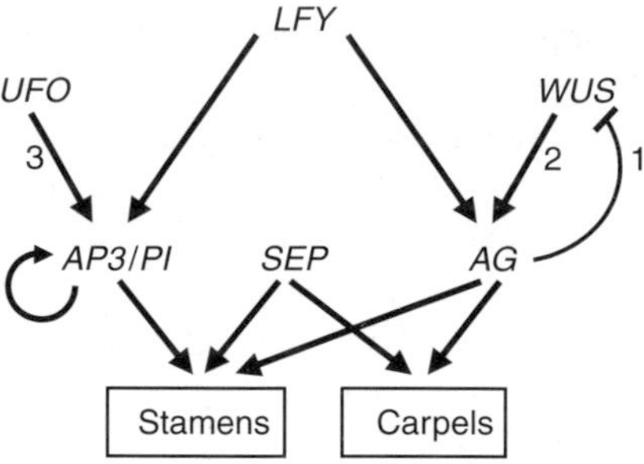

Figure 1.3 Simplified model for the regulation of stamen and carpel identity in *Arabidopsis* (based on Jack, 2004; Lohmann *et al.*, 2001; Parcy *et al.*, 1998). The floral meristem identity gene *LFY* activates both B-class (*AP3* and *PI*) and C-class (*AG*) floral organ identity genes. *UFO* is a coactivator of *AP3* expression, perhaps by targeting a transcriptional activator of *AP3* for protein destruction. *WUS* is a coactivator of *AG* expression at early stages of *Arabidopsis* flower development, but at later stages *AG* negatively regulates *WUS* (either directly or indirectly) resulting in termination of the floral meristem. The B-class proteins (*AP3* and *PI*) function as obligate heterodimers and autoregulate their own expression. The B- and C-class proteins interact with E-class proteins (*SEP*) to promote stamen and carpel development. We hypothesize that three regulatory steps (numbered 1–3) were added relatively late in the evolution of the flower, after the origin of a bisexual axis. 1, The addition of repression of *WUS* by *AG* is hypothesized to have played a role in the compression/determinacy of the floral axis. 2, The dependency of *AG* expression on *WUS* activity is hypothesized to have been a key step in the evolution of a sterile perianth. 3, The dependency of *AP3* expression on *UFO* activity is hypothesized to have been critical in the evolution of the eudicot dimorphic perianth.

activate target genes necessary for stamen development, while quartets consisting of C- and E-class proteins may activate the carpel developmental pathway (Figure 1.2). Although we do not yet know if such quartets exist *in planta*, there is a great deal of experimental evidence suggesting that A-, B-, C- and E-class MADS-box proteins do form higher order protein complexes (Fan *et al.*, 1997; Egea-Cortines *et al.*, 1999; Honma and Goto, 2001; Ferrario *et al.*, 2003).

The second problem that work in *Arabidopsis* has recently shed light on, is the question of why *LFY*, which is expressed throughout the floral meristem (Weigel *et al.*, 1992; Blazquez *et al.*, 1997), only activates C-class genes in the central domain, and B-class genes as a torus. The first breakthrough was the discovery that the gene *UNUSUAL FLORAL ORGANS* (*UFO*) is a coregulator of the B-class gene *AP3*, such that *LFY* can activate *AP3* expression only in the presence of *UFO* (Figure 1.3; Levin and Meyerowitz, 1995; Wilkinson and Haughn, 1995; Lee *et al.*, 1997; Parcy *et al.*, 1998). *UFO* and its homologs, e.g. *FIMBRIATA* (*FIM*) in *Antirrhinum*, encode F-box proteins (Simon *et al.*, 1994; Ingram *et al.*, 1995; Samach *et al.*, 1999). F-box proteins act as components of the SKP1-cullin-F-box (SCF) complex, which target substrates for ubiquitin-mediated protein degradation. Some evidence suggests that *UFO* protein functions in an SCF complex to mediate the destruction of target proteins (Wang *et al.*, 2003). Thus, one plausible model is that *LFY* cannot activate *AP3* expression unless some unknown transcriptional repressor, call it BX, is stripped off an *AP3 cis*-regulatory element through the action of a *UFO*-containing SCF complex (Ingram *et al.*, 1995, 1997; Samach *et al.*, 1999). There is no direct evidence for and some evidence against a role for *UFO* in the regulation of the other B-class gene in *Arabidopsis*, *PI* (Samach *et al.*, 1999; Zhao *et al.*, 2001). However, since the B-class genes *AP3* and *PI* only bind DNA as heterodimers, and functional heterodimers are required for the maintenance of *AP3* and *PI* expression by binding enhancer elements in their respective promoters (Riechmann *et al.*, 1996a, b), *UFO* effectively controls the distribution of both B-class gene products (Figure 1.3). From the point of view of floral evolution, it is noteworthy that *UFO* is expressed in a torus in all shoot meristems, both vegetative and floral (Ingram *et al.*, 1995; Lee *et al.*, 1997; Long and Barton, 1998). It has been suggested, therefore, that *UFO* is a general shoot meristem patterning gene (reviewed in Ng and Yanofsky, 2001; Tooke and Battey, 2003) that interacts with *LFY* to establish the basic pattern of the angiosperm flower: a ring of stamens around the central carpels (Parcy *et al.*, 1998).

When the role of *UFO* was characterized it was hypothesized that there might be another shoot meristem patterning gene that serves a similar coregulatory role to restrict C-class gene expression to the center of the flower meristem (Parcy *et al.*, 1998). This mystery gene was subsequently identified as the homeodomain protein, *WUSCHEL* (*WUS*; Lenhard *et al.*, 2001; Lohmann *et al.*, 2001). *WUS* is expressed in the central zone of all *Arabidopsis* SAMs and is required for the maintenance of a population of meristematic cells (Laux *et al.*, 1996; Mayer *et al.*, 1998). Interestingly, while *WUS* and *LFY* promote C-class (*AG*) expression during early stages of flower development, *AG* represses *WUS* expression at later stages (Lenhard *et al.*, 2001; Lohmann *et al.*, 2001), thereby suppressing the proliferative influence of *WUS*, resulting in floral determinacy (Figure 1.3).

1.6.2 Developmental regulation of the perianth

The regulation of petal versus sepal identity is relatively well studied in a few species that have both organ types (e.g. *Arabidopsis* and *Antirrhinum*). In these species, petal identity only arises in regions of the flower meristem that express B-class genes, *AP3* and *PI* in *Arabidopsis* (or their *Antirrhinum* homologs, *DEFICIENS*, *DEF* and *GLOBOSA*, *GLO*), and the A-gene *APETALA* (*AP1*) in *Arabidopsis* (or its *Antirrhinum* homolog, *SQUAMOSA*, *SQUA*) (Coen and Meyerowitz, 1991; Weigel *et al.*, 1992). By the same token, sepals develop in parts of the flower only expressing A-class genes (Figure 1.2). This model, the classic ABC-model (Coen and Meyerowitz, 1991), can be represented by the shorthand: A = sepals; A + B = petals (Figure 1.2).

An alternative view of the same data can be obtained by recalling that the A-function is antagonistic to the C-function (Coen and Meyerowitz, 1991). Consequently, parts of the flower expressing A-function do not express C-function and vice versa. Thus one can state accurately that petal identity is promoted by B-gene activity and suppressed by C-gene activity, or B − C = petals. Likewise, if sepal identity is the ground state for organs developing within a flower meristem (expressing floral meristem identity, or 'F-genes') then one can assert: F – B – C = sepals. An implication of this view is that 'A-function' disappears, being replaced by two quite distinct classes of genetic functionality, floral meristem identity ('F'), and exclusion of C-function from the outer parts of the flower. Put another way, the control of floral organ identity can be seen as being determined by the intersection of the F-, B- and C-functions (FBC) (Schwarz-Sommer *et al.*, 1990) rather than the more conventional ABC formulation (Figure 1.2). Indeed, besides *Arabidopsis* (and its close relatives), functionally characterized homologs of the A-class gene *AP1*, appear to only establish floral meristem identity and not floral organ identity (Irish and Sussex, 1990; Huijser *et al.*, 1992; Taylor *et al.*, 2002; Litt and Irish, 2003; Fornara *et al.*, 2004).

The difference between the FBC and ABC models is of minor importance when dealing with an extant flower. However, as discussed later, when dealing with the evolutionary origin of the perianth, the FBC model has the advantage of focusing attention on the core problem: achieving a spatial or temporal separation between the onset of F- and C-function.

1.7 Models for the origin of bisexuality

The general presumption has been that a bisexual reproductive axis is derived from an ancestor with unisexual axes (whether dioecious or monoecious). However, this presumption should be reevaluated in light of molecular evidence favoring a sister group relationship between angiosperms and all other living seed plants. Ignoring fossil taxa – and very few extinct taxa have been reconstructed with sufficient detail to rule out the production of bisexual axes – it is equally parsimonious to infer that the seed habit at its inception resulted in bisexual axes, with subsequent evolution of unisexual axes in ancestors of the living gymnosperms. Although this possibility cannot be completely ruled out, the model discussed here assumes that a bisexual (hermaphroditic) strobilus arose from a monoecious ancestor with unisexual axes.

The mostly male (MM) theory of Frohlich and Parker (Frohlich and Parker, 2000; Frohlich, 2002) suggests that the intermediate step between monoecy and hermaphroditism was gynomonoecy, wherein formerly staminate cones became bisexual (Figure 1.4). In the MM model, gynomonoecy evolved when ovule-promoting genes came to be ectopically expressed in the apical region of pollen cones, resulting in the production of ovules on adaxial surfaces of microsporophylls. The ectopic ovules, while initially sterile, subsequently became fertile, at which point the seed cones were lost, resulting in hermaphroditism. The primary evidence in favor of the MM model comes from the inference that the seed-cone-specific *FLO/LFY* homolog, *NEEDLY*, was lost in the stem lineage of the angiosperms, as might be expected to happen when the seed cones were lost (Frohlich, 2002). A variant of the MM hypothesis offered by Albert *et al.* (2002) is

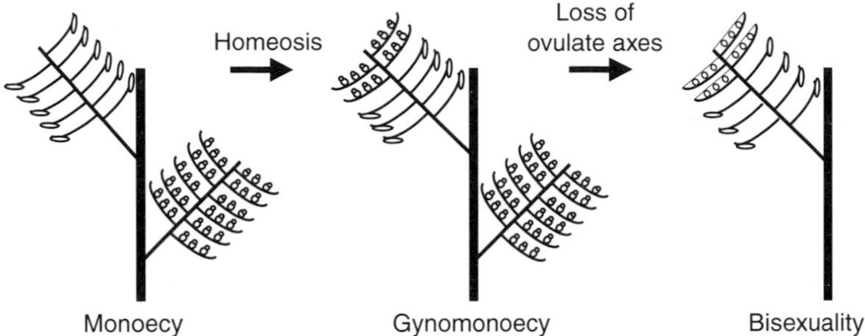

Figure 1.4 A schematic depiction of the transitions to bisexual axes in the ancestor of angiosperms according to the out-of-male, or OOM model (Theissen *et al.*, 2002). The ancestor is taken as being monoecious, having reproductive axes on the same plant that are composed only of megasporophylls (modified ovule-bearing leaves) or microsporophylls (modified pollen-bearing leaves), but not both. The transitional stage involves homeotic conversion distal microsporophylls in pollen-bearing axes to megasporophyll identity. This gynomonoecious intermediate gave rise to the condition found in angiosperms by the loss of fully ovulate cones. The OOM model differs from the 'mostly male model' (Frohlich and Parker, 2000) by invoking homeotic conversion of microsporophylls into megasporophylls rather than ectopic production of ovules on microsporophylls.

similar but shifts attention from the *LFY* and *NEEDLY* proteins to their *cis*-regulatory machinery.

An alternative to the MM model, the out-of-male (OOM) hypothesis, invokes a homeotic rather than ectopic mechanism (Theissen *et al.*, 2002; Theissen and Becker, 2004). Once again the hypothesis involves a gynomonoecious intermediate (Figure 1.4). However, the OOM model envisages the homeotic conversion of distal microsporophylls into megasporophylls (incipient carpels). Further, the OOM model speculates that the reason for homeosis was late repression of B gene activity within developing pollen cones (Theissen and Becker, 2004). The OOM model, however, resembles the MM model in supposing that bisexuality might have been beneficial initially because the megasporophylls would bear pollination droplets that might attract insect pollinators (Theissen and Becker, 2004).

The MM and OOM hypotheses make rather different predictions. The MM model implies that carpels (except for the ovules themselves) are homologous to microsporophylls of extant gymnosperms. Thus, if one found genes expressed differentially in micro- and megasporophylls of, say, cycads, one would expect the angiosperm carpels to have an expression profile resembling microsporophylls rather than megasporophylls (Frohlich and Parker, 2000). Additionally, the MM model would gain circumstantial support from mutations in which ectopic ovules were produced on other organs (especially microsporophylls or stamens) without associated carpel tissue. The OOM model in contrast predicts that homeotic mutations will predominate. Indeed the occasional production of distal ovuliferous organs on the pollen cones of some mutant spruces supports the basic premise (Theissen and Becker, 2004).

While there are positive aspects to the MM theory, we consider the OOM model to be more plausible on the grounds that homeotic mutations are so common in cone/flower development (e.g. De Craene, 2003).

1.8 Apical megasporophyll production on a microsporangiate axis?

If we assume that the presence/absence of B-gene expression acts as a switch between mega- and microsporophyll identity across seed plants, the major unanswered question under the OOM model is of how the B-function is repressed toward the apex of a formally microsporangiate cone? Under the prevailing model for the regulation of stamen and carpel identity in *Arabidopsis* (Parcy *et al.*, 1998; Lohmann *et al.*, 2001; Jack, 2004), the production of microsporophylls (stamens) is dependent upon organ primordia developing within a domain of the floral meristem that expresses *UFO/FIM*. Likewise, the production of carpels requires there to be some portion of the flower that expresses *WUS* (hence C-class genes) but not *UFO*. Because *WUS* and *UFO* maintain stable expression domains relative to the floral shoot apex, and flower meristems do not elongate during development, each reproductive organ primordium develops in a domain that either expresses *UFO* and *WUS* or *WUS* alone. However, such a mechanism would not seem viable in a plant with an elongated reproductive axis. Because of the SAM's continued meristematic activity, sporophylls would all pass through the stable domains of *WUS* and *UFO* expression. Thus, we infer that the role of *WUS* and *UFO* in regulating floral patterning arose only later, perhaps coincident with the switch from an elongated to a compressed floral axis, or even with the evolution of the more fixed floral architecture of eudicots.

Theissen and Becker (2004) hypothesize that the original regulation of B-genes was controlled by an apical-to-basal gradient of an unknown modulator of gene expression (e.g. hormones or other diffusible molecules). We would like to propose an alternative model for the evolution of the bisexual axis, based on the kinetics of gene expression and the current model for the action of MADS-domain proteins in multimeric complexes.

The key to our model is the fact that *LFY*, an important signaling integrator for the transition from vegetative to reproductive development, responds positively to a developmental clock. *LFY* expression levels increase with the age of *Arabidopsis* meristems (Blazquez *et al.*, 1997, 1998; Schmid *et al.*, 2003) and the gibberellin signaling pathway positively regulate *LFY* expression (Blazquez *et al.*, 1998). By extrapolating these findings, we suggest that reproductive shoots in the common ancestor of angiosperms would have gradually accumulated higher and higher levels of *LFY* protein during development.

Next, we hypothesize that the kinetics of C-class gene expression differs from the kinetics of B-class gene expression. Specifically, we suggest that maximal expression levels for C-class genes were much higher than for B-class genes (Figure 1.5). Because both B- and C-class genes are positively regulated by *LFY*, an increase in

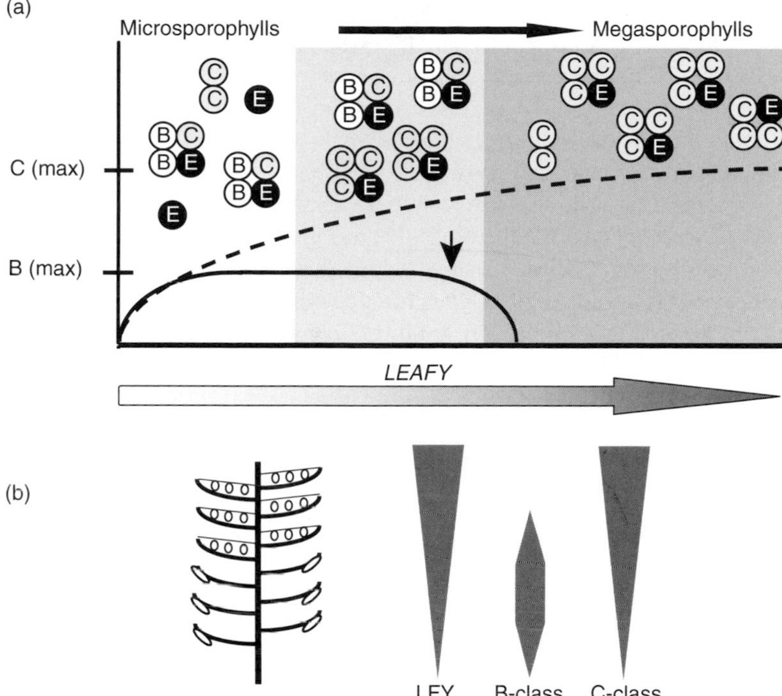

Figure 1.5 Proposed mechanism for the regulation of sexual organ identity on an extended floral axis. We hypothesize that in the ancestor of angiosperms, *LEAFY* (*LFY*) responded to a developmental clock such that *LFY* protein levels increased over ontogenetic time. This is depicted on the Y-axis in (a) and is represented by the increasing width of the gray triangle labeled '*LFY*' in (b). Increase in *LFY* protein levels would initially drive increasing levels of B- and C-class proteins until the B-class gene promoter reached its maximal expression level. At this time the predominant ternary complex would be composed of B-, C- and E-class proteins, thus, promoting microsporophyll/stamen development. After this initial phase, C-class gene expression would continue to rise, and C-class proteins would begin to exclude B-class proteins from active ternary complexes, until the majority of the active ternary complexes were composed of only C- and E-class proteins. We hypothesize that the relative increase in C+E-class ternary complexes over B + C + E-class ternary complexes would prevent the autoregulatory maintenance of B-class gene expression (arrow head). As C + E-class ternary complexes predominate, the reproductive axis would be expected to switch, potentially quite abruptly, from a microsporophyll to a megasporophyll production.

LFY protein over ontogeny could result in C-class proteins becoming progressively more abundant without concomitant increases in B-class proteins (Figure 1.5). If E-class (*SEP*) proteins were in limited supply, the increase in C-class protein levels would tend to displace B-class proteins from active protein complexes (quartets; Figure 1.5). Furthermore, if B-class genes showed positive, transcriptional autoregulation (as they do now; Riechmann *et al.*, 1996b; Hill *et al.*, 1998), the increasing abundance of C + E complexes and the resultant lack of B + C + E complexes would tend to lead to B-class gene expression failing to be maintained (Figure 1.5).

Thus, during the development of a reproductive axis, we imagine that steadily increasing *FLO/LFY* levels would initially promote expression of B- and C-genes, but once C-gene levels hit a certain threshold, activity and then expression of B-genes would be suppressed (arrow in Figure 1.5). A developmental clock of *LFY* protein levels could thereby cause reproductive shoots to shift (potentially quite abruptly) from microsporophyll (stamen) to megasporophyll (carpel) identity (Figure 1.5).

The specification of floral organ identity along the bisexual axis of a derived species such as *Arabidopsis* is likely to be highly modified compared to our hypothetical ancestral angiosperm species. Possible support for our hypothesis comes from the fact that ectopic activation of *LFY*, when fused to the strong VP16 activation domain, results (in the most extreme cases) in flowers consisting only of carpeloid organs (Parcy *et al.*, 2002).

1.9 The compression of the floral axis

Previous workers have not paid great attention to the question of how the floral axis switched from being an elongated stem with temporal control of patterning (e.g. due to a *LFY* developmental clock) to being a compressed, determinate stem with spatial control of patterning (via *WUS* and *UFO*). Here we consider possible genetic models in the hope that this will stimulate attention to this interesting topic.

Extrapolating from what is known in *Arabidopsis*, determinacy of the floral axis seems largely due to the negative regulation of the SAM-proliferation program by C-class genes (Laux *et al.*, 1996; Lenhard *et al.*, 2001; Lohmann *et al.*, 2001). Without a population of actively dividing cells in the SAM, upward progression of the shoot apex ceases. As a result, the floral promordia do not leave the meristematic zone to enter the zone of shoot elongation that subtends the apex, and ultimately the SAM is consumed by floral organs. Therefore, a determinate reproductive axis could have evolved with the evolution of a negative interaction between C-class function and maintenance of the SAM. How could this interaction have evolved?

The hypothesis we propose begins from an ancestor like *Archaefructus* that produces elongated bisexual axes under the regulation of C-protein level, such that B-gene activity is high early but declines with time as C-activity increases (Figure 1.5). We hypothesize that the first step in the evolution of a regulatory interaction between flower patterning genes and SAM-maintenance genes was the random acquisition of a binding site in the *WUS cis*-regulatory sequences that could be bound by a C-class dependent, transcriptional repressor. This repressor could be the C-class protein itself (e.g. *AG*), or a downstream gene product that is upregulated by C-class function. The result would be a graded inhibition of *WUS* by C-proteins. As C-class protein levels increased during shoot maturation, shoot growth would gradually cease due to inhibition of *WUS*, leaving the uppermost carpels congested at the shoot apex (Figure 1.1).

Floral determinacy via *WUS* negative regulation may have contributed to congestion of organs near the flower apex across angiosperm diversity, but the tight spatial control of a highly compressed floral axis (e.g. whorled phyllotaxy in many derived angiosperms compared to spiral phyllotaxy and reduced axis compression in many basal angiosperms) was likely to be refined over evolutionary time. These two critical aspects of floral evolution, meristem determinacy and axis compression, are likely to have evolved through selection for higher reproductive fitness. Reducing the number of carpels (and stamens) per shoot, with a concomitant increase in the total number of reproductive shoots, would have allowed tighter regulation of the time of pollen release and stigmatic receptivity and would permit selection to act more efficiently on stamen versus carpel number. Also, it could have served to bring the microsporophylls (stamens) closer to the megasporophylls (carpels), which may have produced exudates that attracted pollinating insects (Endress, 1996; Hufford, 1996; Frohlich, 2002). Thus, it is not hard to imagine gradual adjustment to the strength of the interactions among *LFY*, C-class gene products B-class gene products and *WUS* resulting in the development of a compressed bisexual axis and the production of a quite stable number of stamens and carpels.

1.10 The evolution of the perianth

Based on the morphology found among ANITA grade plants, it appears that the last common ancestor of the extant angiosperms had a perianth of petaloid organs (Doyle and Endress, 2000; Soltis *et al.*, 2000; Endress, 2001; Zanis *et al.*, 2003). Traditionally, petals were thought to have evolved multiple times (Takhtajan, 1991) – sometimes when modified bracts became added to the flower (bracteopetals) and sometimes when stamens became sterilized (andropetals). Because petal development depends upon floral meristem identity genes, there is good reason to suppose that petals originated from organs that were already within the floral zone, rather than by modification of bracts situated outside the flower. The only floral organs at this time were stamens and carpels, of which one can only assume that the outermost, i.e. stamens, became modified by sterilization and flattening for the purposes of pollinator attraction.

Therefore, the origin of the perianth would seem to require restriction of C-class function to the inner domain of the flower (under our model the floral axis was compressed by this time). This could have been achieved when C-class gene expression came under the control of *WUS*, a gene restricted to the inner region of the floral meristem, most likely by the addition of *WUS* binding sites in the C-class gene *cis*-regulatory regions (Lohmann *et al.*, 2001). Initially, *WUS* binding would simply promote C-class gene expression, but later, it could become necessary for the expression of C-class genes. In the latter condition, one would expect the peripheral region of the flower to be free of C-activity but to contain abundant B-class proteins. These B-class proteins would be expected to form complexes with E-class proteins (*SEP*) and perhaps other MADS-box genes. Whereas the lack of C-class proteins prevents the full activation of the stamen developmental pathway, some of the stamen development genes could have been activated to a sufficient degree so that the organs would develop with mixed leaf/stamen identity, i.e. as andropetals. Thus, we propose that the co-regulation of C-class genes by *WUS* was a key determinant in the origin of the petaloid perianth of ancestral angiosperms.

1.11 The origin of a dimorphic perianth

Within the angiosperms the perianth has been evolutionarily labile. From the inferred ancestral monomorphic, petaloid perianth, three main transitions have occurred repeatedly: modifications to petal form (including shifts to greater sepaloidy), evolution of a dimorphic perianth and perianth loss. The origin of the dimorphic perianth, comprising distinct sepal and petal whorls, deserves special attention because this feature evolved somewhere near the base of the eudicots (Zanis *et al*., 2003) and characterizes *Arabidopsis* and *Antirrhinum*.

Just as the development of petals requires a spatial separation of the zones of F- and C-function, the development of a dimorphic perianth requires a separation of F- and B-function. This is achieved in *Arabidopsis* because the B-gene *AP3* is only expressed in regions expressing both *FLO/LFY* and *UFO*, and *UFO* is absent from the 1st whorl of the flower (Parcy *et al*., 1998). Thus, we speculate that co-regulation of B-activity by *UFO* may be a eudicot-specific phenomenon (postdating the duplication of B-genes into the *DEF/AP3* and *GLO/PI* classes), and may have been the key step in the evolution of the eudicot, dimorphic perianth.

When the dimorphic perianth evolved in the eudicot ancestors of *Arabidopsis*, *UFO* may already have been playing a (partially redundant) role in general meristem functioning, perhaps in the maintenance of meristem size or the regulation of the zone of organ initiation. *UFO* is expressed in a torus in *Arabidopsis* vegetative and reproductive SAMs (Ingram *et al*., 1995; Lee *et al*., 1997; Long and Barton, 1998). This function might have entailed targeting a mystery transcriptional inhibitor, 'BX', for ubiquitin-mediated degradation (Ingram *et al*., 1995, 1997; Samach *et al*., 1999). The ancestral *DEF/AP3* gene may have acquired a *cis*-regulatory binding site for BX meaning that the ancestral *DEF/AP3* gene would only be expressed in the region of *UFO* activity, namely, in a torus around the shoot apex. If it was already necessary for *DEF/AP3* and *GLO/PI* to heterodimerize for proper activation of downstream targets (Riechmann *et al*., 1996b; Hill *et al*., 1998), which is suggested to have originated sometime after the monocot and eudicot lineages diverged (Kramer and Irish, 2000; Winter *et al*., 2002), then *UFO*-dependent transcription of *DEF/AP3* would result in a spatial restriction of B-function (*DEF/AP3* and *GLO/PI*) to the *UFO*-expressing parts of the flower.

Once *UFO* coregulation of B-genes evolved, shifts between petaloidy and sepaloidy and between monomorphic and dimorphic perianths would have been relatively easy developmental transitions, occurring by an adjustment to the zones of expression of *UFO* and by the modulation of the targets of B-gene regulation. If the capacity for floral adaptation to different pollination systems has contributed to the success of angiosperms (e.g. Grant, 1994), then the freedom to regulate B- and

C-function independently of floral meristem identity could have been a critical innovation that explains the remarkable success of the eudicot clade.

1.12 Conclusion

Even in model species like *Arabidopsis* there is still much that we do not know about the development of flowers. For example, what are the downstream targets of the B- and C-class proteins that determine petal, stamen and carpel development? What is the mode of action of *UFO* and does it involve a mystery inhibitor of B-class gene expression that is targeted for degradation by a *UFO*-containing SCF complex? How is the *UFO* expression domain maintained in SAMs? Moreover, we know very little about the function of floral meristem and organ identity genes outside of eudicots, and even less outside of angiosperms. Consequently, the model proposed here is highly speculative and almost certainly false in many regards. Why then take the trouble to spell out all the developmental genetic transitions?

Even if our model is not correct in all its genetic details, certain generalities may nonetheless hold. For example, it could well be that in ancestral species with elongated floral axes, stamen versus carpel identity involved signaling from a developmental clock that specified a transition in organ identity, although the specific mechanism proposed here, limited maximal expression of B-class genes and competition for E-class proteins, may be invalid. Likewise, we will feel vindicated if floral axis compression involved the evolution of cross-talk between SAM genes and floral meristem identity genes even if the series of novel developmental interactions proposed here needs to be adjusted, or if new genetic players are identified.

A further justification for proposing these genetic models is that they may serve to stimulate and focus empirical research on flower development. For example, our model suggests that it would be interesting to construct *Arabidopsis* plants in which SAM genes no longer interacted with the flower developmental program to see if elongated shoots resembling those of *Archaefructus* could be generated. Also, our model makes testable predictions about genetic interactions in non-model plants such as early diverging lineages of angiosperms and other seed plants. Predictions include the following:

(1) In early-diverging angiosperms and non-flowering seed plants, C-gene products are effectively negative regulators of B-function.
(2) *UFO*, *WUS*, and BX are active in SAMs in non-flowering seed plants, but do not control the expression of floral organ identity genes.
(3) The cooption of *UFO* for *DEF/AP3* activation coincided phylogenetically with the origin of a dimorphic perianth in eudicots.
(4) Early diverging angiosperms with a perianth of only petals do not require *UFO* for B-class gene activity, but may require *WUS* for C-class gene activity.

We hope that the fact that such predictions exist in print may serve to stimulate the development of genetic tools like mutant screens and transformation strategies in non-model angiosperm, and non-angiosperm seed plant systems.

Finally, the development of detailed evolutionary developmental models is valuable if they help demystify evolutionary transitions that are otherwise hard to imagine occurring by small sequential steps. Thus, even if the model presented here is completely wrong, it may serve to demonstrate one way in which a complicated developmental system like a flower could have arisen through a series of functional intermediates. Trivial as this may seem to a hardened evolutionary biologist, it should not be forgotten that for many people the origin of evolutionary novelties remains perplexing. Therefore, breaking down evolutionary transitions into small, plausible steps, as we have attempted here, may serve a valuable role even if the detailed theory is ultimately found to be wanting.

Acknowledgements

This chapter arose out of a talk presented by DAB at the symposium *Flowers: Development, Morphology, and Evolution* organized by Peter Linder in Zurich, July 2002, to honor the numerous contributions of Peter Endress. We are grateful for critical discussions and insights from numerous colleagues (many of whom would not subscribe to our conclusions): Paul Berry, James Doyle, Peter Crane, Michael Donoghue, Peter Endress, Michael Frohlich, Detlef Weigel, Elena Kramer, Rebecca Oldham, Jocelyn Hall, Ho-Sung Yoon, Daniel Fulop, Christopher Day, Donna Fernandez, Vivian Irish, Rebecca Lamb, Amy Litt, Eunyoung Chae, Gemma Dimartino, Jens Sundstrom, Moriyah Zik, Naomi Nakayama and Queenie Tan. We gratefully thank Kandis Elliott for assistance with artwork.

References

Albert, V. A., Gustafsson, M. H. G. and Di Laurenzio, L. (1998) Ontogenetic systematics, molecular developmental genetics, and the angiosperm petal, in *Molecular Systematics of Plants* (eds D. E. Soltis, P. S. Soltis and J. J. Doyle), Kluwer Academic Publishers, Boston, pp. 349–374.

Albert, V. A., Oppenheimer, D. G. and Lindqvist, C. (2002) Pleiotropy, redundancy and the evolution of flowers, *Trends in Plant Science*, **7**, 297–301.

Angenent, G. C., Busscher, M., Franken, J., Colombo, L. and Vantunen, A. J. (1993) The homeotic gene Fbp2 regulates floral organogenesis in Petunia and encodes a new class of Mads box proteins, *Journal of Cellular Biochemistry*, **Suppl. 17B**, 13–13.

Arber, E. A. N. and Parkin, J. (1907) On the origin of angiosperms, *Botanical Journal of the Linnean Society*, **38**, 29–80.

Baum, D. A. (1998) The evolution of plant development, *Current Opinion in Plant Biology*, **1**, 79–86.

Blazquez, M. A., Green, R., Nilsson, O., Sussman, M. R. and Weigel, D. (1998) Gibberellins promote flowering of Arabidopsis by activating the LEAFY promoter, *Plant Cell*, **10**, 791–800.

Blazquez, M. A., Soowal, L. N., Lee, I. and Weigel, D. (1997) LEAFY expression and flower initiation in Arabidopsis, *Development*, **124**, 3835–3844.

Bowe, L. M., Coat, G. and dePamphilis, C. W. (2000) Phylogeny of seed plants based on all three genomic compartments: extant gymnosperms are monophyletic and Gnetales' closest relatives are conifers, *Proceedings of the National Academy of Sciences of the United States of America*, **97**, 4092–4097.

Busch, M. A., Bomblies, K. and Weigel, D. (1999) Activation of a floral homeotic gene in Arabidopsis *Science*, **285**, 585–587.

Chaw, S. M., Parkinson, C. L., Cheng, Y. C., Vincent, T. M. and Palmer, J. D. (2000) Seed plant phylogeny inferred from all three plant genomes: monophyly of extant gymnosperms and origin of Gnetales from conifers, *Proceedings of the National Academy of Sciences of the United States of America*, **97**, 4086–4091.

Coen, E. S. and Meyerowitz, E. M. (1991) The war of the whorls – genetic interactions controlling flower development, *Nature*, **353**, 31–37.

Coen, E. S., Romero, J. M., Doyle, S., Elliott, R., Murphy, G. and Carpenter, R. (1990) Floricaula – a homeotic gene required for flower development in *Antirrhinum Majus*, *Cell*, **63**, 1311–1322.

Darwin, F. and Seward, A. C., eds (1903) *More Letters of Charles Darwin*, John Murray, London.

De Craene, L. P. R. (2003). The evolutionary significance of homeosis in flowers: a morphological perspective, *International Journal of Plant Sciences*, **164**, S225–S235.

De Craene, L. P. R., Smets, E. F. and Vanvinckenroye, P. (1998) Pseudodiplostemony, and its implications for the evolution of the androecium in the Caryophyllaceae, *Journal of Plant Research*, **111**, 25–43.

Doyle, J. A. (1994) Origin of the Angiosperm flower – a phylogenetic perspective, *Plant Systematics and Evolution*, **Suppl. 8**, 7–29.

Doyle, J. A. (2000) Paleobotany, relationships, and geographic history of Winteraceae, *Annals of the Missouri Botanical Garden*, **87**, 303–316.

Doyle, J. A. and Endress, P. K. (2000) Morphological phylogenetic analysis of basal angiosperms: comparison and combination with molecular data, *International Journal of Plant Sciences*, **161**, S121–S153.

Doyle, J. A., Hotton, C. L. and Ward, J. V. (1990). Early Cretaceous tetrads, zonasulculate pollen, and Winteraceae. 2. Cladistic – analysis and implications, *American Journal of Botany*, **77**, 1558–1568.

Drews, G. N., Bowman, J. L. and Meyerowitz, E. M. (1991) Negative regulation of the *Arabidopsis* homeotic gene *AGAMOUS* by *APETALA2* product, *Cell*, **65**, 991–1002.

Egea-Cortines, M., Saedler, H. and Sommer, H. (1999) Ternary complex formation between the MADS-box proteins SQUAMOSA, DEFICIENS and GLOBOSA is involved in the control of floral architecture in *Antirrhinum majus*, *EMBO Journal*, **18**, 5370–5379.

Endress, P. K. (1996) Structure and function of female and bisexual organ complexes in gnetales, *International Journal of Plant Sciences*, **157**, S113–S125.

Endress, P. K. (2001) The flowers in extant basal angiosperms and inferences on ancestral flowers, *International Journal of Plant Sciences*, **162**, 1111–1140.

Fan, H. Y., Hu, Y., Tudor, M. and Ma, H. (1997) Specific interactions between the K domains of AG and AGLs, members of the MADS domain family of DNA binding proteins, *Plant Journal*, **12**, 999–1010.

Feild, T. S., Arens, N. C., Doyle, J. A., Dawson, T. E. and Donoghue, M. J. (2004) Dark and disturbed: a new image of early angiosperm ecology, *Paleobiology*, **30**, 82–107.

Feild, T. S., Brodribb, T. and Holbrook, M. (2002) Hardly a relict: freezing and the evolution of vesselless wood in winteraceae, *Evolution*, **56**, 464–478.

Ferrario, S., Immink, R. G. H., Shchennikova, A., Busscher-Lange, J. and Angenent, G. C. (2003) The MADS box gene FBP2 is required for SEPALLATA function in petunia, *Plant Cell*, **15**, 914–925.

Flanagan, C. A. and Ma, H. (1994) Spatially and temporally regulated expression of the Mads-box gene Agl2 in wild-type and mutant Arabidopsis flowers, *Plant Molecular Biology*, **26**, 581–595.

Floyd, S. K. and Friedman, W. E. (2000) Evolution of endosperm developmental patterns among basal flowering plants, *International Journal of Plant Sciences*, **161**, S57–S81.

Fornara, F., Parenicova, L., Falasca, G. *et al* (2004) Functional characterization of OsMADS18, a member of the AP1/SQUA subfamily of MADS box genes, *Plant Physiology*, **135**, 2207–2219.

Friis, E. M., Crane, P. R. and Pedersen, K. R. (1986) Floral evidence for Cretaceous chloranthoid Angiosperms, *Nature*, **320**, 163–164.

Friis, E. M., Doyle, J. A., Endress, P. K. and Leng, Q. (2003) Archaefructus – angiosperm precursor or specialized early angiosperm? *Trends in Plant Science*, **8**, 369–373.

Frohlich, M. W. (2002) The mostly male theory of flower origins: summary and update regarding the Jurassic pteridosperm *Pteroma*, in *Developmental Genetics and Plant Evolution* (eds Q. C. B. Cronk, R. M. Bateman and J. A. Hawkins), Taylor and Francis Inc., London, pp. 85–108.

Frohlich, M. W., and Meyerowitz, E. M. (1997) The search for flower homeotic gene homologs in basal angiosperms and gnetales: a potential new source of data on the evolutionary origin of flowers, *International Journal of Plant Sciences*, **158**, S131–S142.

Frohlich, M. W. and Parker, D. S. (2000) The mostly male theory of flower evolutionary origins: from genes to fossils, *Systematic Botany*, **25**, 155–170.

Givnish, T. (1979) On the adaptive significance of leaf form, in *Topics in Plant Population Biology* (eds O. T. Solbrig, S. Jain, G. B. Johnson and P. H. Raven), Columbia University Press, New York, pp. 375–407.

Goremykin, V. V., Hansmann, S. and Martin, W. F. (1997) Evolutionary analysis of 58 proteins encoded in six completely sequenced chloroplast genomes: revised molecular estimates of two seed plant divergence times, *Plant Systematics and Evolution*, **206**, 337–351.

Goto, K. and Meyerowitz, E. M. (1994) Function and regulation of the Arabidopsis floral homeotic gene Pistillata, *Genes & Development*, **8**, 1548–1560.

Goto, K., Kyozuka, J. and Bowman, J. L. (2001) Turning floral organs into leaves, leaves into floral organs, *Current Opinion in Genetics & Development*, **11**, 449–456.

Graham, S. W. and Olmstead, R. G. (2000) Utility of 17 chloroplast genes for inferring the phylogeny of the basal angiosperms, *American Journal of Botany*, **87**, 1712–1730.

Grant, V. (1971) *Plant Speciation*, 2nd edn, Columbia University Press, New York.

Grant, V. (1994) Modes and origins of mechanical and ethological isolation in angiosperms, *Proceedings of the National Academy of Sciences of the United States of America*, **91**, 3–10.

Hill, T. A., Day, C. D., Zondlo, S. C., Thackeray, A. G. and Irish, V. F. (1998) Discrete spatial and temporal *cis*-acting elements regulate transcription of the Arabidopsis floral homeotic gene APETALA3, *Development*, **125**, 1711–1721.

Honma, T. and Goto, K. (2001) Complexes of MADS-box proteins are sufficient to convert leaves into floral organs, *Nature*, **409**, 525–529.

Hufford, L. (1996) The morphology and evolution of male reproductive structures of gnetales, *International Journal of Plant Sciences*, **157**, S95–S112.

Huijser, P., Klein, J., Lonnig, W. E., Meijer, H., Saedler, H. and Sommer, H. (1992) Bracteomania, an inflorescence anomaly, is caused by the loss of function of the Mads-box gene *Squamosa in Antirrhinum Majus*, *EMBO Journal*, **11**, 1239–1249.

Ingram, G. C., Doyle, S., Carpenter, R., Schultz, E. A., Simon, R. and Coen, E. (1997) Dual role for *fimbriata* in regulating floral homeotic genes and cell division in *Antirrhinum*, *EMBO Journal*, **16**, 6521–6534.

Ingram, G. C., Goodrich, J., Wilkinson, M. D., Simon, R., Haughn, G. W. and Coen, E. S. (1995) Parallels between unusual floral organs and Fimbriata, genes controlling flower development in Arabidopsis and Antirrhinum, *The Plant Cell*, **7**, 1501–1510.

Irish, V. F. and Sussex, I. M. (1990) Function of the Apetala-1 gene during Arabidopsis floral development, *The Plant Cell*, **2**, 741–753.

Jack, T. (2004) Molecular and genetic mechanisms of floral control, *The Plant Cell*, **16**, S1–S17.

Jack, T., Brockman, L. L. and Meyerowitz, E. M. (1992) The homeotic gene *Apetala3* of *Arabidopsis Thaliana* encodes a Mads box and is expressed in petals and stamens, *Cell*, **68**, 683–697.

Kramer, E. M. and Irish, V. F. (2000) Evolution of the petal and stamen developmental programs: evidence from comparative studies of the lower eudicots and basal angiosperms, *International Journal of Plant Sciences*, **161**, S29–S40.

Kuzoff, R. K. and Gasser, C. S. (2000) Recent progress in reconstructing angiosperm phylogeny, *Trends in Plant Science*, **5**, 330–336.

Laux, T., Mayer, K. F. X., Berger, J. and Jurgens, G. (1996) The WUSCHEL gene is required for shoot and floral meristem integrity in Arabidopsis, *Development*, **122**, 87–96.

Lee, I., Wolfe, D. S., Nilsson, O. and Weigel, D. (1997) A LEAFY co-regulator encoded by UNUSUAL FLORAL ORGANS, *Current Biology*, **7**, 95–104.

Lenhard, M., Bohnert, A., Jurgens, G. and Laux, T. (2001) Termination of stem cell maintenance in Arabidopsis floral meristems by interactions between WUSCHEL and AGAMOUS, *Cell*, **105**, 805–814.

Levin, J. Z. and Meyerowitz, E. M. (1995) Ufo – an Arabidopsis gene involved in both floral meristem and floral organ development, *The Plant Cell*, **7**, 529–548.

Litt, A. and Irish, V. F. (2003) Duplication and diversification in the APETALA1/FRUITFULL floral homeotic gene lineage: implications for the evolution of floral development, *Genetics*, **165**, 821–833.

Lohmann, J. U. and Weigel, D. (2002) Building beauty: the genetic control of floral patterning, *Developmental Cell*, **2**, 135–142.

Lohmann, J. U., Hong, R. L., Hobe, M. *et al.* (2001) A molecular link between stem cell regulation and floral patterning in Arabidopsis, *Cell*, **105**, 793–803.

Long, J. A. and Barton, M. K. (1998) The development of apical embryonic pattern in Arabidopsis, *Development*, **125**, 3027–3035.

Mandel, M. A. and Yanofsky, M. F. (1998) The Arabidopsis AGL9 MADS box gene is expressed in young flower primordia, *Sexual Plant Reproduction*, **11**, 22–28.

Mathews, S. and Donoghue, M. J. (1999) The root of angiosperm phylogeny inferred from duplicate phytochrome genes, *Science*, **286**, 947–950.

Mayer, K. F. X., Schoof, H., Haecker, A., Lenhard, M., Jurgens, G. and Laux, T. (1998) Role of WUSCHEL in regulating stem cell fate in the Arabidopsis shoot meristem, *Cell*, **95**, 805–815.

McGonigle, B., Bouhidel, K. and Irish, V. F. (1996) Nuclear localization of the Arabidopsis APETALA3 and PISTILLATA homeotic gene products depends on their simultaneous expression, *Genes & Development*, **10**, 1812–1821.

Ng, M. and Yanofsky, M. F. (2001) Activation of the arabidopsis B class homeotic genes by APETALA1, *Plant Cell*, **13**, 739–753.

Parcy, F., Bomblies, K. and Weigel, D. (2002) Interaction of LEAFY, AGAMOUS and TERMINAL FLOWER1 in maintaining floral meristem identity in Arabidopsis, *Development*, **129**, 2519–2527.

Parcy, F., Nilsson, O., Busch, M. A., Lee, I. and Weigel, D. (1998) A genetic framework for floral patterning, *Nature*, **395**, 561–566.

Pelaz, S., Ditta, G. S., Baumann, E., Wisman, E. and Yanofsky, M. F. (2000) B and C floral organ identity functions require SEPALLATA MADS- box genes, *Nature*, **405**, 200–203.

Pelaz, S., Tapia-Lopez, R., Alvarez-Buylla, E. R. and Yanofsky, M. F. (2001) Conversion of leaves into petals in Arabidopsis, *Current Biology*, **11**, 182–184.

Qiu, Y. L., Lee, J., Bernasconi-Quadroni, F. *et al.* (2000) Phylogeny of basal angiosperms: analyses of five genes from three genomes, *International Journal of Plant Sciences*, **161**, S3–S27.

Qiu, Y. L., Lee, J. H., Bernasconi-Quadroni, F. *et al.* (1999) The earliest angiosperms: evidence from mitochondrial, plastid and nuclear genomes, *Nature* **402**, 404–407.

Regal, P. J. (1977) Ecology and evolution of flowering plant dominance, *Science*, **196**, 622–629.

Riechmann, J. L., Krizek, B. A. and Meyerowitz, E. M. (1996a) Dimerization specificity of Arabidopsis MADS domain homeotic proteins APETALA1, APETALA3, PISTILLATA, and AGAMOUS, *Proceedings of the National Academy of Sciences of the United States of America*, **93**, 4793–4798.

Riechmann, J. L., Wang, M. Q. and Meyerowitz, E. M. (1996b) DNA-binding properties of Arabidopsis MADS domain homeotic proteins APETALA1, APETALA3, PISTILLATA and AGAMOUS, *Nucleic Acids Research*, **24**, 3134–3141.

Samach, A., Klenz, J. E., Kohalmi, S. E., Risseeuw, E., Haughn, G. W. and Crosby, W. L. (1999) The *UNUSUAL FLORAL ORGANS* gene of *Arabidopsis thaliana* is an F-box protein required for normal patterning and growth in the floral meristem, *The Plant Journal*, **20**, 433–445.

Savard, L., Li, P., Strauss, S. H., Chase, M. W., Michaud, M. and Bousquet, J. (1994) Chloroplast and nuclear gene-sequences indicate late Pennsylvanian time for the last common ancestor of extant seed plants, *Proceedings of the National Academy of Sciences of the United States of America*, **91**, 5163–5167.

Savidge, B., Rounsley, S. D. and Yanofsky, M. F. (1995) Temporal Relationship between the Transcription of 2 Arabidopsis Mads Box Genes and the Floral Organ Identity Genes, *The Plant Cell*, **7**, 721–733.

Schmid, M., Uhlenhaut, N. H., Godard, F. *et al.* (2003) Dissection of floral induction pathways using global expression analysis, *Development*, **130**, 6001–6012.

Schwarz-Sommer, Z., Huijser, P., Nacken, W., Saedler, H. and Sommer, H. (1990) Genetic control of flower development by homeotic genes in *Antirrhinum majus*, *Science*, **250**, 931–936.

Simon, R., Carpenter, R., Doyle, S. and Coen, E. (1994) Fimbriata controls flower development by mediating between meristem and organ identity genes, *Cell*, **78**, 99–107.

Soltis, P. S., Soltis, D. E. and Chase, M. W. (1999) Angiosperm phylogeny inferred from multiple genes as a tool for comparative biology, *Nature*, **402**, 402–404.

Soltis, P. S., Soltis, D. E., Zanis, M. J. and Kim, S. (2000) Basal lineages of angiosperms: relationships and implications for floral evolution, *International Journal of Plant Sciences*, **161**, S97–S107.

Sun, G., Dilcher, D. L., Zheng, S. L. and Zhou, Z. K. (1998) In search of the first flower: a Jurassic angiosperm, Archaefructus, from northeast China, *Science*, **282**, 1692–1695.

Sun, G., Ji, Q., Dilcher, D. L., Zheng, S. L., Nixon, K. C. and Wang, X. F. (2002) Archaefructaceae, a new basal angiosperm family, *Science*, **296**, 899–904.

Takhtajan, A. (1991) *Evolutionary Trends in Flowering Plants*, Columbia University Press, New York.

Taylor, S. A., Hofer, J. M. I., Murfet, I. C. *et al.* (2002) PROLIFERATING INFLORESCENCE MERISTEM, a MADS-box gene that regulates floral meristem identity in pea, *Plant Physiology*, **129**, 1150–1159.

Theissen, G. (2001) Development of floral organ identity: stories from the MADS house, *Current Opinion in Plant Biology*, **4**, 75–85.

Theissen, G. and Becker, A. (2004) Gymnosperm orthologues of class B floral homeotic genes and their impact on understanding flower origin, *Critical Reviews in Plant Sciences*, **23**, 129–148.

Theissen, G. and Saedler, H. (2001) Plant biology – floral quartets, *Nature*, **409**, 469–471.

Theissen, G., Becker, A., Winter, K. U., Munster, T., Kirchner, C. and Saedler, H. (2002) How the land plants learned their floral ABCs: the role of MADS-box genes in the evolutionary origin of flowers, in *Developmental Genetics and Plant Evolution* (eds Q. C. B. Cronk, R. M. Bateman and J. A. Hawkins), Taylor and Francis Inc., London, PP. 173–205.

Tooke, F. and Battey, N. (2003) Models of shoot apical meristem function, *New Phytologist*, **159**, 37–52.

Trobner, W., Ramirez, L., Motte, P. *et al.* (1992) Globosa – a homeotic gene which interacts with deficiens in the control of Antirrhinum floral organogenesis, *EMBO Journal*, **11**, 4693–4704.

Vandenbussche, M., Zethof, J., Royaert, S., Weterings, K. and Gerats, T. (2004) The duplicated B-class heterodimer model: whorl-specific effects and complex genetic interactions in *Petunia hybrida* flower development, *The Plant Cell*, **16**, 741–754.

Walker, J. W., Brenner, G. J. and Walker, A. G. (1983) Winteraceous pollen in the lower Cretaceous of Israel – early evidence of a Magnolialean Angiosperm family, *Science*, **220**, 1273–1275.

Wang, X. P., Feng, S. H., Nakayama, N. *et al.* (2003) The COP9 signalosome interacts with SCFUFO and participates in Arabidopsis flower development, *The Plant Cell*, **15**, 1071–1082.

Weigel, D. and Meyerowitz, E. M. (1993) Activation of floral homeotic genes in Arabidopsis, *Science*, **261**, 1723–1726.

Weigel, D., Alvarez, J., Smyth, D. R., Yanofsky, M. F. and Meyerowitz, E. M. (1992) Leafy controls floral meristem identity in Arabidopsis, *Cell*, **69**, 843–859.

Wilkinson, M. D. and Haughn, G. W. (1995) Unusual floral organs controls meristem identity and organ primordia fate in Arabidopsis, *The Plant Cell*, **7**, 1485–1499.

Winter, K. U., Becker, A., Munster, T., Kim, J. T., Saedler, H. and Theissen, G. (1999) MADS-box genes reveal that gnetophytes are more closely related to conifers than to flowering plants, *Proceedings of the National Academy of Sciences of the United States of America*, **96**, 7342–7347.

Winter, K. U., Weiser, C., Kaufmann, K. *et al.* (2002) Evolution of class B floral homeotic proteins: obligate heterodimerization originated from homodimerization, *Molecular Biology and Evolution*, **19**, 587–596.

Yang, Y. Z., Fanning, L. and Jack, T. (2003) The K domain mediates heterodimerization of the Arabidopsis floral organ identity proteins, APETALA3 and PISTILLATA, *The Plant Journal*, **33**, 47–59.

Yanofsky, M. F., Ma, H., Bowman, J. L., Drews, G. N., Feldmann, K. A. and Meyerowitz, E. M. (1990) The protein encoded by the Arabidopsis homeotic gene *Agamous* resembles transcription factors, *Nature*, **346**, 35–39.

Zanis, M. J., Soltis, D. E., Soltis, P. S., Mathews, S. and Donoghue, M. J. (2002). The root of the angiosperms revisited, *Proceedings of the National Academy of Sciences of the United States of America*, **99**, 6848–6853.

Zanis, M. J., Soltis, P. S., Qiu, Y. L., Zimmer, E. and Soltis, D. E. (2003) Phylogenetic analyses and perianth evolution in basal angiosperms, *Annals of the Missouri Botanical Garden*, **90**, 129–150.

Zhao, D. Z., Yu, Q. L., Chen, M. and Ma, H. (2001) The ASK1 gene regulates B function gene expression in cooperation with UFO and LEAFY in Arabidopsis, *Development*, **128**, 2735–2746.

2 Floral induction

Reynald Tremblay and Joseph Colasanti

2.1 Introduction

Plants have evolved elaborate mechanisms to ensure that flowering occurs at an optimal time for reproductive success. The coordination of these regulatory mechanisms requires the integration of diverse signals from the environment with endogenous physiological cues in order for the floral transition to occur at the correct time. Extensive physiological experiments of the last century have established a foundation upon which current genetic models are superimposed in order to reveal the mechanisms underlying the floral transition. These early physiological experiments, however, were performed with a wide variety of species, whereas current genetic models are based almost exclusively on studies with the small flowering plant *Arabidopsis thaliana*. The present challenge is to consolidate the molecular elements of the genetic models with the physiological data.

In *Arabidopsis,* at least four floral inductive pathways have been incorporated into a genetic model of a flowering network (Figure 2.1). This network contains several nodes, each of which represents a site of signal integration. Integration of these pathways is required for coordinated initiation of flowering by the various pathways. The assignment of four floral induction pathways in *Arabidopsis* does not necessarily mean that the same genetic hierarchy exists in all plant species. Indeed, given that the environmental stimuli perceived by plants vary depending on geographical location, it is not unexpected that distinct floral inductive mechanisms may have evolved independently in different plant species. However, the genetic pathways derived from study of *Arabidopsis* mutants define the basic mechanisms common to most plants. This chapter will examine the four major pathways defined in *Arabidopsis* and how the components of each pathway interact. Homologous floral inductive systems in other plant species, and differences between the network in *Arabidopsis* and other species, will be discussed.

2.2 Floral transition is marked by developmental phase changes

The discovery by Garner and Allard (1920) of photoperiodic induction of flowering was the starting point for subsequent physiological studies investigating the floral transition. By defining a controllable external stimulus that could be applied to certain plants to cause flowering, they introduced an experimental system that could be used to dissect and analyse the floral induction process at the physiological level. The numerous studies that followed defined two important features of the floral

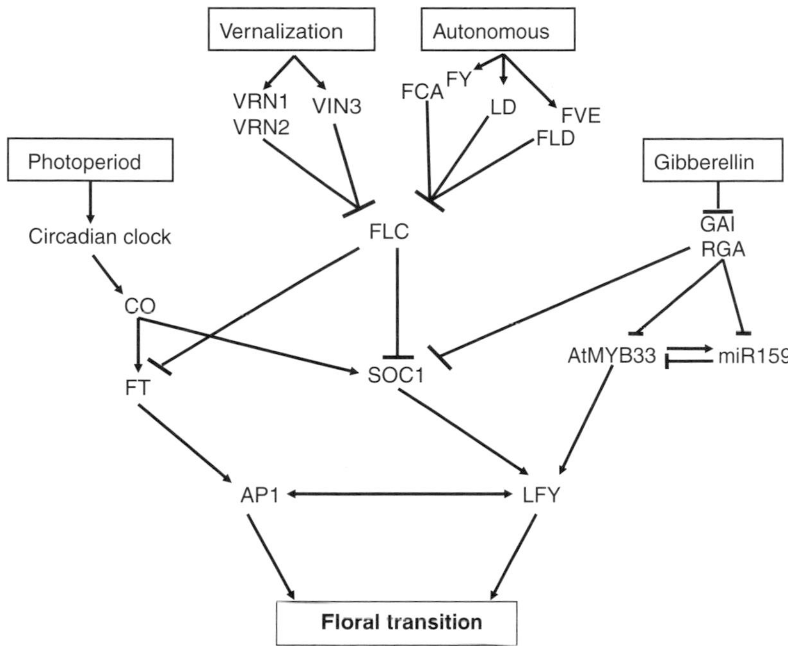

Figure 2.1 Floral inductive pathways in *Arabidopsis thaliana*. A simplified view of flowering pathways in *Arabidopsis* showing induction through two environmental components, photoperiod and vernalization, and two endogenous components, autonomous and gibberellin. Photoperiodic induction involves interpretation of day length changes via the circadian clock.

transition common to higher plants. First, a floral stimulus originates in the leaves and is transported to the shoot apical meristem (SAM). Second, the SAM, as the target of the floral stimulus, must be competent to receive the floral stimulus (Lang, 1965, 1977). The floral transition is manifested at the SAM, a collection of stem cells at the growing apex of the aerial shoot that gives rise to all vegetative and reproductive tissues. The transition from a vegetative to a reproductive stage represents a developmental progression in SAM morphology accompanied by a fundamental shift in gene expression (Smith *et al.*, 2004). The SAM must pass a developmental checkpoint where it progresses from an incompetent state, unable to perceive incoming floral inductive signals, to a competent state, where it is able to interpret floral inductive signals and make the transition to flowering (Poethig, 1990). This transition from an incompetent (juvenile) to a competent (mature) vegetative meristem is marked by changes in organ production and gene expression (Poethig, 1990; Telfer and Poethig, 1998; Haung and Yang, 1998; Aubert *et al.*, 2001; Vega *et al.*, 2002). Once the first developmental checkpoint is passed, the SAM is competent to flower and floral inductive signals generated by external and internal stimuli can evoke floral transition. The genetic components that act in the SAM to specify competence to flower are beginning to be deciphered. A recent study in *Arabidopsis* found that elimination of the function of two paralogous *BEL1*-like (*BELL*) homeobox genes

PENNYWISE (*PNY*) and *POUND-FOOLISH* (*PNF*) resulted in defective SAM that were unable to respond to floral inductive signals and remained in a vegetative state (Smith *et al.*, 2004). This unexpected finding suggests a connection between the architecture of the meristems and their ability to respond to inductive stimuli.

2.3 Floral induction is mediated through multiple pathways

A vegetative SAM generates leaves and shoots, whereas a reproductive SAM will give rise to inflorescences and flowers. Floral inductive signals induce the transformation of a mature vegetative SAM to a reproductive state. External stimuli, generated by environmentally driven changes, such as alterations in day length have been the subject of study for over 100 years. The early studies of Julien Tournois in the early 1900s demonstrated that plants grown with day length alterations, either shorter lengths by shading or longer lengths by incandescent light bulbs, could affect flowering time (Tournois, 1912). Tournois died in World War I, and it was not until Garner and Allard examined a tobacco mutant that required short day lengths to induce flowering that the concept of photoperiodism came to light (Garner and Allard, 1920). In Garner and Allard's experiments, the Maryland Mammoth variety of short-day (SD) tobacco was derived from a normally day-neutral tobacco; i.e. plants that flower at the same time, regardless of day length. Eventually it was established through night break experiments, where the long night period was interrupted by a short exposure to light, that floral induction in SD plants depends on the duration of the night and not on the length of day.

Later experiments using localized shading and lighting of both the SAM and leaves of spinach revealed that inductive signals originate in the leaves (Knott, 1934). Similar experiments in Russia during the 1930s on floral induction in chrysanthemum and *Perilla frutescens* demonstrated that induced leaves (donor) when grafted onto non-induced plants (recipient), could cause early flowering of the recipient (Chailakhyan, 1936). The discovery that leaves are the source of floral inductive signals in photoperiod-induced plants led Chailakhyan to postulate the florigen theory (see later). The hypothetical florigen is a universal, floral inducing substance (or substances) that is produced in the leaves and is transported to the growing apex where it causes the transition to flowering (Chailakhyan, 1936).

As described later, not only the length of light exposure (i.e. quantity) but also quality of light can be an important signal for flowering. Specific photoreceptors enable plants to sense the duration of light exposure, as well as discriminate between different wavelengths of light.

The effect of temperature on the flowering process was established early in the twentieth century where it was known that certain wheat and rye varieties needed to over-winter to induce flowering (Evans, 1969). Gassner (1918) found that exposure to cold temperatures during germination in pots, followed by transfer to soil under normal temperatures, accelerated flowering of the winter variety but had no affect on the spring variety (Gassner, 1918). The term vernalization was first coined by Lysenko, who found that exposure to cold temperatures must be followed by increases in day length, suggesting that plants must pass through different developmental checkpoints in order to flower.

Finally, endogenous cues that relay information about growth status, such as plant size, hormones and nutrient flow, can influence time to flowering (Bernier *et al.*, 1981). Plant species that consistently flower after producing a defined number of leaves, regardless of external environmental factors, may rely exclusively on these endogenous signals to flower. Studies of certain varieties of maize and tobacco that flower only after generating a predictable number of leaves support the existence of an endogenous floral inductive pathway (McDaniel, 1980; Irish and Jegla, 1997).

2.4 Photoperiodic floral induction provides a cue to seasonal changes

Changes in day length provide a reliable and consistent indicator of changes in the environment. These changes portend the onset of a cold period or the commencement of a rainy season and allow plants to coordinate flowering time. SD plants flower as day length shortens and long-day (LD) plants flower as day length increases; day-neutral plants flower irrespective of changes in day length (Lang, 1965). Within different day length responses, further division into facultative and obligate categories is possible. Obligate plants have an absolute requirement for inductive photoperiods and will remain in a vegetative state unless they are induced to flower. The flowering of facultative plants is accelerated under inductive conditions, but even in the absence of an inductive photoperiod, flowering will occur. Comparisons of plant species with different day length requirements for flowering, as well as between different varieties within a given species that vary in day length response, allows for identification of loci and, by extension, the genes required for inductive day length determination. In species such as rice and maize, genomic loci involved in flowering, or quantitative trait loci (QTL), have been identified (Lin *et al.*, 2000; Salvi *et al.*, 2002; Chardon *et al.*, 2004). Dissection and analysis of QTL has facilitated identification and isolation of specific flowering time genes in rice (Yano *et al.*, 2000; Lin *et al.*, 2000, 2002).

2.4.1 Photoreceptors transduce light signals

Plausible candidates for genes that are implicated in the transition to flowering are those that encode photoreceptors. The phytochromes, which perceive red and far-red light, and the cryptochromes, which perceive UV-A and blue light, are two major classes of photoreceptors found in most higher plants and that are involved in flowering. In *Arabidopsis*, the phytochromes A and B, along with the cryptochromes 1 and 2, are involved in day length response (Mouradov *et al.*, 2002). Mutations or misexpression of genes that encode these photoreceptors alters the plant's perception of light and can affect flowering time. In general, far-red and blue light promote flowering in *Arabidopsis* whereas red light inhibits flowering

(Lin, 2000). The transduction of the light signals involves a complex web of interactions between photoreceptors and their corresponding interacting proteins. In terms of floral induction, perception of photoperiod appears to be one of the most important transducers of the plant's environment. An important mechanism used by the plant phytochromes and cryptochromes to communicate photoperiod activity involves the entrainment of the circadian rhythm, a self-reinforcing endogenous clock that allows light/dark coordinated gene expression (see later). Additional inputs into flowering from temperature changes, light quantity and quality, also alter flowering time through elements of the endogenous clock (Putterill *et al.*, 2004; Boss *et al.*, 2004).

Light can have an additional role in transducing information about the quality of light perceived by plants (Simpson and Dean, 2002; Casal *et al.*, 2004). Genetic studies in *Arabidopsis* have discovered one of the components of a potential 'light quality pathway,' the *PHYTOCHROME AND FLOWERING TIME 1* (*PFT1*) gene (Cerdan and Chory, 2003). In this case, the perception of a low red/far-red ratio would activate the 'shade avoidance response' – a potential indicator of competition from neighbouring plants and a signal to flower early.

2.4.2 The circadian clock is self-reinforcing

Day length variation provides plants a predictable mechanism for detecting seasonal changes. Entrainment and subsequent setting of periodicity allows the endogenous clock to detect day length changes through comparison with the light cycle. Tracking changes in light/dark cycles is accomplished by an endogenous mechanism, known as the circadian rhythm, that creates self-reinforcing rhythmic gene expression patterns (Hayama and Coupland, 2003; Millar, 2004). Changes in the periodicity of the light/dark cycle results in the alteration of the endogenous rhythm, which can lead to floral induction by the photoperiodic pathway in day length responsive species. In *Arabidopsis*, three transcription factors, *CIRCADIAN CLOCK ASSOCIATED1* (*CCA1*), *LATE ELONGATED HYPOCOTYL* (*LHY*) and *TIMING OF CAB EXPRESSION1* (*TOC1*), interact to form a negative feedback loop (Somers *et al.*, 1998; Wang and Tobin, 1998; Mizoguchi *et al.*, 2002). *TOC1* belongs to a five-member family of ARABIDOPSIS PSEUDO-RESPONSE REGULATORS (APRR) whose expression is regulated by circadian rhythms. Rice similarly contains a five-member PRR family that shows rhythmic expression under circadian control, suggesting conservation of function amongst monocots and dicots (Murakami *et al.*, 2003). Output from the circadian clock occurs by changing expression of downstream genes to coincide with favorable photoperiodic conditions for flowering.

2.4.3 Key genes integrate photoperiodic induction

The circadian rhythm output to the floral inductive pathway in *Arabidopsis* controls the expression of *CONSTANS* (*CO*), a zinc-finger protein that functions as a transcription factor (Putterill *et al.*, 1995). *CO* was identified in mutant screens where loss of function *co* mutants failed to undergo early flowering when placed in normally inductive LD conditions, but eventually flowered around the same time as plants grown in non-inductive SD (Putterill *et al.*, 1995). *CO* is specifically involved in photoperiodic induction under LD conditions, as mutant plants retain their ability to flower. Plants that over-express *CO* flower early and are insensitive to changes in day length (Onouchi *et al.*, 2000). *CO* expression is therefore sufficient to induce flowering; it shows a diurnal pattern with peaks at dawn and 16 h after light in LD conditions, linking *CO* to the circadian clock (Suarez-Lopez *et al.*, 2001). Expression under SD conditions is similar to LD, but with a narrower window and lower peaks of expression corresponding to changes in day length. Previous physiological experiments revealed that the length of day in LD plants appears to be important in generating a floral inductive signal, and the finding of increased levels of *CO* expression under LD conditions further supports this concept (Deitzer, 1984; Suarez-Lopez *et al.*, 2001). The finding that light stabilizes *CO* expression and that, in dark conditions, CO protein undergoes proteasome-mediated degradation, links the circadian clock to the timing of *CO* expression (Valverde *et al.*, 2004).

The CO protein contains two regions that are required for proper function: a zinc-finger motif that is similar to protein–protein interacting zinc-fingers found in animals, and a CCT region, which is found in *CO*, *CO-LIKE* and *TOC1* genes and is required for the nuclear localization of these proteins (Robson *et al.*, 2001). Recent work has shown that while *CO* expression is found in many cell types, phloem companion cell-specific expression of *CO* results in early flowering, while restriction of *CO* expression to the SAM does not (An *et al.*, 2004). Similarly, Ayre and Turgeon (2004) found that expression of *CO* under the control of the galactinol synthase promoter, such that its expression was confined to minor vein companion cells, caused early flowering in non-inductive SD conditions. Further, they reported that leaves expressing *CO* in minor veins when grafted to *co* mutant receptor plants were able to accelerate flowering. These results suggest that *CO* acts by mediating the generation of a photoperiod-induced floral stimulus from the leaves that travels through the phloem to the SAM. As with earlier physiological studies, further research is needed to uncover the nature of the signaling molecule. In addition, while expression from phloem-specific promoters confirms that *CO* has a function in forming the floral stimulus in leaves, the wide-ranging expression pattern suggests that *CO* may have other roles apart from integrating flower-inducing light signals.

A direct downstream target of *CO* is *FLOWERING LOCUS T* (*FT*), a RAF-kinase inhibitor-like protein. Late flowering *ft* mutants were first isolated in a genetic screen of early flowering *CO* over-expressing lines (Samach *et al.*, 2000). This places *FT* downstream of *CO*, as it represses the over-expression phenotype, and also indicates the existence of another target gene that acts in parallel with *CO*, as flowering still occurred sooner than in *co* knockouts. Mutation in another direct target of *CO*, *SUPRESSOR OF OVEREXPRESSION OF CO 1* (*SOC1*), a MADS-box transcription factor, was also found to suppress the early flowering phenotype in *CO* over-expressing lines (Samach *et al.*, 2000). Mutation in *soc1* results in delayed transition to flowering, as with *ft* mutants and, like *ft*, flowering is only delayed slightly relative to *co* mutants. However, when functions of both *FT* and *SOC1* genes are eliminated by mutation, *CO* over-expressing plants are delayed to nearly the same degree as *co* mutants. This suggests that *FT* and *SOC1* are two key downstream targets of *CO*, and function in parallel, partially redundant, inductive pathways (Figure 2.1). Both *FT* and *SOC1* up-regulate the floral meristem identity genes *LEAFY* (LFY) and *APETALA1 (AP1),* as described later.

2.4.4 CO *and* FT *gene function is conserved in other plant species*

The discovery of *CO* and *FT* homologues in both monocot and dicot species suggests that they play important roles in controlling flowering in other plants. *CO* from *Brassica napus* and *Pharbitis nil* were able to complement *co* loss of function mutants in *Arabidopsis* (Robert *et al.*, 1998; Liu *et al.*, 2001). Recent studies have identified functional homologs of these genes within monocot species, with the majority of work focused on rice, a facultative SD plant. In rice, QTL associated with flowering time mutants, called *Heading date*, were found to encode genes that are related to *CO* and *FT*. *Heading Date 1* (*Hd1*), a *CO* homologue, was found to influence early transition to flowering in inductive SD conditions and to inhibit flowering under LD conditions (Yano *et al.*, 2000). Like *CO* in *Arabidopsis*, *Hd1* shows a diurnal pattern of expression, with peak levels at dawn and during the night under LD conditions (Hayama *et al.*, 2003). Under SD conditions, mRNA expression is similar, but was much lower at dawn than under LD conditions. These expression patterns coincide with the output pattern from the rice circadian clock expression (Figure 2.2). Circadian clock regulation of the *CO-like* family genes in both monocots and dicots supports an evolutionarily conserved mechanism for photoperiodic induction.

Further evidence of this conserved function was found in the analysis of rice *hd1* mutant lines. The late flowering *hd1* mutants were found to have lower expression levels of an important flowering time gene *Heading date 3a* (*Hd3a*), which has high sequence similarity to the *Arabidopsis FT* gene (Kojima *et al.*, 2002). Mutant *hd3a* plants flower much later under SD than wild-type plants. Therefore CO/FT interaction appears to be conserved in rice, even though rice flowering is induced by SD rather than LD conditions. The level of *Hd3a* gene expression is highest during the day under SD conditions but no expression was found under LD treatments (Hayama *et al.*, 2003). Evidence suggests that *Hd1* functions to inhibit expression of *Hd3a* under LD conditions and to induce *Hd3a* under inductive SD conditions (Figure 2.2).

Comparison of the photoperiod inducibility of flowering in rice and *Arabidopsis* suggest that a large part of this pathway, from photoreceptor perception to the creation of endogenous circadian rhythm, is conserved. Changes in circadian rhythms due to changes in day-length from a non-inductive to an inductive period have been found to result in the activation of *CO-LIKE* genes in these diverse species

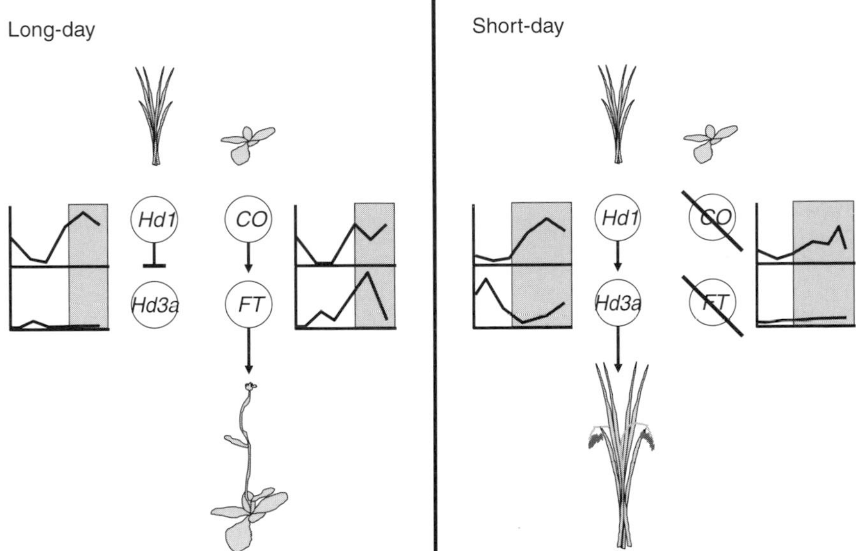

Figure 2.2 *CO* operates in a conserved pathway for photoperiod induction of flowering. *CO* expression is regulated by the circadian rhythm, which establishes a diurnal pattern of expression. The correlation of peaks under long-day (LD) conditions with light allows *CO* to up-regulate *FT* expression. *CO* expression under short days (SD) peaks after dark, with no resulting increase in *FT* expression. In rice, inductive photoperiods are reversed. Under SD, the *CO* orthologue *Hd1* is expressed and induces the expression of the *FT* orthologue *Hd3a*. Under LD conditions, *Hd1* represses *Hd3a*.

(Hayama *et al.*, 2003). These *CO-LIKE* regulators in turn act upon *FT-LIKE* genes in a linear cascade that can then induce the transition to flowering (see Figure 2.1). While conservation of key components like *CO* and *FT* may exist in various species, other independent parallel mechanisms have evolved to regulate photoperiodic floral transition. Given the complexity of the transition to flowering process and the importance of imparting information about seasonal growing conditions, it might be expected that independent mechanisms have evolved to transmit information about photoperiod conditions.

2.4.5 *Photoperiod induction through* CO*-independent pathways*

Photoperiodic floral induction through non-*CO* pathways exists in different species and appears to have arisen independently (Doi *et al.*, 2004). For instance, while *Hd1* acts in rice to promote flowering under SD and repress it under LDs, an alternative pathway, defined by another rice flowering time locus, *EARLY HEADING DATE 1* (*EHD1*) was found to initiate flowering in an *Hd1*-independent manner (Doi *et al.*, 2004). *Ehd1* encodes a B-type response regulator with, based on sequence analysis, no functional orthologue in *Arabidopsis*. *Ehd1* may represent a unique evolutionary adaptation where a gene has been co-opted to induce photoperiodic flowering despite the presence of an already conserved linear pathway. In lines deficient for a functional *hd1* gene, *Ehd1* is sufficient to induce early flowering in SD. Unlike *Hd1*, *Ehd1* does not appear to repress flowering under LD conditions (Doi *et al.*, 2004). The existence of a *CO*-independent pathway demonstrates that adaptation to local environments, i.e. different day lengths, can result in subtle changes to an existing pathway. Incorporation of new genes into a pre-existing regulatory network, as is the case for *Ehd1*, represents a fine-tuning process to local environmental conditions.

2.5 Autonomous pathway

Most plants, even in the absence of external inductive signals, will flower eventually; plants with obligate requirements for external flower-inducing signals are rare. This suggests that other factors intrinsic to plant growth must provide signals that cause flowering. These autonomous, or constitutive, signals are most likely derived from the physiological outputs from the plant that reflect readiness to flower, such as plant size, age or leaf number. One possibility is that these endogenous cues may be directly related to the amount of resources accumulated in the plant (Bernier and Perilleux, 2005). In *Arabidopsis*, a number of mutants that flower late in both inductive and non-inductive photoperiods define an autonomous pathway, separate from the photoperiod pathway (Figure 2.1). However, many of the mutant genes respond to elements of the vernalization pathway (see later), so clearly the two pathways intersect at various points.

The nature of the specific physiological elements that combine to create an autonomous inductive signal has yielded much speculation. It is generally assumed that a combination of nutrients, most notably sucrose, and phytohormones, such as gibberellins or cytokinins, moves from leaves (and possibly roots) to the SAM to elicit flowering (Bernier and Perilleux, 2005). Other factors found in phloem sap, such as small proteins and RNA, could also make up components of the signal.

So far, genetic studies have not unambiguously revealed the nature of these components. However, one trend that has emerged from studies of *Arabidopsis* autonomous pathway genes is that they may encode factors involved in RNA processing and in maintenance of the epigenetic state of key regulatory genes (Simpson, 2004). The FLC gene in particular is a common component of both the autonomous and vernalization pathways in *Arabidopsis* and thus appears to be an important node for signal integration in this species. Although the universal function of FLC as a global flowering repressor is uncertain given that functional equivalents have not been detected in other species, it does provide a useful model for understanding the molecular nature of floral inductive signals.

2.5.1 FLOWERING LOCUS C *integrates different floral inductive pathways*

Genes involved in the autonomous pathway in *Arabidopsis* were first identified as mutants that delayed the flowering regardless of day length; i.e. in both SD and LD conditions (Martinez-Zapater and Somerville, 1990; Koornneef *et al.*, 1991). The first isolated flowering time gene, *LUMINIDEPENDENS* (*LD*), was found to play a role in autonomous flowering, where mutant *ld* plants were delayed in floral transition regardless of light exposure (Lee *et al.*, 1994). *LD* encodes a homeodomain protein that may bind DNA or possibly RNA. The molecular function of *LD* is unknown.

Other genes were found to affect flowering by altering the expression levels of a known floral inhibitor *FLOWERING LOCUS C* (*FLC*). *FLC* functions in a dosage dependent manner, with multiple positive and negative regulators affecting its transcription rate (Koornneef *et al.*, 1998a,b). *FLC*, which encodes a MADS box protein, is involved in maintaining the vegetative state through repression of several floral induction genes, including *SOC1* and *FT* (Michaels and Amasino, 1999; Samach *et al.*, 2000; Hepworth *et al.*, 2002). FLC maintains both of these genes in a repressed state, such that the elimination of FLC function or up-regulation of inducers *SOC1* and *FT* will expedite the floral transition. Many regulators have been identified that control *FLC* expression. Two other autonomous pathway genes, *FCA* and *FY*, were found to interact with each other to repress *FLC* expression (Simpson *et al.*, 2003). FCA contains an RNA-binding domain and is thought to interact directly with FY, which shares homology to known 3′-end RNA-processing proteins from *Saccharomyces cerevisiae*. The FCA–FY complex may be involved in premature polyadenylation of the *FLC* mRNA during intron splicing (Figure 2.3a). This would result in the premature termination of the *FLC* transcript, although this has yet to be verified, possibly due to the instability of the transcript produced (Simpson, 2004).

Another way autonomous pathway genes may induce flowering is through epigenetic control. The genes *FLD* and *FVE* have been shown to act by repressing *FLC* through transcriptional silencing (Figure 2.3b). FLD is homologous to components

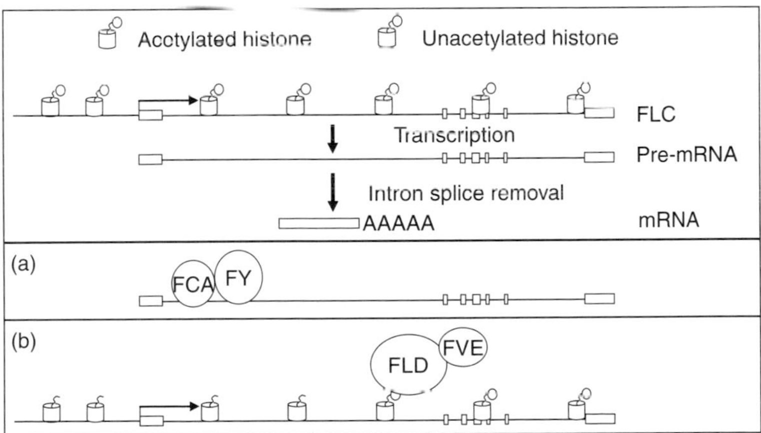

Figure 2.3 Potential models for *FLC* regulation by the autonomous pathway. (a) FCA/FY regulates mRNA splicing and polyadenylation, through mis-splicing of the *FLC* pre-mRNA. (b) FLD/FVE, as components of the histone deacetylase complex, mediate localized transcriptional inactivation.

of histone deacetylase complexes (HDAC) found in animal systems (He *et al.*, 2003). Histone deacetylation is associated with altered histone packaging and is an important regulator of euchromatic gene silencing. The FVE protein resembles animal and yeast factors that make up HDAC, which are involved in chromatin assembly (Ausin *et al.*, 2004). *FLD* and *FVE* function to lower *FLC* expression by interfering with transcriptional initiation, although a low level of *FLC* expression is still observed. *FVE*, *FLD*, *FY* and *FCA* also are associated with activities, such as histone methylation and polyadenylation, and appear to act in parallel to inactivate *FLC* (Simpson, 2004).

Overall, autonomous regulation of *FLC* levels by multiple mechanism supports its role as a node in floral induction in *Arabidopsis*. Regulation of gene expression through histone modifications that cause silencing and the potential mis-splicing of mRNA transcripts represent independent regulatory mechanisms that control expression of the FLC floral repressor. While the information represented here is limited to a single plant family, as*FLC* activity has only been described in the *Brassicaceae*, the mechanisms described earlier for transcriptional/translational inactivation could be used to regulate flowering nodes in other species.

2.6 Vernalization

Vernalization is a checkpoint through which many plants must pass, where flowering is facilitated by a prolonged exposure to cold temperature and subsequently induced by other inductive pathways (Sung and Amasino, 2004). Grafting leaves from plants induced to flower by vernalization to non-induced recipients did not accelerate flowering time, indicating that the site of cold perception is the SAM and not the leaves (Bernier *et al.*, 1981). Flowering only occurs after a prolonged exposure to cold temperatures, and many plants that require vernalization require a subsequent photoperiodic response in order to flower. This prevents precocious flowering due to a short warm period before the onset of winter conditions.

Vernalization is a somatically heritable state; once vernalization has occurred, daughter cells derived from induced SAM cells retain the state induced by extended cold exposure (Sung and Amasino, 2004). In species that require photoperiodic floral induction following vernalization, if non-inductive photoperiods follow vernalization, normal vegetative growth continues (Bernier *et al.*, 1981). However, plants exposed to inductive photoperiods will flower after an extended period after the vernalization event. Flowering of some varieties of wheat and *Arabidopsis* is accelerated by vernalization although the genes involved in mediating vernalization may be different for each species (Yan *et al.*, 2004a).

2.6.1 Mediation of vernalization in Arabidopsis by FLC *repression*

Vernalization in *Arabidopsis* involves several steps that culminate in the stable repression of *FLC* (Figure 2.4). First, plants must distinguish between short cold

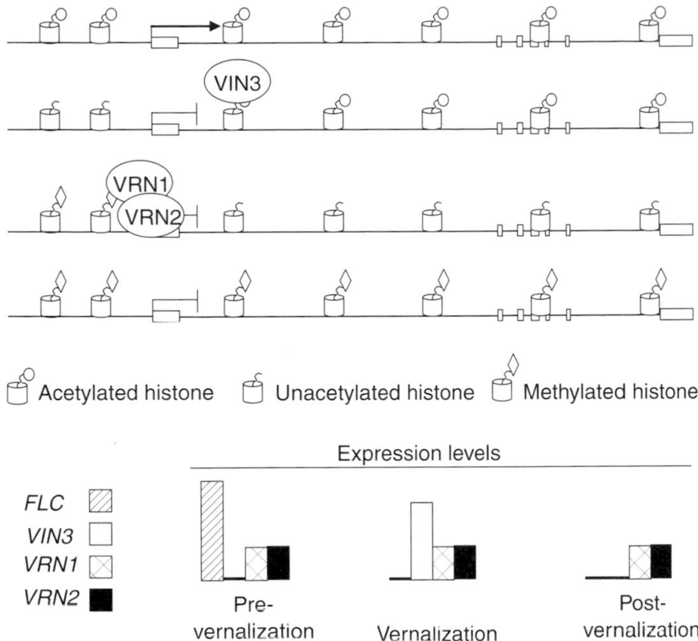

Figure 2.4 Proposed mechanism for silencing of *FLC* by vernalization. Extended cold periods induce VIN3 expression. VIN3 mediates histone deacetylation that results in silencing of *FLC*. Stable silencing of FLC is mediated by VRN1 and VRN2. Histone methylation mediates stable transcriptional silencing. Expression of both *VRN1* and *VRN2* is constant throughout plant growth, while *VIN3* expression is maintained only during vernalization.

periods and prolonged periods such as winter; different mechanisms are present to respond to either period, with cold acclimation to short periods and vernalization to longer periods. Experiments in *Arabidopsis* have shown that during extended cold periods, inactivation of the *FLC* locus is mitotically heritable (Sheldon *et al.*, 2000). Mutations that disrupt the vernalization response reveal that epigenetic silencing of *FLC* occurs through a multistep process. First, *VIN3*, a plant homeodomain (PHD) protein, hypoacetylates histones at the *FLC* locus (Sung and Amasino, 2004). PHD proteins play a role in chromatin remodeling, possibly through protein–protein interactions (Aasland *et al.*, 1995). Detection of *VIN3* mRNA only occurs after an extended exposure to cold temperatures, and expression levels decline following a return to warmer conditions. In mutant *vin3* plants, histone acetylation patterns in the *FLC* promoter remain unchanged during cold exposure, so *FLC* expression is unaffected. *VIN3* is involved in the initial inactivation of *FLC* but not in the repression following removal from cold conditions, as *VIN3* expression occurs only during the cold period.

Stable repression of *FLC* occurs through the activity of two other genes, *VERNALIZATION 1* and *2* (*VRN1* and *VRN2*). In *vrn1* and *vrn2* mutants, *FLC* is repressed after exposure to vernalizing conditions, but fails to maintain this repression once plants are returned to warm temperatures (Gendall *et al.*, 2001; Levy *et al.*, 2002). The lack of stable repression following vernalization suggests that both genes are important in the maintenance of the repressed state, and therefore may function in propagating the 'memory' of cold exposure. *VRN1* encodes a nuclear localized protein that binds to DNA in a sequence un-specific manner (Levy *et al.*, 2002). *VRN1* is expressed in many developing tissues before, during and after vernalization. Over-expression analysis has identified a vernalization-dependent and independent pathway for *VRN1* activity, as over-expression lines are not vernalized and show an accelerated floral transition with no changes in *FLC* levels (Levy *et al.*, 2002). *VRN1* over-expression activates *FT* and, as mentioned earlier, this in turn can initiate the floral transition (Levy *et al.*, 2002). *VRN2* encodes a nuclear localized zinc-finger protein that shows similarity to polycomb proteins found in *Arabidopsis* and *Drosophila*. *Polycomb* genes are involved in the stable repression of genes in *Drosophila*, but not in the initiation of the repressed state, as is the case with *VRN2* (Francis and Kingston, 2001). Recent work has shown that *polycomb* genes in *Arabidopsis* function like those in *Drosophila*, interacting with each other and forming multimeric complexes that function as a histone methyltransferase, although the specific binding partners for *VRN2* have not been identified yet (Chanvivattana *et al.*, 2004). In *vrn2* mutants, *FLC* is hypoacetylated during vernalization but there is no subsequent methylation of the histones and *FLC* expression increases upon return to normal temperatures. In *vrn1* mutants, only methylation of lysine residue 27 of histone H3 is observed, but *FLC* repression is likewise not maintained (Sung and Amasino, 2004).

Overall, vernalization in *Arabidopsis* involves the epigenetic inactivation of *FLC* that is initiated through histone hypoacetylation by *VIN3*. *VRN1* and *VRN2* function to stabilize *FLC* silencing through methylation of the histones, which in turn leads to a stable, heritable change that maintains the cells exposed to the prolonged cold period in a state of competence to respond to subsequent floral inductive signals.

2.6.2 Vernalization in cereals

Breeding studies examining the flowering times of winter and spring wheat varieties uncovered two QTL that are associated with differences in vernalization response. Variants of the *VRN-1* QTL (not related to *Arabidopsis VRN1*) account for the majority of vernalization responses between winter and spring varieties, with the dominant form of the gene in spring varieties (Yan *et al.*, 2004a). A gene within the *VRN-1* QTL shares a high degree of identity to *AP1* in *Arabidopsis*, and represents a potential node in the integration of signaling pathways (Murai *et al.*, 2003; Trevasksis *et al.*, 2003). This gene, *WAP1*, shows increased expression following vernalization in winter varieties of wheat, supporting its role as a possible vernalization gene that defines the *VRN-1* QTL. A second QTL involved in vernalization in wheat is *VRN-2* (different from *Arabidopsis VRN2*). Recent work has identified a gene within this QTL, *ZCCT1*, which contains a CCT domain similar to the one found in *CO*

(Yan *et al.*, 2004b). Dominant *ZCCT1* is associated with winter varieties, and a marked reduction in *ZCCT1* expression occurs following vernalization, similar to the results seen in floral repression by *FLC* in *Arabidopsis*.

Understanding how *WAP1* and *ZCCT1* interact will provide important insights into the evolution of vernalization within various plant species. Vernalization in wheat could be a relatively new and unique evolutionary innovation, as wheat evolved from sub-tropical grasses that would not have required a cold period to induce flowering (Clayton and Renvoize, 1986; Yan *et al.*, 2004b). Comparison of vernalization in *Arabidopsis* and wheat shows that while both appear to utilize different nodes in the floral inductive pathway (i.e. no *FLC* gene has been found in wheat), the inactivation of floral repressors appears to play an important role in cold responsiveness.

2.7 Hormones and other factors

The transmission of hormones, nutrients and other endogenous compounds to the SAM during floral transition is associated with the induction of flowering in many species (Sachs and Hackett, 1983; Chandler and Dean, 1994). However, no single hormone or compound has been shown to have floral inductive activity in all higher plants. As described later, universality is one of the criteria for a putative florigenic substance; i.e. like other phytohormones, it is common to all species. The complexities of some models that attempt to integrate physiological and genetic data are in early stages of development (Bernier and Perilleux, 2005).

2.7.1 Nutrient diversion theory of floral induction

Nutrient content at the SAM changes at the time of floral transition, induced by changes in transport of sucrose and other sugars to the apex (Bodson, 1977). The nutrient diversion theory of floral transition is based on changes in sucrose concentration during floral transition, with higher sucrose levels at the SAM following floral inductive treatments in species like *Sinapis alba* (Bodson, 1977; Sachs and Hackett, 1983; Bodson and Outlaw Jr., 1985; Lejeune *et al.*, 1993). Changes in the relative ratio of sugar and nitrogen transport to the apex are also implicated in the nutrient diversion theory (Raper *et al.*, 1988). No genetic evidence for this model exists to date, but analysis of carbohydrate to nitrogen ratios prior to and during floral transition show that under vegetative conditions, the ratio is consistent but at time of flowering nitrogen transport becomes proportionally greater to the apex (Rideout *et al.*, 1992). However, more recent work has shown that increases in sucrose levels at the apex change at the time of flowering in *Fuchsia hybida* regardless of whether or not an inductive photoperiod occurred (King and Ben-Tal, 2001). It is possible that changes in sucrose concentrations, or other nutrients, are a result of floral transition and not the signal that induces flowering. Indeed, when LD inductive signals at low light levels are applied, the amount of sucrose at the apex did not increase, while plants still flowered in response to the day length changes (King and Ben-Tal, 2001). This suggests that sucrose marks photosynthetic activity, and that the photoinduction pathway does not simply require photosynthesis to carry out induction. While these results indicate that nutrient changes at the apex are not required for floral evocation, it is still not known if these changes can induce flowering.

2.7.2 Gibberellin

The role of gibberellic acid (GA) in affecting flowering time in *Arabidopsis* is clear. Application of GA results in early flowering in SD and LD, and can bypass the late flowering phenotype of many mutants from the other three pathways (Chandler and Dean, 1994). One of the first steps in the GA pathway involves the generation of a signal by *GA1*, which converts geranylgeranyl pyrophosphate to copalyl pyrophosphate, although the signals that induce *GA1* are unknown. The production of GA is important, as demonstrated in GA biosynthesis deficient mutants that do not flower under SD conditions due to the absence of active GA (Wilson *et al.*, 1992). The role of GA as a possible long distance floral inductive signal has been considered. Mutational analysis of genes in the GA pathway reveals that generation of the signal, and not GA itself, results in floral induction (Dill and Sun, 2001).

GA molecular targets were first described in barley, with the *GAMYB* gene being an integral part of GA signal transduction. Over-expression of GAMYB protein mimics the effects of GA application (Gubler *et al.*, 1995; Cercos *et al.*, 1999). GAMYB binds to a specific DNA sequence, called the GA-response element (GARE), which is present in promoter regions of genes known to be up-regulated by GA application. *Arabidopsis* has three genes that are similar to barley *GAMYB*, with *AtMYB33* showing the highest degree of similarity (Gocal *et al.*, 2001). This factor has been implicated in flowering in *Arabidopsis*, and can bind *in vitro* to the GARE sequence found upstream of the *LEAFY* (*LFY*) floral meristem identity gene (Blazquez *et al.*, 1998; Blazquez and Weigel, 1999). *AtMYB33* may act in concert with other inducers of *LFY*, such as *SOC1*, to regulate this critical step in floral transition.

GA affects two floral transition repressors, *GIBBERELLIC ACID INSENSITIVE* (*GAI*) and *RGA* (*Repressor of ga1-3*), both of which are *DELLA* class proteins (Cheng *et al.*, 2004). DELLA proteins are characterized by having an *N*-terminal DELLA domain, which may act as the target for GA-directed ubiquitination followed by subsequent proteasome-mediated destruction. *GAI* and *RGA* are important in the inhibition of flowering, as recent results indicate that *GAI* acts to repress expression of *SOC1*, although there does not appear to be a direct effect on *LFY* (Achard *et al.*, 2004). Both *GAI* and *RGA* function to repress the expression of the microRNA miR159, a 21-bp sequence with significant identity to the *AtMYB* genes (Gocal *et al.*, 2001; Reinhart *et al.*, 2002). MicroRNAs have been shown to have an important role in the transcriptional silencing of target genes, either by binding mRNA transcripts and initiating the RNA interference response, or by binding to mRNA and preventing translation (Baulcombe, 2004). Over-expression of miR159 results in lower levels of *AtMYB33* in leaves, which in turn delays flowering significantly. The lower levels of *AtMYB33* are likely the result of miRNA-mediated destruction. Levels of miR159 appear to fluctuate with changes in the GA signaling pathway, as well as in response to other hormones like auxin and ethylene, which can affect DELLA protein levels (Achard *et al.*, 2004). *AtMYB33*, which contains a putative GARE motif, positively regulates miR159, suggesting the possibility of a negative feedback loop that prevents over-expression, and therefore premature flowering.

In *Arabidopsis* GA is thought to overcome *SOC1* repression by targeting the *DELLA* proteins for ubiquitination, which leads to protein degradation in the proteasome (Achard *et al.*, 2004). Degradation of *GAI* and *RGA* alleviates floral repression and allows for induction of *LFY* through both *SOC1* and *AtMYB33* promotion (Figure 2.5). The pathway of DELLA protein repression, GA mediated destruction of DELLA proteins and microRNA regulation has been found in other developmental pathways as well as other plant species, suggesting that the conserved regulatory pathway has been co-opted to serve different developmental purposes.

Changes in concentrations of hormones or nutrients can influence floral evocation (Rideout *et al.*, 1992). GA has a major influence on floral transition in *Arabidopsis*, with a complex balance of gene products that promote and repress flowering. Other hormones in *Arabidopsis* may play a role in flowering, as mutants for ethylene and ABA signaling show flowering time defects. Although GA can also promote flowering in species as different from *Arabidopsis* as the grass *Lolium* (King and Evans, 2003), the flower inducing effect of GA is not universal to all plants. Cytokinins also

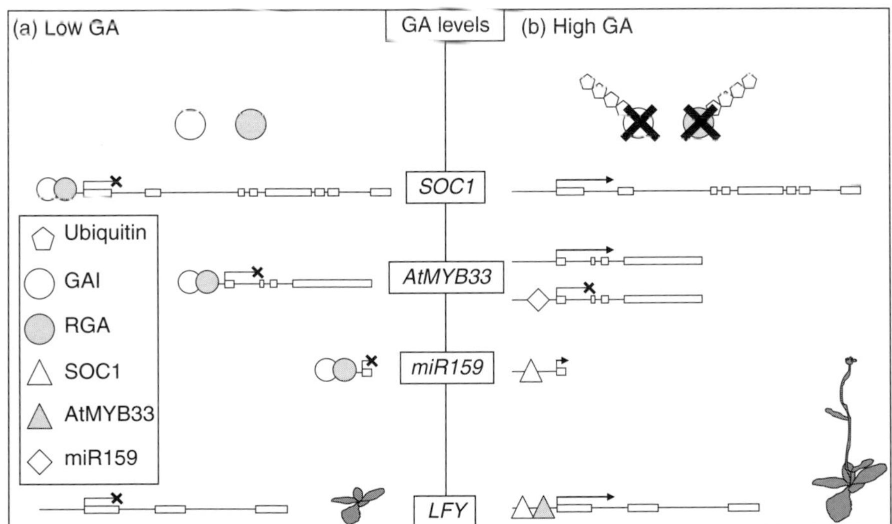

Figure 2.5 Gibberellin-mediated regulation of floral induction in *Arabidopsis*.

have a floral inductive effect in many species, and an elaborate model that postulates the movement of a combination of cytokinins with sucrose to the apex to induce flowering has been proposed (Bernier and Perilleux, 2005).

2.7.3 Long-distance floral inductive signals

The discovery that floral inductive signals move from the leaves to the shoot apex was made at a time when different aspects of plant growth were found to be controlled by a handful of relatively simple compounds common to all plants – the phytohormones. This convergence of discoveries may have lead to the idea that the floral inductive signal was a universal phytohormone as well. A hypothetical flower-inducing substance, dubbed florigen by Chailakhyan (1936), was believed to have specific features: (i) it is produced in leaves and moves to the shoot apex, (ii) it is transported via the phloem and cannot pass through non-living tissue, (iii) it has a measurable rate of movement and (iv) like other phytohormones, it serves the same function in all plants, i.e. to induce transition to flowering. However, a defined biochemical substance that meets all these criteria has not been found. A fifth criterion might be that florigen cannot be isolated by standard biochemical means.

Genetic and molecular evidence has so far not yielded many insights into the nature of long-distance flower-inducing signals. However, recent findings, such as the phloem-specific activity of the *Arabidopsis* CO gene described earlier, are beginning to make headway in this effort. Another example of a genetic approach that may provide some insight into long-distance floral inductive signals comes from research in maize. The maize *INDETERMINATE1* gene *ID1* is an important regulator of flowering time that is expressed exclusively in developing leaves (Colasanti *et al.*, 1998). Severe *id1* mutants remain in a prolonged vegetative state and eventually produce inflorescences with vegetative characteristics. ID1 is the only flowering time regulator discovered so far that accumulates exclusively in leaf tissue. This unique spatial and temporal expression pattern suggests that ID1 could be involved in the generation or transmission of a leaf-derived floral inductive signal (Colasanti *et al.*, 1998; Colasanti and Sundaresan, 2000). Similar to many of the flowering time genes isolated from other plant species, the ID1 protein encodes a zinc-finger protein with DNA-binding activity (Kozaki *et al.*, 2004). Genes with a high level of similarity to maize ID1 are found in all higher plants, although none of them has been found to control flowering. Identification of ID1 target genes should shed some light on the molecular and biochemical nature of the floral stimulus.

One intriguing possibility for the floral stimulus has become apparent with the discovery of the role of microRNAs in controlling plant development. A recent study reports the isolation of small RNA molecules in phloem sap – presumably these molecules could be transmitting information from one part of the plant to another (Yoo *et al.*, 2004). The idea that the microRNAs could carry floral inductive information from leaf to SAM opens up some interesting speculation on the nature of florigen (Corbesier and Coupland, 2005). An miRNA-based floral stimulus also might fit most of the criteria for florigen described earlier.

2.7.4 Integration and commencement of the floral transition

Regardless of the origin of the signal, the ultimate purpose of floral induction is to convert the SAM from a vegetative to a reproductive state (Perilleux and Bernier, 2002). In *Arabidopsis*, this process is genetically defined, with several genes identified that are important for floral transition. The genes described here function at the threshold between floral induction and the development events that give rise to the flowers. The accompanying review by Kramer (2005) describes the molecular basis of flower development.

2.7.4.1 LEAFY *and* APETALA1

In severe mutants of Arabidopsis *LEAFY* (*LFY*) and *APETALA1* (*AP1*) genes floral apices are replaced by organs with vegetative characteristics (Irish and Sussex, 1990; Schultz and Haughn, 1991). *LFY* is a transcription factor unique to higher plants that directly targets floral meristem identity genes like *AP1* and the *AP1*-related factor *CAULIFLOWER* (*CAL*), as well as other downstream targets (William *et al.*, 2004). *LFY* has a role in the integration of signals from different floral inductive pathways and activation of floral meristem identity genes, such as *AP1* (William *et al.*, 2004).

2.7.4.2 TERMINAL FLOWER 1

TERMINAL FLOWER 1 (*TFL1*) maintains the SAM in a vegetative state. *TFL1* is expressed in the shoot apex during the juvenile period, and this expression continues during flower development (Alvarez *et al.*, 1992). Mutant *tfl1* plants flower early and, instead of maintaining an indeterminate inflorescence meristem, produce a terminal flower. *TFL1* inhibits the expression of *AP1* and *LFY*, as demonstrated by knockout *tfl1* plants showing ectopic *AP1* and *LFY* expression, and in over-expression experiments, where *AP1* and *LFY* mRNA levels are greatly reduced (Liljegren *et al.*, 1999). *AP1* and *LFY* knockout and over-expression analyses show similar changes in *TFL1* expression. Unlike many of the genes involved in meristem identity, *TFL1* does not encode a transcription factor, although it may act in a signal transduction pathway by inhibiting a mitogen-activated protein kinase pathway.

2.7.4.3 Conservation of LFY *function in higher plants*

The identification of *LFY* homologues with similar roles in the floral transition and specifying flower structure in diverse species shows a conservation of function (Veit *et al.*, 2004). The discovery that *LFY* homologues in maize, *ZFL1* and *ZFL2,* have roles in both transition to flowering and in proper organization of floral meristems confirms the broad extent of LFY conserved function (Bomblies *et al.*, 2003).The discovery of LFY genes with similar functions in both monocot and dicot species supports the conserved role of this gene family in integrating floral inductive signals and regulating reproductive meristem architecture.

2.8 Perspective

The immobile lifestyle of plants requires that they be exquisitely responsive to both internal and external influences to insure optimal flowering time. Integration of different floral inductive signals provides a stable way to assess both external and internal conditions and allow for reproductive success. Functional genomics studies have found that some components of floral evocation are conserved in many dicot and monocot species, implying that the process of transition is ancient. On the other hand, floral induction pathways in some plants are controlled by genes that do not perform the same function in other species. This suggests that the transition mechanism can take diverse forms and depends on the specific developmental requirements of the plant. Current research is focused on understanding the nature of floral inductive signals and their connection to the molecular machinery that brings about flower formation.

References

Aasland, R., Gibson, T. J. and Stewart, A. F. (1995) *Trends in Biochemical Sciences*, **20**, 56–59.
Achard, P., Herr, A., Baulcombe, D. and Harberd, N. P. (2004) *Development*, **131**, 3357–3365.
Alvarez, J., Guli, C. L., Yu, X.-H. and Smyth, D. R. (1992) *The Plant Journal*, **2**, 103–116.
An, H., Roussot, C., Suarez-lopez, P. *et al.* (2004) *Development*, **131**, 3615–3626.
Aubert, D., Chen, L. J., Moon, Y. H. *et al.* (2001) *The Plant Cell*, **13**, 1865–1875.
Ausin, I., Alonso-Blanco, C., Jarillo, J. A., Ruiz-Garcia, L. and Martinez-Zapater, J. M. (2004) *Nature Genetics*, **36**, 162–166.
Ayre, B. G. and Turgeon, R. (2004) *Plant Physiology*, **135**, 2271–2278.
Baulcombe, D. (2004) *Nature*, **431**, 356–363.
Bernier, G. and Perilleux, C. (2005) *Plant Biotechnology Journal*, **3**, 3–16.
Bernier, G., Kinet, J.-M. and Sachs, R. M. (1981) *The Physiology of Flowering*, CRC Press, Boca Raton, Florida.
Blazquez, M. A. and Weigel, D. (1999) *Plant Physiology*, **120**, 1025–1032.
Blazquez, M. A., Green, R., Nilsson, O., Sussman, M. R. and Weigel, D. (1998) *The Plant Cell*, **10**, 791–800.
Bodson, M. (1977) *Planta*, **135**, 19–23.
Bodson, M. and Outlaw Jr., W. H. (1985) *Plant Physiology*, **79**, 420–424.
Bomblies, K., Wang, R. L., Ambrose, B. A., Schmidt, R. J., Meeley, R. B. and Doebley, J. (2003) *Development*, **130**, 2385–2395.
Boss, P. K., Bastow, R. M., Mylne, J. S. and Dean, C. (2004) *The Plant Cell*, **16 (Suppl.)**, S18–S31.
Casal, J. J., Fankhauser, C., Coupland, G. and Blazquez, M. A. (2004) *Trends in Plant Science*, **9**, 309–314.
Cercos, M., Gomez-Cadenas, A. and Ho, T. H. D. (1999) *The Plant Journal*, **19**, 107–118.
Cerdan, P. D. and Chory, J. (2003) *Nature*, **423**, 881–885.
Chailakhyan, M. K. (1936) *Comptes Rendus (Dokl.) Acad. Sci. U.R.S.S.*, **13**, 79–83.
Chandler, J. and Dean, C. (1994) *Journal of Experimental Botany*, **278**, 1279–1288.
Chanvivattana, Y., Bishopp, A., Schubert, D. *et al.* (2004) *Development*, **131**, 5263–5276.
Chardon, F., Virlon, B., Moreau, L. *et al.* (2004) *Genetics*, **168**, 2169–2185.
Cheng, H., Qin, L., Lee, S. *et al.* (2004) *Development*, **131**, 1055–1064.
Clayton, W. D. and Renvoize, S. A. (1986) *Genera Graminum: Grasses of the World*, Royal Botanic Gardens, Kew, London.

Colasanti, J. and Sundaresan, V. (2000) *Trends in Biochemical Sciences*, **25**, 236–240.
Colasanti, J., Yuan, Z. and Sundaresan, V. (1998) *Cell*, **93**, 593–603.
Corbesier, L. and Coupland, G. (2005) *Plant Cell and Environment*, **28**, 54–66.
Deitzer, G. F. (1984) In *Light and the Flowering Process* (ed., K. E. Cockshull), Academic Press Inc., Orlando, Florida, pp. 51–63.
Dill, A. and Sun, T. (2001) *Genetics*, **159**, 777–785.
Doi, K., Izawa, T., Fuse, T. *et al.* (2004) *Genes & Development*, **18**, 926–936.
Evans, L. T. (1969) In *The Induction of Flowering – Some Case Histories* (ed., L. T. Evans), Cornell University Press, Ithaca, New York, pp. 1–13.
Francis, N. J. and Kingston, R. E. (2001) *Nature Reviews Molecular Cell Biology*, **2**, 409–421.
Garner, W. W. and Allard, H. A. (1920) *Journal of Agricultural Research*, **18**, 553–606.
Gassner, G. (1918) *Zeitschr Bot.*, **10**, 417–480.
Gendall, A. R., Levy, Y. Y., Wilson, A. and Dean, C. (2001) *Cell*, **107**, 525–535.
Gocal, G. F. W., Sheldon, C. C., Gubler, F. *et al.* (2001) *Plant Physiology*, **127**, 1682–1693.
Gubler, F., Kalla, R., Roberts, J. K. and Jacobsen, J. V. (1995) *The Plant Cell*, **7**, 1879–1891.
Haung, M. D. and Yang, C. H. (1998) *Plant and Cell Physiology*, **39**, 382–393.
Hayama, R. and Coupland, G. (2003) *Current Opinion in Plant Biology*, **6**, 13–19.
Hayama, R., Yokoi, S., Tamaki, S., Yano, M. and Shimamoto, K. (2003) *Nature*, **422**, 719–722.
He, Y. H., Michaels, S. D. and Amasino, R. M. (2003) *Science*, **302**, 1751–1754.
Hepworth, S. R., Valverde, F., Ravenscroft, D., Mouradov, A. and Coupland, G. (2002) *EMBO Journal*, **21**, 4327–4337.
Irish, E. and Jegla, D. (1997) *The Plant Journal*, **11**, 63–71.
Irish, V. F. and Sussex, I. M. (1990) *The Plant Cell*, **2**, 741–753.
King, R. W. and Ben-Tal, Y. (2001) *Plant Physiology*, **125**, 488–496.
King, R. W. and Evans, L. T. (2003) *Annual Review of Plant Biology*, **54**, 307–328.
Knott, J. E. (1934) *Proceedings of the American Society of Horticultural Science*, **31**, 152–154.
Kojima, S., Takahashi, Y., Kobayashi, Y. *et al.* (2002) *Plant and Cell Physiology*, **43**, 1096–1105.
Koornneef, M., Alonso-Blanco, C., Blankestijn-de Vries, H., Hanhart, C. J. and Peeters, A. J. (1998a) *Genetics*, **148**, 885–892.
Koornneef, M., Alonso-Blanco, C., Peeters, A. J. M. and Soppe, W. (1998b) *Annual Review of Plant Physiology and Plant Molecular Biology*, **49**, 345–370.
Koornneef, M., Hanhart, C. J. and van den Veen, J. H. (1991) *Molecular General Genetics*, **229**, 57–66.
Kozaki, A., Hake, S. and Colasanti, J. (2004) *Nucleic Acids Research*, **32**, 1710–1720.
Lang, A., ed. (1965) *Physiology of Flower Initiation, Differentiation and Development*, Springer, Berlin.
Lang, A. (1977) *Proceedings of the National Academy of Sciences of the United States of America*, **74**, 2412–2416.
Lee, I., Aukerman, M. J., Gore, S. L. *et al.* (1994) *The Plant Cell*, **6**, 75–83.
Lejeune, P., Bernier, G., Requier, M. C. and Kinet, J. M. (1993) *Planta*, **190**, 71–74.
Levy, Y. Y., Mesnage, S., Mylne, J. S., Gendall, A. R. and Dean, C. (2002) *Science*, **297**, 243–246.
Liljegren, S. J., Gustafson-Brown, C., Pinyopich, A., Ditta, G. S. and Yanofsky, M. F. (1999) *The Plant Cell*, **11**, 1007–1018.
Lin, C. (2000) *Plant Physiology*, **123**, 39–50.
Lin, H. X., Ashikari, M., Yamanouchi, U., Sasaki, T. and Yano, M. (2002) *Breeding Science*, **52**, 35–41.
Lin, H. X., Yamamoto, T., Sasaki, T. and Yano, M. (2000) *Theoretical and Applied Genetics*, **101**, 1021–1028.
Liu, J. Y., Yu, J. P., McIntosh, L., Kende, H. and Zeevaart, J. A. D. (2001) *Plant Physiology*, **125**, 1821–1830.
Martinez-Zapater, J. M. and Somerville, C. R. (1990) *Plant Physiology*, **92**, 770–776.
McDaniel, C. N. (1980) *Planta*, **148**, 462–467.
Michaels, S. D. and Amasino, R. M. (1999) *The Plant Cell*, **11**, 949–945.
Millar, A. J. (2004) *Journal of Experimental Botany*, **55**, 277–283.
Mizoguchi, T., Wheatley, K., Hanzawa, Y. *et al.* (2002) *Developmental Cell*, **2**, 629–641.

Mouradov, A., Cremer, F. and Coupland, G. (2002) *The Plant Cell*, **14**, S111–S130.
Murai, K., Miyamae, M., Kato, H., Takumi, S. and Ogihara, Y. (2003) *Plant and Cell Physiology*, **44**, 1255–1265.
Onouchi, H., Igeno, M. I., Perilleux, C., Graves, K. and Coupland, G. (2000) *The Plant Cell*, **12**, 885–900.
Perilleux, C. and Bernier, G. (2002) In *Plant Reproduction, Annual Plant Reviews*, Vol. 6 (ed., J. A. Roberts), Sheffield Academic Press, Sheffield, England, pp. 1–32.
Poethig, R. S. (1990) *Science*, **250**, 923–930.
Putterill, J., Laurie, D. A. and Macknight, R. (2004) *Bioessays*, **26**, 363–373.
Putterill, J., Robson, F., Lee, K., Simon, R. and Coupland, G. (1995) *Cell*, **80**, 847–857.
Raper, C. D. J., Thomas, J. F., Tolley-Henry, L. and Rideout, J. W. (1988) *Botanical Gazette*, **149**, 289–294.
Reinhart, B. J., Weinstein, E. G., Rhoades, M. W., Bartel, B. and Bartel, D. P. (2002) *Genes & Development*, **16**, 1616–1626.
Rideout, J. W., Raper, C. D. J. and Miner, G. S. (1992) *International Journal of Plant Sciences*, **153**, 78–88.
Robert, L. S., Robson, F., Sharpe, A., Lydiate, D. and Coupland, G. (1998) *Plant Molecular Biology*, **37**, 763–772.
Robson, F., Costa, M. M. R., Hepworth, S. R. *et al.* (2001) *The Plant Journal*, **28**, 619–631.
Sachs, R. M. and Hackett, W. P. (1983) In *Strategies of Plant Reproduciton* (ed., W. J. Meudt), Allenheld-Osmun, Totowa, New Jersey.
Salvi, S., Tuberosa, R., Chiapparino, E. *et al.* (2002) *Plant Molecular Biology*, **48**, 601–613.
Samach, A., Onouchi, H., Gold, S. E. *et al.* (2000) *Science*, **288**, 1613–1616.
Schultz, E. A. and Haughn, G. W. (1991) *The Plant Cell*, **3**, 771–781.
Sheldar, C. C., Rouse, D. T., Finnegan, E. J., Peacock, W. J. and Dennis, E. S. (2000) *Proceedings of the National Academy of Sciences of the United States of America*, **97**, 3753–3758.
Simpson, G. (2004) *Current Opinion in Plant Biology*, **7**, 1–5.
Simpson, G., Dijkwel, P., Quesada, V., Henderson, I. and Dean, C. (2003) *Cell*, **113**, 777–787.
Simpson, G. G. and Dean, C. (2002) *Science*, **296**, 285–289.
Smith, H. M. S., Campbell, B. C. and Hake, S. (2004) *Current Biology*, **14**, 812–817.
Somers, D. E., Webb, A. A., Pearson, M. and Kay, S. A. (1998) *Development*, **125**, 485–494.
Suarez-Lopez, P., Wheatley, K., Robson, F., Onouchi, H., Valverde, F. and Coupland, G. (2001) *Nature*, **410**, 1116–1120.
Sung, S. B. and Amasino, R. M. (2004) *Current Opinion in Plant Biology*, **7**, 4–10.
Telfer, A. and Poethig, R. S. (1998) *Development*, **125**, 1889–1898.
Tournois, J. (1912) *Comptes Rendus de l' Academie Sciences de Paris*, **155**, 297–300.
Trevasksis, B., Bagnall, D. J., Ellis, M. H., Peacock, W. J. and Dennis, E. S. (2003) *Proceedings of the National Academy of Sciences of the United States of America*, **100**, 13099–13104.
Valverde, F., Mouradov, A., Soppe, W., Ravenscroft, D., Samach, A., and Coupland, G. (2004) *Science*, **303**, 1003–1006.
Vega, S. H., Sauer, M., Orkwiszewski, J. A. J. and Poethig, R. S. (2002) *The Plant Cell*, **14**, 133–147.
Veit, J., Wagner, E. and Albrechtova, J. T. (2004) *Plant Physiology and Biochemistry*, **42**, 573–578.
Wang, Z. Y. and Tobin, E. M. (1998) *Cell*, **93**, 1207–1217.
William, D. A., Su, Y. H., Smith, M. R., Lu, M., Baldwin, D. A. and Wagner, D. (2004) *Proceedings of the National Academy of Sciences of the United States of America*, **101**, 1775–1780.
Wilson, R. N., Heckman, J. W. and Somerville, C. R. (1992) *Plant Physiology*, **100**, 403–408.
Yan, L., Helguera, M., Kato, K., Fukuyama, S., Sherman, J. and Dubcovsky, J. (2004a) *Theoretical and Applied Genetics*, **109**, 1677–1686.
Yan, L., Loukoianov, A., Blechl, A. *et al.* (2004b) *Science*, **303**, 1640–1644.
Yano, M., Katayose, Y., Ashikari, M. *et al.* (2000) *The Plant Cell*, **12**, 2473–2483.
Yoo, B. C., Kragler, F., Varkonyi-Gasic, E. *et al.* (2004) *The Plant Cell*, **16**, 1979–2000.

3 Floral patterning and control of floral organ formation

Elena M. Kramer

3.1 Introduction

The flower is the most recognizable feature of the angiosperms, which are the dominant form of plant life on land today. The majority of flowers possess four types of floral organs: two outer whorls of sterile organs, the sepals and petals (also known as the perianth), and two inner whorls of fertile organs, the male stamens and female carpels, with the carpels positioned centrally. Although this organization is strictly adhered to in the major dicot model species (Figure 3.1), great variation, affecting both organ position and type, is observed in other taxa. Before we can consider this diversity, however, we need to understand the fundamental genetic program controlling floral organ identity. Luckily, work conducted over the last 15 years, particularly in the model species *Arabidopsis* and *Antirrhinum*, has greatly enhanced our understanding of these pathways. This research has focused on the so-called ABC model, which can be thought of as the central unifying principle of floral development (Jack, 2004). This chapter will outline the present state of knowledge on the function, regulation and conservation of the genes that participate in the ABC program as they have been studied in model species, particularly *Arabidopsis* and *Antirrhinum*. What will become apparent is that although the ABC model is striking in its simplicity, the genetic mechanisms underlying the program are quite complex.

3.2 The ABC model of floral organ identity

During the late 1980s, forward mutagenesis studies of *Arabidopsis* and *Antirrhinum* in several laboratories uncovered an intriguing series of homeotic floral mutants (Komaki *et al.*, 1988; Bowman *et al.*, 1989; Coen *et al.*, 1991). In both taxa, the mutants appeared to fall into similar classes: mutations that affected sepal and petal identity were placed into what was termed the 'A' class; those that affected petal and stamen identity, the 'B' class; and those that affected stamen and carpel identity, the 'C' class. For instance, B mutants exhibited the transformation of petals into sepals and stamens in carpels (Bowman *et al.*, 1989; Carpenter and Coen, 1990). Analysis of double and triple mutants (Bowman *et al.*, 1991) led to the proposition of a simple and elegant model that explained the major aspects of genetic interactions among the loci; this became known as the ABC model (Figure 3.2; Coen and

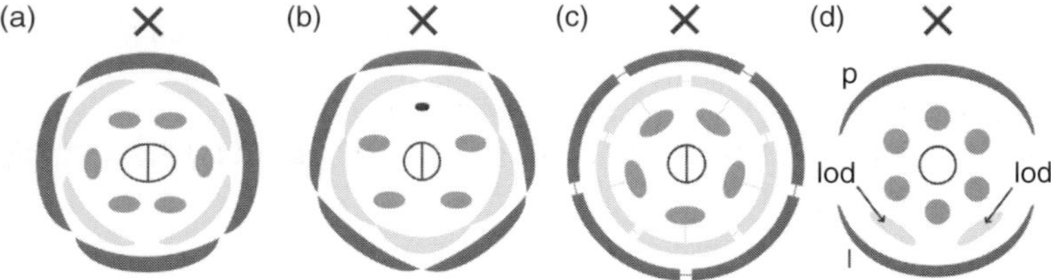

Figure 3.1 Floral diagrams indicating the arrangement and number of organs in the flowers of *Arabidopsis* (a), *Antirrhinum* (b), *Petunia* (c) and *Oryza* (d). In (a–c), the first whorl represents sepals; the second whorl, petals; the third whorl, stamens; and the fourth, carpels. In (d), the palea (p) and lemma (l) are external to the lodicules (lod), stamens and carpels. The X indicates the position of the inflorescence axis.

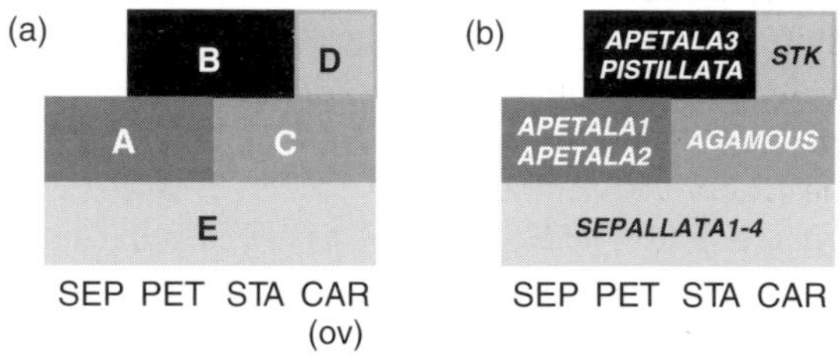

Figure 3.2 (a) Schematic representing the ABC program of floral organ identity, with addition of D and E class genes as suggested by Theissen (2001). (b) *Arabidopsis* representatives for the A, B, C and E gene classes.

Meyerowitz, 1991). Fundamentally, the ABC model holds that the overlapping domains of three classes of gene activity, referred to as A, B and C, produce a combinatorial code that determines floral organ identity in successive whorls of the developing flower. Another critical component of the ABC program is that A and C functions are mutually exclusive (Bowman *et al.*, 1991), such that elimination of C gene activity causes the A domain to expand and vice versa (Drews *et al.*, 1991; Gustafson-Brown *et al.*, 1994). Additional studies of mutants and over-expressing lines have largely confirmed the model and demonstrated the completely homeotic nature of this developmental program (Bowman *et al.*, 1991; Mizukami and Ma, 1992; Krizek and Meyerowitz, 1996a). As our understanding of the genes involved in the program has grown, the model has expanded as well to what is now referred to as the ABCDE model (Theissen, 2001). 'D' class genes were proposed as ovule identity genes based on work done in *Petunia* (Colombo *et al.*, 1995) while 'E' class genes function broadly across the floral meristem to facilitate the function of many of the original ABC loci (Pelaz *et al.*, 2000; Ditta *et al.*, 2004).

3.2.1 The major genetic players in the ABC model

Molecular identification of the genes corresponding to the primary ABC mutants revealed that genetically homologous loci generally control similar aspects of floral identity in *Arabidopsis* and *Antirrhinum* (Table 3.1; reviewed by Weigel and Meyerowitz, 1994; Lohmann and Weigel, 2002). In *Arabidopsis*, the A class genes are represented by *APETALA1* (*AP1*) and *APETALA2* (*AP2*), which have early roles in the determination of floral meristem identity (see Chapter 2) but are thought to be required for sepal and petal identity as well. The B class genes, *APETALA3* (*AP3*) and *PISTILLATA* (*PI*), are responsible for the establishment of petal and stamen identity in the second and third whorls, respectively. *AGAMOUS* (*AG*) – the C class gene – is necessary for stamen and carpel identity, but is also required to specify the determinacy of the floral meristem. The D class in *Arabidopsis* corresponds to the gene *SEEDSTICK* (*STK*, see below; Pinyopich *et al.*, 2003) while the E class is comprised of a set of four paralogs known as *SEPALLATA1*-4 (*SEP1*-4) (Pelaz *et al.*, 2000; Ditta *et al.*, 2004). In *Antirrhinum*, the functions of the C class gene *PLENA* (*PLE*) and the B class genes *DEFICIENS* (*DEF*) and *GLOBOSA* (*GLO*) are very similar to those of their *Arabidopsis* homologs (reviewed Irish and Kramer, 1998). Although mutant forms of *SEP*-like genes have not yet been isolated in *Antirrhinum*, their genetic homologs have similar patterns of gene expression and protein interaction (Davies *et al.*, 1996). The A class genes do not exhibit direct functional equivalency, however. *LIP1* and *LIP2*, which are the *Antirrhinum* homologs of *AP2*, contribute to organ identity but have not yet been ascribed a role in floral meristem identity (Keck *et al.*, 2003). Conversely, the *Antirrhinum AP1*-like gene *SQUAMOSA* (*SQUA*) clearly functions as a determinant of floral meristem identity but appears to be dispensable for organ identity (Huijser *et al.*, 1992). Much less is known about the conservation of function among homologs of *STK*, which remain to be described in *Antirrhinum*.

3.2.2 Members of the MADS-box transcription factor family

With the exception of *AP2*, all of the organ identity genes identified to date are members of the pan-eukaryotic MADS transcription factor family (reviewed by Becker and Theissen, 2003; Messenguy and Dubois, 2003). More specifically, they are type II MADS proteins, which are characterized by a distinct domain structure (Figure 3.3a; Alvarez-Buylla *et al.*, 2000). Our understanding of the biochemical nature of ABC gene function is based on the roles of each of these domains. DNA binding at sequence elements known as CArG boxes is controlled by the highly conserved *N*-terminal MADS domain (Riechmann *et al.*, 1996b). This only occurs, however, when the proteins are dimerized, which is primarily mediated by the adjacent I and K domains (Riechmann *et al.*, 1996a). Different dimerization preferences are observed between proteins and this appears to be important for determining functional specificity (Riechmann *et al.*, 1996a; Krizek and Meyerowitz, 1996b). The AP3 and PI gene products are known to function as obligate heterodimers while AP1 and AG have broader interaction potentials (Riechmann *et al.*, 1996a; Davies *et al.*, 1996). The current model is that AP1/SEP and AG/SEP are the critical heterodimers functioning in floral organ identity (Honma and Goto, 2001; Theissen and Saedler, 2001). These various dimer combinations are now thought to associate

Table 3.1 ABCDE Gene Homologs of the Major Model Species

Gene class	*Arabidopsis*	*Antirrhinum*	*Petunia*	*Oryza*	Gene function*
A class genes	*APETALA1* *APETALA2*	*SQUAMOSA* *LIP1, LIP2*	*FBP26, FBP29* *PhAp2A*	*RAP1* *OSJNBa0010D21.13*	Floral meristem identity, sepal and petal identity
B class genes	*APETALA3* *PISTILLATA*	*DEFICIENS* *GLOBOSA*	*PhDEF, PhTM6* *PhGLO1, PhGLO2*	*SUPERWOMAN1* *OsMADS2, OsMADS4*	Petal and stamen identity
C class genes	*AGAMOUS* *SHP1/2*	*FARINELLI* *PLENA*	*pMADS3* *FBP6*	*OsMADS3*	Stamen and carpel identity, floral meristem determinacy
D class genes	*SEEDSTICK*	nk	*FBP7, FBP11*	*OsMADS13*	Ovule identity
E class genes	*SEPALLATA1* *SEPALLATA2* *SEPALLATA3*	*DEFH49* *DEFH72*	*FBP2, FBP4,* *FBP5, FBP9,* *FBP23, pMADS12*	*LHS1, OsMADS34* *OsMADS5,*	Facilitation of ABC gene function

nk = not known.

*Gene function has not been established for all of these loci. The listed gene function represents the functional repertoire that appears to be conserved for members of these lineages (see text for references).

Note It is important to remember that in some cases the evolutionary relationships among homologs are such that simple genetic orthology does not exist. This is particularly true in comparisons between the core eudicot and monocot model species. Similarly, morphological homology has been inferred between the core eudicot petal and grass lodicule, but it should be noted that the phylogenetic distance in this comparison is such that the assignment of 'sameness' may be an over-simplification.

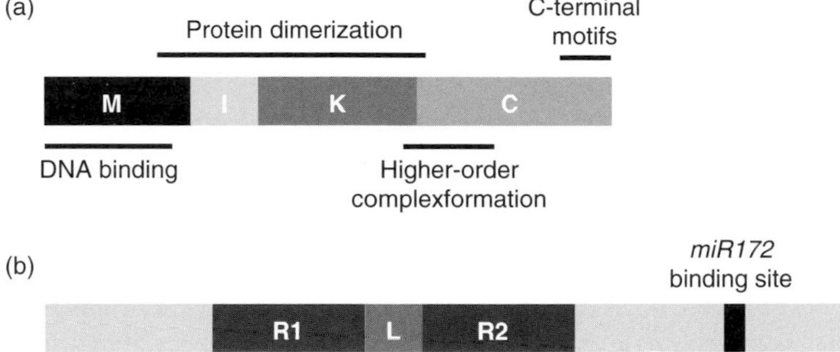

Figure 3.3 (a) Schematic representation of a typical MIKC-type MADS-box gene. The general functional role of each protein domain is noted above and below the schematic. (b) Schematic representation of a typical AP2 subfamily member. R1 and R2 are the AP2 domain repeats which are joined by the Linker region (L). The evolutionarily conserved *miR172* binding site is indicated. See text for references on both gene types.

in larger complexes, an interaction mediated at least in part by the C-terminal domain of the proteins (Egea-Cortines *et al.*, 1999; Honma and Goto, 2001). This region exhibits much lower levels of overall sequence conservation (Purugganan *et al.*, 1995), but has previously been shown to be essential for proper gene function (Krizek and Meyerowitz, 1996b). The so-called 'quartet' model holds that tetramers, consisting of two MADS protein dimers, are responsible for the specification of organ identity in each whorl (Theissen and Saedler, 2001). For instance, in the second whorl, AP3/PI dimers would associate with AP1/SEP dimers to control petal identity. Presumably, differentiation of organ identities would result from the distinct DNA-binding specificities of each complex (Egea-Cortines *et al.*, 1999). It is important to note that although this 'quartet' model is very attractive, it is currently supported by limited direct data (Jack, 2004). The C domain is also significant due to the presence of short, highly conserved motifs that are lineage specific. These sequences have been implicated in transcriptional activation in some cases (Moon *et al.*, 1999; Honma and Goto, 2001) and aspects of functional specificity in others (Lamb and Irish, 2003).

3.2.2.1 Redundant and complex functions among the floral MIKC MADS-box genes

As we have learned more about the MADS-box gene family and the many roles they play in floral development, it has become clear that their functions are often not as straightforward as the ABC model would suggest. This is due in large part to multiple levels of functional redundancy, often between evolutionarily related genes. In *Arabidopsis* for instance, *AG* shares a role in carpel and ovule identity with its paralogs *SHATTERPROOF1*/2 (SHP1/2) and *SEEDSTICK* (*STK*) (Pinyopich *et al.*, 2003). The closely related *SHP1* and *2* loci are completely redundant with each other but double *shp1 shp2* mutants exhibit defects in valve margin differentiation and fruit dehiscence (Liljegren *et al.*, 2000). Similarly, the phenotype of *stk* mutants would suggest that the gene has a very limited role, in this case controlling the abscission of seeds from the mature fruit (Pinyopich *et al.*, 2003). Combinations of these mutants, however, reveal that all of the loci contribute to ovule identity and development. Furthermore, they are capable of promoting carpel identity, even in the absence of *AG*. In *Antirrhinum*, paralogs of *AG*-like genes have also been found to display some functional redundancy. *PLE*, the original C function gene in *Antirrhinum*, shares stamen identity functions with the closely related gene *FARINELLI* (*FAR*) (Davies *et al.*, 1999). It is even possible that *FAR* is the primary stamen identity gene in *Antirrhinum* since the *ple* mutant shows reduction of *FAR* expression, thereby reflecting loss of function in both genes. The four *SEP* loci in *Arabidopsis* present similar patterns of significant overlap in their functional repertoires combined with some unique functions (Pelaz *et al.*, 2000; Ditta *et al.*, 2004). The complex evolutionary relationships among the loci in the *SEP* and *AG* subfamilies (Zachgo *et al.*, 1995; Kramer *et al.*, 2004) serve to highlight the variability that can be introduced into an otherwise conserved genetic program when gene duplication has occurred at many phylogenetic levels.

In this context, it is important to consider some complicated findings regarding *AP1* function. The difficulty in understanding *AP1*'s contribution to organ development is a result of several factors, including its dual role in separate stages of floral meristem development, and the lack of one-to-one correspondence between 'A' gene functions in *Arabidopsis* and *Antirrhinum* (Gutierrez-Cortines and Davies, 2000). In *ap1* mutants, the early flowers are transformed into inflorescence meristems while later flowers show a mixed inflorescence/floral identity with bracts subtending axillary meristems in the outer whorls and normal stamens and carpels in the inner whorls (Irish and Sussex, 1990). Consistent with this phenotype, *AP1* is expressed throughout the early developing floral meristem but becomes restricted to the outer two whorls soon after the initiation of the sepal primordia (Mandel *et al.*, 1992). For some time it has been known that *AP1* has partial functional redundancy with a very closely related paralog, the *CAULIFLOWER* (*CAL*) gene, which contributes to floral meristem identity and exhibits an expression pattern similar to that of *AP1* (Bowman *et al.*, 1993). It was not surprising then to find that a more distantly related *AP1*-like gene *FRUITFULL* (*FUL*) can also promote the establishment of floral meristem identity (Ferrandiz *et al.*, 2000). Interestingly, however, expression of *FUL* is not normally overlapping with that of *AP1*, being initially detected in the inflorescence meristem and only appearing in presumptive fourth whorl of the developing floral meristem (Mandel and Yanofsky, 1995). In fact, *FUL* is only expressed in the early floral meristem in the *ap1* mutant background, suggesting that *AP1* normally functions to restrict the expression of *FUL*. In the wild-type background, therefore, *AP1* and *FUL* may be controlling floral meristem identity via separate genetic pathways, rather than the simple functional redundancy seen with *AP1* and *CAL*.

Further insight into the complex *ap1* phenotype and the apparent differences between the *AP1* and *SQUA* functions has come from recent work in *Arabidopsis*. Analysis of *AGL24*, a MIKC-type MADS gene involved in flowering time control

(Yu *et al.*, 2002; Michaels *et al.*, 2003), indicates that aspects of the *ap1* phenotype are actually due to over-expression of *AGL24* (Yu *et al.*, 2004). *AP1* is required in the early floral meristem to limit expression of *AGL24*, which is thought to promote flowering and inflorescence meristem identity (Yu *et al.*, 2002). Double *agl24 ap1* mutants show a considerably improved phenotype relative to *ap1*, including significant rescue of organ identity defects. These findings indicate that *AP1* itself may not be absolutely required for sepal or petal identity, similar to what is observed with *SQUA* in *Antirrhinum* (Huijser *et al.*, 1992). One caveat to this suggestion is that any role for *AP1* in organ identity could be complemented by the presence of the *SEP* genes, which have some functional equivalency with *AP1* (Honma and Goto, 2001). Thus we see that due to the highly ramified nature of the MIKC-type MADS gene phylogeny (Becker and Theissen, 2003), understanding gene function is often not as simple as examining single gene mutants. It is necessary to examine multiple combinations of mutant alleles in all related loci and consider how changes in the expression of other genes may produce the phenotypes in question.

3.2.3 Members of the AP2/EREBP transcription factor family

In contrast to the detailed understanding of the MIKC-type genes, we know relatively little about the specific functions of *AP2*, which is a member of the AP2/EREBP family of transcription factors (Okamuro *et al.*, 1997a; Riechmann and Meyerowitz, 1998). The main characteristic of this family is the highly conserved AP2 domain, which is present as a single domain in the EREBP genes but as a tandem repeat in the AP2 subfamily (Figure 3.3b; Weigel, 1995). Recently, it was shown that this domain may have been derived by horizontal gene transfer from a cyanobacterial endonuclease (Magnani *et al.*, 2004). Limited phylogenetic analyses of the family indicate that it is ancient, with pre-seed plant divergences (~170 mya) even among the lineages of the AP2 subfamily (Shigyo and Ito, 2004). Unusually long DNA-binding sites have been identified for members of the subfamily and it appears that each AP2 domain repeat (R1 and R2) as well as the intervening linker region (L) contributes to DNA binding (Figure 3.3b; Nole-Wilson and Krizek, 2000; Kuhn, 2001).

Our understanding of *AP2* function has been slow to develop for a number of reasons, including the complexity of the gene family and the functional redundancies that exist among its members in *Arabidopsis* and *Antirrhinum*. In the case of *Arabidopsis*, many aspects of *AP2* function overlap with those of another subfamily member, the gene *AINTEGUMENTA* (*ANT*) (Elliott *et al.*, 1996). Both *AP2* and *ANT* are now known to contribute to floral meristem and petal identity, repression of *AG* in the outer whorls of the flower, and ovule/seed development (Drews *et al.*, 1991; Jofuku *et al.*, 1994; Krizek *et al.*, 2000). In contrast to the case with *AP2* and *ANT*, which are fairly ancient paralogs (Shigyo and Ito, 2004), the functionally redundant *LIP1* and *LIP2* of *Antirrhinum* are closely related (Keck *et al.*, 2003), raising the possibility that additional *Antirrhinum* homologs share functions with these two loci. A further complicating factor in understanding *AP2* function is that although the gene is known to be necessary for the repression of *AG* in the first two whorls of the flower (Drews *et al.*, 1991), its expression pattern is much broader (Jofuku *et al.*, 1994), suggesting post-transcriptional gene regulation. Along these lines, several co-factors have been identified as important for *AP2*-dependent repression of *AG*, including *LEUNIG* (*LEU*) and *SEUSS* (*SEU*) (Liu and Meyerowitz, 1995; Franks *et al.*, 2002; Sridhar *et al.*, 2004). In addition, it has recently been demonstrated that *AP2* function is restricted to the first two whorls via repression by a microRNA that is expressed in whorls 3 and 4 (Aukerman and Sakai, 2003; Chen, 2004). Given the fact that the recognition site for this microRNA, *miR172*, is conserved across divergent representatives of the AP2 subfamily in *Arabidopsis* (Aukerman and Sakai, 2003), it seems likely that this regulatory mechanism is similarly conserved across angiosperms.

3.3 Regulating the expression of the floral organ identity genes

Unlike the situation with *AP2*, the gene expression patterns of the floral MADS-box genes largely correspond to their domain of function: *AP1* is expressed throughout the early floral meristem but rapidly becomes restricted to whorls 1 and 2; *AP3* and *PI* are primarily expressed in whorls 2 and 3; and *AG* expression is limited to whorls 3 and 4 (reviewed Jack, 2004). Although these loci are unified by their common activation via the floral meristem identity gene *LEAFY* (*LFY*) (Parcy *et al.*, 1998), each gene has a distinct set of genetic controls. In the case of *AP1*, the gene is quantitatively dependent on *LFY* for activation such that eliminating *LFY* function significantly reduces but does not eliminate *AP1* expression (Wagner *et al.*, 1999). This activation by *LFY* does not seem to depend on other co-factors (Parcy *et al.*, 1998), which may be related in part to the fact that the LFY protein has very high affinity for its binding site in the *AP1* promoter (Maizel *et al.*, 2005). *LFY* and *AP1* are actually involved in a positive feedback loop where each gene enhances the expression of the other (Bowman *et al.*, 1993; Weigel and Nilsson, 1995), consistent with their cooperative role in determining floral meristem identity. Other components of the floral-promotion pathway may be responsible for up-regulation of *AP1* in the absence of *LFY* (Okamuro *et al.*, 1997b; Hempel *et al.*, 1998). Following the initiation of *AP1* expression across the floral meristem, the ensuing restriction of the *AP1* domain to the first and second whorls is due to the activity of *AG* in the inner whorls (Gustafson-Brown *et al.*, 1994). It remains to be determined whether the repressive effect of *AG* on *AP1* is due to a direct interaction. Similarly, we do not know if the negative miRNA-based regulation of the other A class gene, *AP2*, is genetically downstream of *AG*.

This relatively simple activation by *LFY* cannot be the case with *AP3*, *PI* or *AG* since each of these loci has its own distinct expression pattern. The expression of the B genes *AP3* and *PI* in the second and third whorls is traditionally separated into two phases: an initiation phase that is *LFY*-dependent and a maintenance phase that results from an autoregulatory feedback loop. During the initiation period, *LFY* up-regulates *AP3* with the assistance of the F-box protein *UNUSUAL FLORAL ORGANS* (*UFO*) (Lee *et al.*, 1997; Parcy *et al.*, 1998). In contrast to the broad expression of *LFY*, the expression of *UFO* is limited to the second and third whorls at the developmental stage when B gene expression first occurs (Ingram *et al.*, 1995; Lee *et al.*, 1997). Genetic and biochemical evidence suggests that UFO targets a repressor of *AP3* for degradation via the SKP1-cullin-F-box (SCF) complex (Lee *et al.*, 1997; Wang *et al.*, 2003), although no candidates for this repressor have been identified to date. *AP1* also contributes to the activation of *AP3* in a UFO-dependent manner (Ng and Yanofsky, 2001a). The genetic basis for *PI* activation is poorly understood by comparison. In particular, it remains unclear as to whether *LFY*/*UFO* directly activate *PI* or whether their requirement for *PI* expression is indirect (Honma and Goto, 2000; Chen *et al.*, 2000). Given the fact that the initial expression patterns of *AP3* and *PI* differ to some degree, with weak *AP3* expression in the first whorl and *PI* expression in the fourth whorl (Jack *et al.*, 1992; Goto and Meyerowitz, 1994), it would appear that there are genetic differences in their early regulation. What is certain is that the maintenance regulatory phase is dependent on the AP3/PI heterodimer itself, resulting in the restriction of *AP3* and *PI* expression to the same domain (Jack *et al.*, 1992; Goto and Meyerowitz, 1994). The dimer binds directly to the *AP3* promoter but acts indirectly on *PI* (Hill *et al.*, 1998; Honma and Goto, 2000). Further integration of *AP3* and *PI* function takes place post-translationally since the proteins are not stable *in vivo* as monomers (Jack *et al.*, 1994; Jenik and Irish, 2001) and also require dimerization for nuclear localization (McGonigle *et al.*, 1996).

The establishment of *AG* expression in whorls 3 and 4 similarly depends on *LFY* but requires an additional positive co-factor in the form of the homeodomain gene *WUSCHEL* (*WUS*) (Lohmann *et al.*, 2001; Lenhard *et al.*, 2001). *WUS* primarily functions to promote meristematic stem-cell identity and is expressed in the center of early floral meristems (Mayer *et al.*, 1998). At this stage, WUS has been shown to activate *AG* expression in the presumptive third and fourth whorls by binding to the *AG* promoter in conjunction with LFY (Lohmann *et al.*, 2001; Lenhard *et al.*, 2001). Apparently, once *AG* is activated, its expression becomes *WUS*-independent since *AG* mediates the subsequent down-regulation of *WUS* in the floral meristem. This negative feedback is a major component of *AG*'s role in promoting floral determinacy and explains the indeterminate nature of *ag* flowers. As discussed earlier, the expression of *AG* in the floral meristem is also regulated by the repressive activities of *AP2*, *ANT*, *LEU* and *SEU* in the first two whorls (Drews *et al.*, 1991; Liu and Meyerowitz, 1995; Elliott *et al.*, 1996; Conner and Liu, 2000; Franks *et al.*, 2002). Thus *LFY* is able to activate all the major components of the ABC model in distinct expression domains through cooperation with multiple cofactors. The resultant activities of the genes themselves further refine the pattern. This process, which remains in agreement with the original formation of the ABC model, has recently been computationally modeled to demonstrate that it represents a robust developmental module with predictable behavior (Espinoza-Soto *et al.*, 2004).

3.4 Conservation and modification of the ABC program

The detailed functional characterization of the MIKC-type organ identity genes in *Arabidopsis* and *Antirrhinum*, together with the genes' high degree of sequence conservation have made these loci prime targets for comparative studies of floral morphology across the angiosperms (reviewed Theissen *et al.*, 2000). Putting aside the noted difficulties regarding the A class genes, most aspects of *AP3*, *PI*, *AG* and *SEP* homolog function seem to be well conserved across the core eudicots (reviewed Becker and Theissen, 2003). In the two major grass model species, considerable conservation is also observed (Kang *et al.*, 1998; Kyozuka *et al.*, 2000; Kyozuka and Shimamoto, 2002), leading to the suggestion that the ABC model is applicable to all angiosperms (Ma and dePamphilis, 2000). Such analyses have tended to highlight conserved aspects of the program, but many complicating factors have also been uncovered. These include the very common occurrence of gene duplications, independent patterns of functional evolution, changes in aspects of protein biochemistry and gene regulation, and shifts in gene expression patterns. While these considerations may seem like annoyances, they have actually provided rich insight into the complexities of the evolution of floral developmental programs. In order to illustrate this point, it is useful to consider the extensive work that has been done in two alternative model species – the core eudicot *Petunia* and the derived monocot grasses, most notably *Oryza* (rice).

3.4.1 Floral organ identity gene function in Petunia

A full complement of floral organ identity gene homologs has been identified from *Petunia hybrida* (Table 3.1; Maes *et al.*, 2001; Immink *et al.*, 2003). The functional analysis of these loci has been advanced by the development of an effective insertional mutagenesis system that is being used for forward and reverse genetics (Vandenbussche *et al.*, 2003b). In addition to these tools, *Petunia* has advantages related to the gene lineage evolution of the floral organ identity loci. Extensive analyses of the MIKC-type MADS-box genes have shown that numerous gene duplications have occurred at every phylogenetic level (Kramer *et al.*, 1998, 2003, 2004; Theissen *et al.*, 2002; Munster *et al.*, 2002; Becker and Theissen, 2003; Litt and Irish, 2003; Malcomber and Kellogg, 2004; Stellari *et al.*, 2004). One cluster of duplication events that has received considerable attention is a group that occurred in the *AG*, *AP1*, *SEP* and *AP3* lineages close to the base of the core eudicots (Figure 3.4; Kramer *et al.*, 1998, 2004; Litt and Irish, 2003; Zahn *et al.*, 2005). These are of particular interest because the core eudicot radiation was a critical event in flowering plant evolution, giving rise to approximately 75% of all angiosperms species (Magallon *et al.*, 1999). Furthermore, following the duplications in the *AP3* and *AP1* lineage, the resultant paralogs underwent unusual patterns of sequence evolution (Kramer and Hu, unpublished data; Litt and Irish, 2003; Vandenbussche *et al.*, 2003a). This process involved frameshift mutations in the C-terminal domain that remodeled otherwise highly conserved sequence motifs.

In the *AP3* lineage, the ancestral paleoAP3 motif was retained in one of the paralogous core eudicot lineages, known as the *TM6* lineage, while in the other gene lineage a new conserved sequence was formed, the euAP3 motif (Kramer *et al.*, 1998). A similar pattern is observed in the *AP1* lineage: *Arabidopsis AP1* is a member of the divergent eu*AP1* lineage and *FUL* represents the eu*FUL* lineage, which is relatively unchanged from the ancestral *FUL*-like genes (Litt and Irish, 2003). Motif-swapping experiments in *Arabidopsis* have demonstrated that the paleoAP3 and euAP3 motifs are not functionally equivalent, indicating that the sequence change does reflect a shift in biochemical function (Lamb and Irish, 2003). The functional significance of the euAP1 motif is yet to be determined, although one tantalizing possibility is that the changes in eu*AP3* and eu*AP1* represent co-evolutionary processes (Vandenbussche *et al.*, 2003a; Litt and Irish, 2003). In contrast to the rather dramatic changes seen in eu*AP3* and eu*AP1*, the *SEP* and *AG* duplications are not associated with major modification of gene sequence, and the paralogs seem to have been primarily retained due to sub-functionalization (Kramer *et al.*, 2004; Zahn *et al.*, 2005). One advantage of *Petunia* is that unlike *Arabidopsis*, which has lost its *TM6* ortholog, it has retained both the eu*AP3* and *TM6* ortholog, as well as all other paralogs dating from the ancient core eudicot duplications, thereby facilitating the study of this phenomenon. Of course, subsequent duplications have also occurred, but this caveat is true of essentially all comparative studies of the angiosperms.

Even taking this complexity into consideration, genetic studies of floral organ identity gene homologs in *Petunia* suggest an overall conservation of function. For the genetically redundant *SEP*-like genes *FBP2* and *FBP5*, it has been found that the proteins can form similar higher-order protein complexes and they are required for B, C and D function (Angenent *et al.*, 1994; Ferrario *et al.*, 2003; Vandenbussche *et al.*, 2003b) *AP1* and *AP2* homologs have been identified in *Petunia* (Maes *et al.*, 2001; Immink *et al.*, 2003; Vandenbussche *et al.*, 2003b), but have not been functionally analysed to date. The situation with the B gene homologs is complex due to the duplications described earlier, but has provided insight into how the core eudicot paralogs have been evolutionarily retained. In addition to orthologs of the ancient eu*AP3* and *TM6* lineages, *Petunia* possesses two *PI*-like genes, *PhGLO1* and *PhGLO2*, which are the result of a recent duplication (Kramer *et al.*, 1998). Genetic analysis has demonstrated that while *PhDEF*, the eu*AP3* representative, promotes petal and stamen identity, *PhTM6* contributes only to stamen identity (van der Krol *et al.*, 1993; Angenent *et al.*, 1995a; Vandenbussche *et al.*, 2004). Furthermore, PhGLO1 and PhGLO2 have become biochemically subfunctionalized such that although they are expressed in whorls 2 and 3, the latter shows a preference for interacting with the stamen-specific PhTM6 (Vandenbussche *et al.*, 2004). Since the *Petunia PI* duplication occurred much later than the eu*AP3*/*TM6* event (Kramer *et al.*, 1998), this represents a recent co-evolutionary process. It remains unclear as to whether the stamen-specific function of *PhTM6* is representative of the ancestral condition of the paleo*AP3* lineage or whether this reflects a more recent subfunctionalization.

The *Petunia AG* paralogs *FBP6* and *pMADS3*, also the products of an ancient duplication, appear to have relatively redundant functions that are typical for C class genes (Kater *et al.*, 1998; Kapoor *et al.*, 2002). However, a clear distinction between *Petunia* and *Arabidopsis* is seen in the function of their *STK* homologs. As mentioned earlier, ovule-specific D function was first defined in *Petunia* as a result of functional analyses of the closely related paralogs *FBP7* and *FBP11* (Angenent *et al.*, 1995b; Colombo *et al.*, 1995). When the functions of both these genes are disrupted, ovule identity is lost. Furthermore, over-expression of *FBP11* results in the ectopic production of ovules, suggesting that the loci represent ‘master control’ genes for ovule identity. In contrast, although the *Arabidopsis* homolog *STK* contributes to ovule identity, its function is redundant with the *AG* and *SHP* genes (Pinyopich *et al.*, 2003), which are orthologous to *pMADS3* and *FBP6*, respectively (Kramer *et al.*, 2004). What these findings indicate is that in *Petunia*, *FBP7/11* play a distinct role in ovule identity but in *Arabidopsis* this function is shared among all the *AG* subfamily members. Further comparative studies will be required in order to determine whether a dissociable D function exists in most other taxa or if the situation in *Petunia* is an unusual case, possibly related to the central axile placentation of the *Petunia* ovary (Colombo *et al.*, 1996).

3.4.2 Floral organ identity gene function in Oryza

While *Petunia* has provided useful additional data points within the core eudicots, comparisons to the grass model species represent a much larger evolutionary divergence of at least 130 mya (Chaw *et al.*, 2004). Although these monocots theoretically contain representatives of all the ancestral gene lineages predating the core eudicot duplications (Figure 3.4), the grasses have undergone several rounds of independent gene duplications (Gaut and Doebley, 1997; Schmidt and Ambrose, 1998; Munster *et al.*, 2001; Malcomber and Kellogg, 2004). Thus models such as *Oryza* and *Zea* (corn) give us both a measure of deep evolutionary conservation as well as a separate system for studying the impact of gene duplications. Similar to *Petunia*, homologs of all the major floral organ identity genes have been identified in *Oryza* (Lopez-Dee *et al.*, 1999; Kyozuka *et al.*, 2000; Malcomber and Kellogg, 2004) and functional studies have been pursued using transgenic techniques. The *SEP*-like genes are perhaps the best example of lineage proliferation via gene duplication in the grasses. These taxa possess at least three separate lineages of *SEP* homologs that are independently derived relative to those of the core eudicots (Malcomber and Kellogg, 2004). Their gene expression patterns are diverse and highly variable, and functional data has been difficult to obtain, most likely due to extensive redundancy. The *Oryza SEP*-like gene *LEAFY HULL STERILE* (*LHS1*) does appear to contribute to the identity of the palea and lemma as well as to meristem determinacy and the structure of the inflorescence (Jeon *et al.*, 2000, Malcomber and Kellogg, 2004). A broad comparative study of expression patterns of *LHS1* orthologs across the grasses has revealed a high degree of variability, both within and between florets (Malcomber and Kellogg, 2004). These results highlight the variability that can be induced by

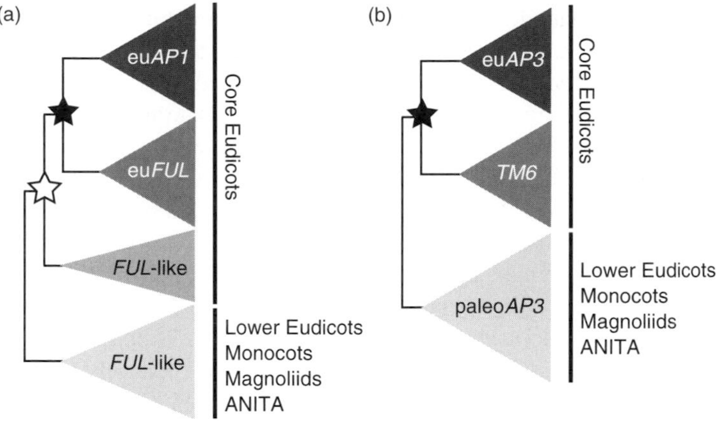

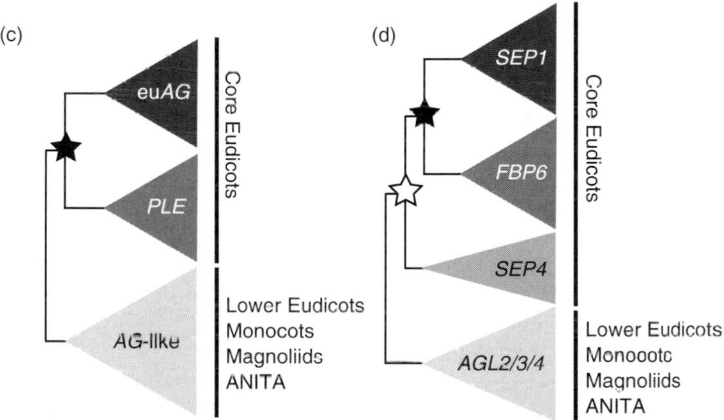

Figure 3.4 Simplified phylogenies of the *AP1* (a), *AP3* (b), *AG* (c) and *SEP* (d) gene lineages (Kramer *et al.*, 1998, 2004; Litt and Irish 2003; Zahn *et al.*, 2005). Black stars indicate gene duplication events that occurred at the base of the core eudicots while white stars indicate other major duplication events. The angiosperm groups that have representatives of each lineage are indicated to the right of each phylogeny.

the presence of multiple gene copies and suggest that orthologs in different grass subfamilies are following independent trajectories of functional evolution. Unfortunately, very little is known about the *AP1* and *AP2* homologs of *Oryza*, although homologs have been identified and the expression of *RAP1*, the rice *AP1*-like gene, is strikingly similar to that of *Arabidopsis AP1* (Kyozuka *et al.*, 2000; Feng *et al.*, 2002).

An added complication in interpreting the conservation of floral organ identity gene function in the grasses is their highly derived floral morphology. In particular, members of the family commonly possess a novel form of floral organ, the lodicule, which is positioned outside of the stamen whorl. This structure varies considerably in aspects of its morphology and has been proposed to be homologous to a wide array of organs, as well as being suggested to be a completely novel organ (Clifford, 1987). Functional studies in *Oryza* of *SUPERWOMAN1* (*SPW1*), a paleo*AP3* ortholog, and the *PI* homologs *OsMADS2* and *OsMADS4* indicate that the genes are necessary for lodicule and stamen identity (Kang *et al.*, 1998; Nagasawa *et al.*, 2003). Similar results have been obtained from genetic analysis of the *Zea* paleo*AP3* gene *SILKY* (Ambrose *et al.*, 2000). These findings are taken as confirmation that the lodicule represents a modified petal (Dahlgren and Clifford, 1982; Clifford, 1987) and, given the ancient divergence of core eudicots and grasses, that the role of *AP3* and *PI* homologs in petal identity is deeply conserved (Ma and dePamphilis, 2000; Ng and Yanofsky, 2001b). Although *AP3* and *PI* homologs clearly contribute to the identity of the lodicule, it is important to note that the structure is not petaloid in nature. This suggests that a novel identity pathway has evolved by modification of a pre-existing organ identity program, a phenomenon that seems likely to have occurred in other angiosperms.

The *AG*-like genes in the grasses duplicated at some point close to the base of the family (Schmidt and Ambrose, 1998; Kramer *et al.*, 2004). One of these paralogs has been functionally characterized in *Oryza*, the gene *OsMADS3* (also called *RAG*) (Kang *et al.*, 1995; Kyozuka *et al.*, 2000). This locus is strongly expressed in the developing stamens and carpels, and can promote floral meristem determinacy in addition to the identity of these organs (Kang *et al.*, 1998; Kyozuka and Shimamoto, 2002). Functional analyses of two *AG*-like genes in *Zea* has revealed that *ZMM2*, which is orthologous to *OsMADS3*, and *ZAG1* exhibit sub-functionalization (Mena *et al.*, 1996). *ZMM2* is primarily expressed in stamens while *ZAG1* is strongly transcribed in the carpels (Schmidt *et al.*, 1993; Mena *et al.*, 1996). Moreover, *zag1* mutants exhibit indeterminacy in only their female flowers (Mena *et al.*, 1996). These results indicate that some redundancy exists between the genes, although studies in rice have provided yet another candidate for carpel identity. In *Arabidopsis*, *AG* is the major determinant of carpel identity but *CRABS CLAW* (*CRC*), a member of the YABBY gene family, also contributes (Alvarez and Smyth, 1999; Bowman and Smyth, 1999). In *Oryza* by contrast, a *CRC* homolog known as *DROOPING LEAF* (*DL*) appears to be the more critical player (Nagasawa *et al.*, 2003; Yamaguchi *et al.*, 2004), raising the possibility that organ identity functions can shift between non-homologous loci. Last, the putative *STK* ortholog in *Oryza*, *OsMADS13*, does show ovule-specific expression but its function has not been determined (Lopez-Dee *et al.*, 1999).

3.5 Sex determination as a modification of floral organ identity

Although the primitive condition in angiosperms is thought to be hermaphroditism, diverse forms of sex determination have evolved many times independently (Westergaard, 1958). Most dioecious or monoecious plants go through a hermaphroditic stage early in flower development, followed by differential abortion or arrest of sex organs (Ainsworth, 2000). Examples of this type of unisexual flower development include the most commonly studied dioecious plants, such as *Silene* and *Rumex*, as well as monoecious species such as maize and cucumber. On the other hand, some dioecious plants like *Spinacia oleraceae* (spinach), *Mercurialis annua* (mercury) and *Cannabis sativa* (hemp) have unisexual flowers with no vestiges of organs of the opposite sex. The genetic pathways underlying sex determination have been a subject of study literally since the time of Darwin, but still remain somewhat elusive. From the standpoint of floral architecture, analyses of various model species have revealed divergent systems controlling organ abortion. In cucumber, for instance, developmental arrest appears to be largely position dependent (Kater *et al.*, 2001), while in maize, abortion relies on both identity and positional cues (Ambrose *et al.*, 2000). A possible role for the organ identity genes has been investigated in several taxa (reviewed Ainsworth, 2000), but has not been found to be of great significance. This is not especially surprising, however, since in all of these cases sex determination occurs after organ identity is established. It remains to be seen whether developmental programs that produce flowers which are entirely male or female from inception are acting upstream of organ identity (Di Stilio *et al.*, 2004; Pfent *et al.*, 2005).

3.6 Future perspectives

It is reassuring to find that the ABC model as first described 15 years ago has survived largely unchanged after extensive scrutiny. While we now understand much more about the finer aspects of the model, such as the mechanisms underlying the mutual repression of A and C genes, the broader picture remains the same. Clearly, components such as the E class genes could not have been recovered in the initial forward mutagenesis screens due to their highly redundant functions. The one feature that requires further analysis is the nature of A function, particularly the relative contribution of the *AP1* genes. Comparative studies of the organ identity program across diverse angiosperms have revealed a complex pattern of conservation and divergence. It seems likely that a program relatively similar to that first described in *Arabidopsis* and *Antirrhinum* was functioning in the ancestors of extant flowering plants. However, considerable developmental system drift has occurred over time, giving rise to new gene combinations and novel forms of floral organs. Teasing apart the complex interactions among gene duplication, functional evolution, and shifts in gene expression will require considerably more research, but promises to be very illuminating.

Acknowledgements

I would like to thank Jocelyn Hall for comments on the manuscript and NSF award IBN-03139103 for providing funding for some of the work described here.

References

Ainsworth, C. (2000) Boys and girls come out to play: the molecular biology of dioecious plants, *Annals of Botany*, **86**, 211–221.

Alvarez-Buylla, E. R., Pelaz, S., Liljegren, S. J. *et al.* (2000) An ancestral MADS-box gene duplication occurred before the divergence of plants and animals, *Proceedings of the National Academy of Sciences of the United States of America*, **97**, 5328–5333.

Alvarez, J. and Smyth, D. R. (1999) 'CRABS CLAW and SPATULA, two Arabidopsis genes that control carpel development in parallel with AGAMOUS', *Development*, **126**, 2377–2386.

Ambrose, B. A., Lerner, D. R., Ciceri, P., Padilla, C. M., Yanofsky, M. F. and Schmidt, R. J. (2000) Molecular and genetic analyses of the *Silky1* gene reveal conservation in floral organ specification between eudicots and monocots, *Molecular Cell*, **5**, 569–579.

Angenent, G. C., Busscher, M., Franken, J., Dons, H. J. M. and van Tunen, A. J. (1995a) Functional interaction between the homeotic genes *fbp1* and *pMADS1* during Petunia floral organogenesis, *The Plant Cell*, **7**, 507–516.

Angenent, G. C., Franken, J., Busscher, M. *et al.* (1995b) A novel class of MADS box genes is involved in ovule development in Petunia, *Plant Cell*, **7**, 1569–1582.

Angenent, G. C., Franken, J., Bussher, M., Weiss, D. and van Tunen, A. J. (1994) Co-suppression of the petunia homeotic gene fbp2 affects the identity of the generative meristem, *The Plant Journal*, **5**, 33–44.

Aukerman, M. J. and Sakai, H. (2003) Regulation of flowering time and floral organ identity by a MicroRNA and its APETALA2-like target genes, *The Plant Cell*, **15**, 2730–2741.

Becker, A. and Theissen, G. (2003) The major clades of MADS-box genes and their role in the development and evolution of flowering plants, *Molecular Phylogenetics and Evolution*, **29**, 464–489.

Bowman, J. L. and Smyth, D. R. (1999) CRABS CLAW, a gene that regulates carpel and nectary development in Arabidopsis, encodes a novel protein with zinc finger and helix-loop-helix domains, *Development*, **126**, 2387–2396.

Bowman, J. L., Alvarez, J., Weigel, D., Meyerowitz, E. M. and Smyth, D. R. (1993) Control of flower development in Arabidopsis thaliana by APETALA1 and interacting genes, *Development*, **119**, 721–743.

Bowman, J. L., Smyth, D. R. and Meyerowitz, E. M. (1989) Genes directing flower development in *Arabidopsis*, *The Plant Cell*, **1**, 37–52.

Bowman, J. L., Smyth, D. R. and Meyerowitz, E. M. (1991) Genetic interactions among floral homeotic genes of *Arabidopsis*, *Development*, **112**, 1–20.

Carpenter, R. and Coen, E. S. (1990) Floral homeotic mutations produced by transposon-mutagenesis in *Antirrhinum majus*, *Genes & Development*, **4**, 1483–1493.

Chaw, S. M., Chang, C. C., Chen, H. L. and Li, W. H. (2004) Dating the monocot–dicot divergence and the origin of core eudicots using whole chloroplast genomes, *Journal of Molecular Evolution*, **58**, 424–441.

Chen, X. (2004) A microRNA as a translational repressor of APETALA2 in Arabidopsis flower development, *Science*, **303**, 2022–2025.

Chen, X., Riechmann, J. L., Jia, D. and Meyerowitz, E. M. (2000) Minimal regions in the *Arabidopsis PISTILLATA* promoter responsive to the *APETALA3/PISTILLATA* feedback control do not contain a CArG box, *Sexual Plant Reproduction*, **13**, 85–94.

Clifford, H. T. (1987) Spiklet and floral morphology, in *Grass Systematics* (eds T. R. Soderstrom, K. W. Hilu, C. S. Campbell and M. E. Barkworth), Smithsonian Institution Press, Washington, D.C., pp. 21–30.

Coen, E. S. and Meyerowitz, E. M. (1991) The war of the whorls: genetic interactions controlling flower development, *Nature*, **353**, 31–37.

Coen, E. S., Doyle, S., Romero, J. M., Elliot, R., Magrath, R. and Carpenter, R. (1991) Homeotic genes controlling flower development in *Antirrhinum*, *Development*, **1** (Suppl.), 149–156.

Colombo, L., Franken, J., Koetje, E. *et al.* (1995) The petunia MADS box gene FBP11 determines ovule identity, *The Plant Cell*, **7**, 1859–1868.

Colombo, L., van Tunen, A. J., Dons, H. J. M. and Angenent, G. C. (1996) Molecular control of flower development in *Petunia hybrida*, in *Advances in Botanical Research*, Vol. 26 (ed. J. A. Callow), Academic Press, New York, pp. 229–250.

Conner, J. and Liu, Z. (2000) LEUNIG, a putative transcriptional corepressor that regulates *AGAMOUS* expression during flower development, *Proceedings of the National Academy of Sciences of the United States of America*, **97**, 12902–12907.

Dahlgren, R. M. T. and Clifford, H. T. (1982) *The Monocotyledons: A comparative Study*, Springer Verlag, New York.

Davies, B., Egea-Cortines, M., de Andrade Silva, E., Saedler, H. and Sommer, H. (1996) Multiple interactions amongst floral homeotic MADS box proteins, *EMBO Journal*, **15**, 4330–4343.

Davies, B., Motte, P., Keck, E., Saedler, H., Sommer, H. and Schwarz-Sommer, Z. (1999) PLENA and FARINELLI: redundancy and regulatory interactions between two Antirrhinum MADS-box factors controlling flower development, *EMBO Journal*, **18**, 4023–4034.

Di Stilio, V. S., Kramer, E. M. and Baum, D. A. (2004) Floral MADS box genes and the evolution of homeotic gender dimorphism in meadow rues (Thalictrum, Ranunculaceae), *The Plant Journal*, **41**, 755–766.

Ditta, G., Pinyopich, A., Robles, P., Pelaz, S. and Yanofsky, M. (2004) The SEP4 gene of *Arabidopsis thaliana* functions in floral organ and meristem identity, *Current Biology*, **14**, 1935–1940.

Drews, G. N., Bowman, J. L. and Meyerowitz, E. M. (1991) Negative regulation of the Arabidopsis homeotic gene AGAMOUS by the APETELA2 product, *Cell*, **65**, 991–1002.

Egea-Cortines, M., Saedler, H. and Sommer, H. (1999) Ternary complex formation between the MADS-box proteins SQUAMOSA, DEFICIENS and GLOBOSA is involved in the control of floral architecture in *Antirrhinum majus*, *EMBO Journal*, **18**, 5370–5379.

Elliott, R. C., Betzner, A. S., Huttner, E. *et al.* (1996) *AINTEGUMENTA*, an *APETALA2*-like gene of *Arabidopsis* with pleiotropic roles in ovule development and floral organ growth, *The Plant Cell*, **8**, 155–168.

Espinoza-Soto, C., Padilla-Longoria, P. and Alvarez-Buylla, E. R. (2004) A gene regulatory network model for cell-fate determination during *Arabidopsis thaliana* flower development that is robust and recovers experimental gene expression profiles, *The Plant Cell*, **16**, 2923–2939.

Feng, Q., Zhang, Y. J., Hao, P. *et al.* (2002) Sequence and analysis of rice chromosome 4, *Nature*, **420**, 316–320.

Ferrandiz, C., Gu, Q., Martienssen, R. and Yanofsky, M. F. (2000) Redundant regulation of meristem identity and plant architecture by FRUITFULL, APETALA1 and CAULIFLOWER, *Development*, **127**, 725–734.

Ferrario, S., Immink, R. G. H., Shchennikova, A., Busscher-Lange, J. and Angenent, G. C. (2003) The MADS box gene *FBP2* is required for SEPALLATA function in Petunia, *The Plant Cell*, **15**, 914–925.

Franks, R. G., Wang, C. X., Levin, J. Z. and Liu, Z. C. (2002) SEUSS, a member of a novel family of plant regulatory proteins, represses floral homeotic gene expression with LEUNIG, *Development*, **129**, 253–263.

Gaut, B. S. and Doebley, J. (1997) DNA sequence evidence for the segmental allotetraploid origin of maize, *Proceedings of the National Academy of Sciences of The United States of America*, **94**, 6809–6814.

Goto, K. and Meyerowitz, E. M. (1994) Function and regulation of the Arabidopsis floral homeotic gene PISTILLATA, *Genes & Development*, **8**, 1548–1560.

Gustafson-Brown, C., Savidge, B. and Yanofsky, M. F. (1994) Regulation of the Arabidopsis floral homeotic gene APETALA1, *Cell*, **76**, 131–143.

Gutierrez-Cortines, M. E. and Davies, B. (2000) Beyond the ABCs: ternary complex formation in the control of floral organ identity, *Trends in Plant Sciences*, **5**, 471–476.

Hempel, F. D., Zambryski, P. C. and Feldman, L. J. (1998) Photoinduction of flower identity in vegetatively biased primordia, *The Plant Cell*, **10**, 1663–1675.

Hill, T., Day, C. D., Zondlo, S. C., Thackeray, A. G. and Irish, V. F. (1998) Discrete spatial and temporal *cis*-acting elements regulate transcription of the *Arabidopsis* floral homeotic gene *APETALA3*, *Development*, **125**, 1711–1721.

Honma, T. and Goto, K. (2000) The Arabidopsis floral homeotic gene PISTILLATA is regulated by discrete *cis*-elements responsive to induction and maintenance signals, *Development*, **127**, 2021–2030.

Honma, T. and Goto, K. (2001) Complexes of MADS-box proteins are sufficient to convert leaves into floral organs, *Nature*, **409**, 525–529.

Huijser, P., Klein, J., Lonnig, W.-E., Meijer, H., Saedler, H. and Sommer, H. (1992) Bractomania, an inflorescence anomaly, is caused by the loss of function of the MADS-box gene *squamosa* in *Antirrhinum majus*, *EMBO Journal*, **11**, 1239–1249.

Immink, R. G. H., Ferrario, S., Busscher-Lange, J., Kooiker, M., Busscher, M. and Angenent, G. C. (2003) Analysis of the petunia MADS-box transcription factor family, *Molecular Genetics & Genomics*, **268**, 598–606.

Ingram, G. C., Goodrich, J., Wilkinson, M. D., Simon, R., Haughn, G. W. and Coen, E. S. (1995) Parallels between UNUSUAL FLORAL ORGANS and FIMBRIATA, genes controlling flower development in Arabidopsis and Antirrhinum, *The Plant Cell*, **7**, 1501–1510.

Irish, V. F. and Kramer, E. M. (1998) Genetic and molecular analysis of angiosperm flower development, in *Advances in Botanical Research*, Vol. 28 (ed. J. A. Callow), Academic Press, New York, pp. 197–231.

Irish, V. F. and Sussex, I. M. (1990) Function of the *apetela-1* gene during *Arabidopsis* floral development, *The Plant Cell*, **2**, 741–753.

Jack, T. (2004) Molecular and genetic mechanisms of floral control, *The Plant Cell*, **16**, S1–S17.

Jack, T., Brockman, L. L. and Meyerowitz, E. M. (1992) The homeotic gene APETALA3 of *Arabidopsis thaliana* encodes a MADS box and is expressed in petals and stamens, *Cell*, **68**, 683–697.

Jack, T., Fox, G. L. and Meyerowitz, E. M. (1994) Arabidopsis homeotic gene APETALA3 ectopic expression: transcriptional and posttranscriptional regulation determine floral organ identity, *Cell*, **76**, 703–716.

Jenik, P. D. and Irish, V. F. (2001) The Arabidopsis floral homeotic gene APETALA3 differentially regulates intercellular signaling required for petal and stamen development, *Development*, **128**, 13–23.

Jeon, J.-S., Jang, S., Lee, S. *et al.* (2000) *leafy hull sterile* is a homeotic mutation in a rice MADS box gene affecting rice flower development, *The Plant Cell*, **12**, 871–884.

Jofuku, K. D., den Boer, B. G. W., Van Montague, M. and Okamuro, J. K. (1994) Control of Arabidopsis flower and seed development by the homeotic gene APETALA2, *The Plant Cell*, **6**, 1211–1225.

Kang, H.-G., Jeon, J.-S., Lee, S. and An, G. (1998) Identification of class B and class C floral organ identity genes from rice plants, *Plant Molecular Biology*, **38**, 1021–1029.

Kang, H.-G., Noh, Y.-S., Chung, Y.-Y., Costa, M. A., An, K. and An, G. (1995) Phenotypic alterations of petal and sepal by ectopic expression of a rice MADS box gene in tobacco, *Plant Molecular Biology*, **29**, 1–10.

Kapoor, M., Tsuda, S., Tanaka, Y. *et al.* (2002) Role of petunia *pMADS3* in determination of floral organ meristem identity, as revealed by its loss of function, *The Plant Journal*, **32**, 115–127.

Kater, M. M., Colombo, L., Franken, J. *et al.* (1998) Multiple *AGAMOUS* homologs from Cucumber and Petunia differ in their ability to induce reproductive organ fate, *The Plant Cell*, **10**, 171–182.

Kater, M. M., Franken, J., Carney, K. J., Colombo, L. and Angenent, G. C. (2001) Sex determination in the monoecious species cucumber is confined to specific floral whorls, *The Plant Cell*, **13**, 481–493.

Keck, E., McSteen, P., Carpenter, R. and Coen, E. (2003) Separation of genetic functions controlling organ identity in flowers, *EMBO Journal*, **22**, 1058–1066.

Komaki, M., Okada, K., Nishino, E. and Shimura, Y. (1988) Isolation and characterization of novel mutants of *Arabidopsis thaliana* defective in flower development, *Development*, **104**, 195–203.

Kramer, E. M., Di Stilio, V. S. and Schluter, P. (2003) Complex patterns of gene duplication in the APETALA3 and PISTILLATA lineages of the Ranunculaceae, *International Journal of Plant Science*, **164**, 1–11.

Kramer, E. M., Dorit, R. L. and Irish, V. F. (1998) Molecular evolution of genes controlling petal and stamen development: duplication and divergence within the *APETALA3* and *PISTILLATA* MADS-box gene lineages, *Genetics*, **149**, 765–783.

Kramer, E. M., Jaramillo, M. A. and Di Stilio, V. S. (2004) Patterns of gene duplication and functional evolution during the diversification of the AGAMOUS subfamily of MADS-box genes in angiosperms, *Genetics*, **166**, 1011–1023.

Krizek, B. A. and Meyerowitz, E. M. (1996a) The Arabidopsis homeotic genes APETALA3 and PISTILLATA are sufficient to provide the B class organ identity function, *Development*, **122**, 11–22.

Krizek, B. A. and Meyerowitz, E. M. (1996b) Mapping the protein regions responsible for the functional specificities of the Arabidopsis MADS domain organ-identity proteins, *Proceedings of the National Academy of Sciences of the United States of America*, **93**, 4063–4070.

Krizek, B. A., Prost, V. and Macias, A. (2000) AINTEGUMENTA promotes petal identity and acts as a negative regulator of AGAMOUS, *The Plant Cell*, **12**, 1357–1366.

Kuhn, E. (2001) From library screening to microarray technology: strategies to determine gene expression profiles and to identify differentially regulated genes in plants, *Annals of Botany*, **87**, 139–155.

Kyozuka, J. and Shimamoto, K. (2002) Ectopic expression of *OsMADS3*, a rice ortholog of *AGAMOUS*, caused a homeotic transformation of lodicules to stamens in transgenic rice plants, *Plant and Cell Physiology*, **43**, 130–135.

Kyozuka, J., Kobayashi, T., Morita, M. and Shimamoto, K. (2000) Spatially and temporally regulated expression of rice MADS box genes with similarity to Arabidopsis class A, B and C genes (In Process Citation), *Plant and Cell Physiology*, **41**, 710–718.

Lamb, R. S. and Irish, V. F. (2003) Functional divergence within the APETALA3/PISTILLATA floral homeotic gene lineages, *Proceedings of the National Academy of Sciences of the United States of America*, **100**, 6558–6563.

Lee, I., Wolfe, D. S., Nilsson, O. and Weigel, D. (1997) A *LEAFY* co-regulator encoded by *UNUSUAL FLORAL ORGANS*, *Current Biology*, **7**, 95–104.

Lenhard, M., Bohnert, A., Jurgens, G. and Laux, T. (2001) Termination of stem cell maintenance in Arabidopsis floral meristems by interactions between WUSCHEL and AGAMOUS, *Cell*, **105**, 805–814.

Liljegren, S. J., Ditta, G. S., Eshed, Y., Savidge, B., Bowman, J. L. and Yanofsky, M. F. (2000) *SHATTERPROOF* MADS-box genes control seed dispersal in *Arabidopsis*, *Nature*, **404**, 766–770.

Litt, A. and Irish, V. F. (2003) Duplication and diversification in the *APETALA1/FRUITFULL* floral homeotic gene lineage: implications for the evolution of floral development, *Genetics*, **165**, 821–833.

Liu, Z. and Meyerowitz, E. M. (1995) LEUNIG regulates AGAMOUS expression in Arabidopsis flowers, *Development*, **121**, 975–991.

Lohmann, J. U. and Weigel, D. (2002) Building beauty: the genetic control of floral patterning, *Developmental Cell*, **2**, 135–142.

Lohmann, J. U., Hong, R. L., Hobe, M. *et al.* (2001) A molecular link between stem cell regulation and floral patterning in Arabidopsis, *Cell*, **105**, 793–803.

Lopez-Dee, Z. P., Wittich, P., Pe, M. E. *et al.* (1999) *OsMADS13*, a novel rice MADS-box gene expressed during ovule development, *Developmental Genetics*, **25**, 237–244.

Ma, H. and dePamphilis, C. (2000) The ABCs of Floral Evolution, *Cell*, **101**, 5–8.

Maes, T., Van de Steene, N., Zethof, J. *et al.* (2001) Petunia Ap2-like genes and their role in flower and seed development, *The Plant Cell*, **13**, 229–244.

Magallon, S., Crane, P. R. and Herendeen, P. S. (1999) Phylogenetic pattern, diversity, and diversification of eudicots, *Annals of the Missouri Botanical Garden*, **86**, 297–372.

Magnani, E., Sjolander, K. and Hake, S. (2004) From endonucleases to transcription factors: evolution of the AP2 DNA binding domain in plants, *The Plant Cell*, **16**, 2265–2277.

Maizel, A., Busch, M. A., Tanahashi, T. *et al.* (2005) The floral regulator LEAFY evolves by substitutions in the DNA binding domain, *Science*, **308**, 260–263.

Malcomber, S. T. and Kellogg, E. A. (2004) Heterogeneous expression patterns and separate roles of the SEPALLATA gene LEAFY HULL STERILE1 in Grasses, *The Plant Cell*, **16**, 1692–1706.

Mandel, M. A. and Yanofsky, M. F. (1995) The Arabidopsis AGL8 MADS box gene is expressed in inflorescence meristems and is negatively regulated by APETALA1, *The Plant Cell*, **7**, 1763–1771.

Mandel, M. A., Gustafson-Brown, C., Savidge, B. and Yanofsky, M. F. (1992) Molecular characterization of the *Arabidopsis* floral homeotic gene *apetela1*, *Nature*, **360**, 274–277.

Mayer, K. F., Schoof, H., Haecker, A., Lenhard, M., Jurgens, G. and Laux, T. (1998) Role of WUSCHEL in regulating stem cell fate in the Arabidopsis shoot meristem, *Cell*, **95**, 805–815.

McGonigle, B., Bouhidel, K. and Irish, V. F. (1996) Nuclear localization of the Arabidopsis APETALA3 and PISTILLATA homeotic gene products depends on their simultaneous expression, *Genes & Development*, **10**, 1812–1821.

Mena, M., Ambrose, B. A., Meeley, R. B., Briggs, S. P., Yanofsky, M. F. and Schmidt, R. J. (1996) Diversification of C-function activity in Maize flower development, *Science*, **274**, 1537–1540.

Messenguy, F. and Dubois, E. (2003) Role of MADS box proteins and their cofactors in combinatorial control of gene expression and cell development, *Gene*, **316**, 1–21.

Michaels, S. D., Ditta, G., Gustafson-Brown, C., Pelaz, S., Yanofsky, M. and Amasino, R. M. (2003) AGL24 acts as a promoter of flowering in Arabidopsis and is positively regulated by vernalization, *The Plant Journal*, **33**, 867–874.

Mizukami, Y. and Ma, H. (1992) Ectopic expression of the floral homeotic gene *agamous* in transgenic arabidopsis plants alters floral organ identity, *Cell*, **71**, 119–131.

Moon, Y.-W., Kang, H.-G., Jung, J.-Y., Jeon, J.-S., Sung, S.-K. and An, G. (1999) Determination of the motif responsible for interaction between the rice APETALA1/AGAMOUS-LIKE9 family proteins using a yeast two-hybrid system, *Plant Physiology*, **120**, 1193–1203.

Munster, T., Deleu, W., Wingen, L. U. *et al.* (2002) Maize MADS-box genes galore, *Maydica*, **47**, 287–301.

Munster, T., Wingen, L. U., Faigl, W., Werth, S., Saedler, H. and Theissen, G. (2001) Characterization of three *GLOBOSA*-like MADS-box genes from maize: evidence for ancient paralogy in one class of floral homeotic B-function genes of grasses, *Gene*, **262**, 1–13.

Nagasawa, N., Miyoshi, M., Sano, Y. *et al.* (2003) *SUPERWOMAN1* and *DROOPING LEAF* genes control floral organ identity in rice, *Development*, **130**, 705–718.

Ng, M. and Yanofsky, M. F. (2001a) Activation of the arabidopsis B class homeotic genes by apetala1, *The Plant Cell*, **13**, 739–754.

Ng, M. and Yanofsky, M. F. (2001b) Function and evolution of the plant MADS-box gene family, *Nature Review Genetics*, **2**, 186–195.

Nole-Wilson, S. and Krizek, B. A. (2000) DNA binding properties of the Arabidopsis floral development protein AINTEGUMENTA, *Nucleic Acids Research*, **28**, 4076–4082.

Okamuro, J. K., Caster, B., Villarroel, R., van Montagu, M. and Jofuku, K. D. (1997a) The AP2 domain of *APETALA2* defines a large new family of DNA binding proteins in Arabidopsis, *Proceedings of the National Academy of Sciences of the United States of America*, **94**, 7076–7081.

Okamuro, J. K., Szeto, W., Lotys-Prass, C. and Jofuku, K. D. (1997b) Photo and hormonal control of meristem identity in the Arabidopsis flower mutants apetala2 and apetala1, *The Plant Cell*, **9**, 37–47.

Parcy, F., Nilsson, O., Busch, M. A., Lee, I. and Weigel, D. (1998) A genetic framework for floral patterning, *Nature*, **395**, 561–565.

Pelaz, S., Ditta, G. S., Baumann, E., Wisman, E. and Yanofsky, M. (2000) B and C floral organ identity functions require SEPALLATA MADS-box genes, *Nature*, **405**, 200–203.

Pfent, C., Pobursky, K. J., Sather, D. N. and Golenberg, E. M. (2005) Characterization of SpAPETALA3 and SpPISTILLATA, B class floral identity genes in *Spinacia oleracea*, and their relationship to sexual dimorphism, *Development Genes and Evolution*, **215**, 132–142.

Pinyopich, A., Ditta, G. S., Savidge, B. *et al.* (2003) Assessing the redundancy of MADS-box genes during carpel and ovule development, *Nature*, **424**, 85–88.

Purugganan, M. D., Rounsley, S. D., Schmidt, R. J. and Yanofsky, M. F. (1995) Molecular evolution of flower development: diversification of the plant MADS-box regulatory gene family, *Genetics*, **140**, 345–356.

Riechmann, J. L. and Meyerowitz, E. M. (1998) The AP2/EREBP family of plant transcription factors, *Biological Chemistry*, **379**, 633–646.

Riechmann, J. L., Krizek, B. A. and Meyerowitz, E. M. (1996a) Dimerization specificity of Arabidopsis MADS domain homeotic proteins APETALA1, APETALA3, PISTILLATA, and AGAMOUS, *Proceedings of the National Academy of Sciences of the United States of America*, **93**, 4793–4798.

Riechmann, J. L., Wang, M. and Meyerowitz, E. M. (1996b) DNA-binding properties of Arabidopsis MADS domain homeotic proteins APETALA1, APETALA3, PISTILLATA and AGAMOUS, *Nucleic Acids Research*, **24**, 3134–3141.

Schmidt, R. J. and Ambrose, B. A. (1998) The blooming of grass flower development, *Current Opinion in Plant Biology*, **1**, 60–67.

Schmidt, R. J., Veit, B., Mandel, M. A., Mena, M., Hake, S. and Yanofsky, M. F. (1993) Identification and molecular characterization of *ZAG1*, the maize homolog of the *Arabidopsis* floral homeotic gene *AGAMOUS*, *The Plant Cell*, **5**, 729–737.

Shigyo, M. and Ito, M. (2004) Analysis of gymnosperm two-AP2-domain-containing genes, *Development Genes and Evolution*, **214**, 105–114.

Sridhar, V. V., Surendrarao, A., Gonzalez, D., Conlan, R. S. and Liu, Z. C. (2004) Transcriptional repression of target genes by LEUNIG and SEUSS, two interacting regulatory proteins for Arabidopsis flower development, *Proceedings of the National Academy of Sciences of the United States of America*, **101**, 11494–11499.

Stellari, G. M., Jaramillo, M. A. and Kramer, E. M. (2004) Evolution of the *APETALA3* and *PISTILLATA* lineages of MADS-box containing genes in basal angiosperms, *Molecular Biology Evolution*, **21**, 506–519.

Theissen, G. (2001) Development of floral organ identity: stories from the MADS house, *Current Opinion in Plant Biology*, **4**, 75–85.

Theissen, G. and Saedler, H. (2001) Floral quartets, *Nature*, **409**, 469–471.

Theissen, G., Becker, A., Di Rosa, A. *et al.* (2000) A short history of MADS-box genes in plants, *Plant Molecular Biology*, **42**, 115–149.

Theissen, G., Becker, A., Winter, K. U., Munster, T., Kirchner, C. and Saedler, H. (2002) How the land plants learned their floral ABCs: the role of MADS-box genes in the evolutionary origin of flowers, in *Developmental Genetics and Plant Evolution* (eds Q. C. B. Cronk, R. M. Bateman and J. A. Hawkins), Taylor and Francis, London, pp. 173–205.

van der Krol, A. R., Brunelle, A., Tsuchimoto, S. and Chua, N. H. (1993) Functional analysis of Petunia floral homeotic MADS box gene PMADS1, *Genes & Development*, **7**, 1214–1228.

Vandenbussche, M., Theissen, G., Van de Peer, Y. and Gerats, T. (2003a) Structural diversification and neo-functionalization during floral MADS-box gene evolution by C-terminal frameshift mutations, *Nucleic Acids Research*, **31**, 4401–4409.

Vandenbussche, M., Zethof, J., Royaert, S., Weterings, K. and Gerats, T. (2004) The duplicated B-class heterodimer model: whorl-specific effects and complex genetic interactions in Petunia hybrida flower development, *The Plant Cell*, **16**, 741–754.

Vandenbussche, M., Zethof, J., Souer, E. *et al.* (2003b) Toward the analysis of the Petunia MADS box gene family by reverse and forward transposon insertion mutagenesis approaches: B, C, and D function organ identity functions require SEPALLATA-like MADS box genes in Petunia, *The Plant Cell*, **15**, 2680–2693.

Wagner, D., Sablowski, R. W. M. and Meyerowitz, E. M. (1999) Transcriptional activation of *APETALA1* by *LEAFY*, *Science*, **285**, 582–584.

Wang, X., Feng, S., Nakayama, N. *et al.* (2003) The COP9 signalosome interacts with SCF UFO and participates in Arabidopsis flower development, *The Plant Cell*, **15**, 1071–1082.

Weigel, D. (1995) The APETALA2 domain is related to a novel type of DNA binding domain, *The Plant Cell*, **7**, 388–389.

Weigel, D. and Meyerowitz, E. M. (1994) The ABCs of floral homeotic genes, *Cell*, **78**, 203–209.

Weigel, D. and Nilsson, O. (1995) A developmental switch sufficient for flower initiation in diverse plants, *Nature*, **377**, 495–500.

Westergaard, M. (1958) The mechanisms of sex determination in dioecious flowering plants, *Advances in Genetics*, **9**, 217–281.

Yamaguchi, T., Nagasawa, N., Kawasaki, S., Matsuoka, M., Nagato, Y. and Hirano, H.-Y. (2004) The YABBY gene *DROOPING LEAF* regulates carpel specification and midrib development in *Oryza sativa*, *The Plant Cell*, **16**, 500–509.

Yu, H., Ito, T., Wellmer, F. and Meyerowitz, E. M. (2004) Repression of AGAMOUS-LIKE 24 is a crucial step in promoting flower development, *Nature Genetics*, **36**, 157–161.

Yu, H., Xu, Y. F., Tan, E. L. and Kumar, P. P. (2002) AGAMOUS-LIKE 24, a dosage-dependent mediator of the flowering signals, *Proceedings of the National Academy of Sciences of the United States of America*, **99**, 16336–16341.

Zachgo, S., de Andrade Silva, E., Motte, P., Trobner, W., Saedler, H. and Schwarz-Sommer, Z. (1995) Functional analysis of the Antirrhinum floral homeotic Deficiens gene in vivo and in vitro by using a temperature-sensitive mutant, *Development*, **121**, 2861–2875.

Zahn, L. M., Kong, H., Leebens-Mack, J. H. *et al.* (2005) The evolution of the SEPALLATA subfamily of MADS-box genes: a preangiosperm origin with multiple duplications throughout angiosperm history, *Genetics*, **169**, 2209–2223.

4 The genetic control of flower size and shape

Lynette Fulton, Martine Batoux, Ram Kishor Yadav and Kay Schneitz

4.1 Introduction

The diversity and beauty of flowers has fascinated mankind for thousands of years. The molecular basis of floral development, however, has only recently been addressed. Work performed in the last 15–20 years has accumulated a tremendous body of knowledge concerning the specification of flowers and floral organ identity (Coen and Meyerowitz, 1991; Weigel and Meyerowitz, 1994; Ng and Yanofsky, 2000; Lohmann and Weigel, 2002). Despite this concentrated effort, several key issues remain to be explored. For example, what factors regulate the size and shape of flowers and floral organs? Curiously, little is known about the intrinsic mechanisms coordinating cell proliferation and growth at the organ level in plants (Mizukami, 2001; Fleming, 2002; Beemster *et al.*, 2003). The situation in plants and animals is comparable in this regard as organ-level growth regulation in animals also remains enigmatic (Conlon and Raff, 1999; Potter and Xu, 2001; Hafen and Stocker, 2003). It is therefore of particular interest to study the molecular basis of these processes. Several significant factors should facilitate this task in plants. First, the size and shape of plant organs is achieved mainly through the control of the number and patterns of cell division as well as through the regulation of cell expansion (Meyerowitz, 1997). Second, cells do not move relative to each other during development, as in animals. Finally, only a limited number of different cell types exist in any given plant organ.

Where do plant organs originate from during development? In a typical higher plant, aerial organs are produced post-embryonically and ultimately derive from a small group of cells, the shoot apical meristem (SAM), located at the very tip or apex of the main growth axis or shoot (Steeves and Sussex, 1989). During the vegetative phase of the life cycle the SAM produces leaves and axillary meristems. Upon floral induction, the plant enters the reproductive phase of development. The SAM converts to an inflorescence meristem and produces bracts, secondary inflorescences and flowers. Flowers, by virtue of their own floral meristem, generate sepals, petals, stamens and carpels bearing ovules, the ultimate female reproductive organs. The apical meristem originates during embryogenesis and represents a zone of continuous cell proliferation. Within the meristem, a small group of stem cells in the central zone (CZ) divides infrequently and is maintained in an undifferentiated state. Some progeny of these cells will populate the peripheral zone (PZ), where they divide at a higher frequency. Organ primordia form in a regular spatial fashion at the flanks of the meristem (phyllotaxy). Progeny cells of the stem cells will also be found in the rib meristem, an interior region located beneath the central zone. There is a second level of organization overlaying the zonation described earlier. Meristems are also organized into clonally distinct layers (Satina and Blakeslee, 1941) that are maintained by virtue of oriented cell divisions. The L1 and L2 layers are the epidermal and uppermost sub-epidermal layers, respectively. Cells in these two layers divide anticlinally. The L3 lies beneath the L2 and cells in the L3 divide in an essentially random fashion. Thus, cells can be members of the PZ and, e.g. the L2 layer and are as such clonally distinct from the neighbouring PZ cells in the L1 or L3 layers. The outer L1 and L2 layers and the more interior L3 and deeper cell layers are often also referred to as tunica and corpus, respectively (Steeves and Sussex, 1989).

While the apical and floral meristems share many basic aspects regarding their development and organization, several principal features distinguish the two types of meristems. The floral meristem is a lateral or axillary meristem that generates floral organs rather than leaves or whole flowers. Floral organs are produced in a whorled phyllotaxy. Finally, the floral meristem is determinate; i.e. in contrast to the SAM, the stem cell population in the flower is maintained only transiently and, upon production of the ovule-bearing carpels, the floral meristem does not generate additional floral organs.

The identity of floral organs is regulated by a set of floral homeotic genes. In short, the ABC model of floral homeotic gene function states that the combinatorial action of homeotic genes, encoding mostly MADS-domain transcription factors (Parenicova *et al.*, 2003), regulates floral organ identity at the whorl level (Schwarz-Sommer *et al.*, 1990; Coen and Meyerowitz, 1991; Weigel and Meyerowitz, 1994; Ng and Yanofsky, 2000; Lohmann and Weigel, 2002). A distinct combination of active MADS-box genes defines an individual whorl identity. A-function gene activity alone establishes the outermost whorl, or whorl 1, which develops sepals. A- and B-function genes together establish whorl 2, B- and C-function genes, whorl 3 and genes of the C-function group alone, whorl 4, which generates carpels. Further complexities of the ABC model have become apparent with the recent identification of the *SEPALLATA* (*SEP*)-class of MADS-box genes. These genes act in concert with ABC genes to regulate whorl identity (Pelaz *et al.*, 2000, 2001; Honma and Goto, 2001; Ditta *et al.*, 2004). Furthermore, little is known about the direct target genes of the homeotic genes and how these target genes contribute to the morphogenesis of individual floral organs (Sablowski and Meyerowitz, 1998; Ito *et al.*, 2004).

Are floral homeotic genes master regulatory genes that ultimately regulate all aspects of floral organogenesis? The answer is no. Interestingly, and perhaps counter to intuition, there exist two types of cell proliferation patterns during floral organ ontogenesis in Arabidopsis (Hill and Lord, 1989; Crone and Lord, 1994; Bossinger and Smyth, 1996; Jenik and Irish, 2000). Early cell proliferation events, such as initiation and early outgrowth of floral organs, depend not on floral organ identity but on the relative position of organ primordia in the floral meristem. Floral homeotic gene activity does not influence those patterns. Conversely, cell proliferation patterns during advanced floral organogenesis depend on organ identity and thus, the floral homeotic genes (Hill and Lord, 1989; Crone and Lord, 1994, Bossinger and Smyth, 1996; Jenik and Irish, 2000).

This chapter focuses on the genetic control of size and shape of flowers and floral organs. For general discussions, regarding the cellular aspects underlying growth control in all plant organs, we recommend a number of excellent recent reviews (Meijer and Murray, 2001; Mizukami, 2001; Fleming, 2002; Beemster *et al.*, 2003; Menand and Robaglia, 2004).

4.2 Flower primordium outgrowth

Flowers develop from lateral meristems arising at the flanks of the inflorescence apex with a regular and usually spiral phyllotaxy. The process of how lateral meristems acquire floral identity is well understood. A complex regulatory network integrating the action of extrinsic signals, such as temperature and photoperiod, with the action of intrinsic factors, such as floral meristem genes, is responsible for lateral meristems adopting a floral fate (Lohmann and Weigel, 2002; Mouradov *et al.*, 2002; Simpson and Dean, 2002). Much less is known regarding the mechanisms governing the radial positioning, initiation and the outgrowth of floral primordia and floral organ primordia. A combination of results based on physiology, micromanipulations, genetics and molecular cell biology provides clear evidence for a prominent role of auxin and polar auxin transport in these processes (Estelle, 1998; Friml and Palme, 2002; Friml, 2003).

A major advance was the analysis of *pin-formed* (*pin*) and *pinoid* (*pid*) mutants in Arabidopsis (Goto *et al.*, 1987; Okada *et al.*, 1991; Bennett *et al.*, 1995). The *pin* and *pid* mutants show pleiotropic effects but are also characterized by "naked" inflorescences that mostly fail to form flowers and end in cone-like tips. Wild-type plants cultured in the presence of inhibitors of polar auxin transport, such as naphthylphthalamic acid, mimic the pin-like phenotypes of *pin* or *pid* mutants. Sometimes both mutants are able to produce flowers that exhibit variable defects regarding floral organ number, size and shape (Okada *et al.*, 1991; Bennett *et al.*, 1995). Marker gene expression studies indicated that the basic organization of the apical meristems of *pin* and *pid* mutants is not noticeably disturbed (Christensen *et al.*, 2000; Vernoux *et al.*, 2000). Lateral organ markers or a marker for the organ boundary were expressed in a ring-like pattern in *pin1-1* plants (Vernoux *et al.*, 2000). The combined results suggest that *PIN* and *PID* are normally required for the outgrowth of floral primordia and for organ positioning and separation. Interestingly, exogenous application of the auxin IAA to the apices of *pin1-1* mutants induced the formation of flower primordia at the site of auxin application (Reinhardt *et al.*, 2000). Further experiments provided convincing evidence that polar auxin transport is a central regulator of phyllotaxis (Reinhardt *et al.*, 2003b).

Auxin and polar auxin transport also contribute to the regulation of lateral organ outgrowth and size (see later). Micromanipulation experiments indicate that the amount of auxin present at the site of the budding primordium correlates with its size. For example, the application of increasingly higher concentrations of auxin to tomato apices cultured in the presence of a polar auxin transport inhibitor resulted in the formation of correspondingly larger leaf primordia (Reinhardt *et al.*, 2000). In accordance with a role of auxin in organ formation, a dynamic auxin gradient, with the maximum at the organ tip, is present in all developing organ primordia of Arabidopsis (Benková *et al.*, 2003). The *PIN* and *PID* genes appear to be central players in the formation of such auxin gradients. *PIN* is part of a gene family that encodes putative auxin efflux carriers (Gälweiler *et al.*, 1998; Luschnig *et al.*, 1998; Müller *et al.*, 1998). PIN proteins are involved in a range of auxin-related activities ranging from processes regulating patterning in the root meristem or early embryogenesis, tropisms and lateral organ formation (Benková *et al.*, 2003; Friml, 2003). The PINs are considered to be important auxin efflux regulators despite the absence of direct proof that they actually function as auxin transporters (Palme and Gälweiler, 1999; Friml and Palme, 2002; Friml, 2003). PIN proteins are localized in the plasma membrane in a polarized fashion correlating with the direction of auxin flux. PIN proteins cycle between the plasma membrane and endosomes (Geldner *et al.*, 2001, 2003). Such a mechanism allows for the rapid relocation of PIN proteins and thus for plasticity in the directions of auxin fluxes. *PID* controls the polarity of this cycling process within a cell (Friml *et al.*, 2004). *PID* encodes a protein kinase that affects polar auxin transport (Christensen *et al.*, 2000; Benjamins *et al.*, 2001) by acting as a binary switch regulating the apical–basal distribution of PIN within a cell (Friml *et al.*, 2004).

The PIN-dependent redirections of polar auxin transport and the transient local auxin gradients constitute a central developmental module required for organ formation in plants (Benková *et al.*, 2003; Reinhardt *et al.*, 2003b). According to the 'reverse fountain' model for the development of aerial organs, auxin accumulates at the site of organ initiation stimulating cell division and primordium outgrowth commences. During the early development of the primordium, the *PIN*-dependent polar auxin transport machinery supplies auxin to the tip through the outer cell layers generating a new auxin gradient, with its maximum at the primordium tip. This mechanism inherently determines the growth axis of the primordium. Auxin from the tip is transported downward through the interior of the primordium, allowing the differentiation of vascular tissue (Benková *et al.*, 2003).

4.3 Regulating flower meristem size

The size of a flower depends on factors such as the number of floral organs and floral organ size, but also on floral meristem size which is influenced by the extent of primordium outgrowth (described in the previous section). But there are other influences, some relating to pattern formation and others relating to more basic cellular aspects of floral meristem development. As outlined in the introduction, several distinct cell populations, either organized in cell layers or zones involving several cell layers, make up the typical floral meristem. It seems reasonable to suggest that some mechanism coordinates the development of these cell populations to allow the formation of a floral primordium of the correct size. Formally, differences in the development of any individual cell population, or a combination thereof, should affect meristem size. For example, a bigger floral meristem may be obtained by increasing the size of the PZ, through elevating the number of PZ cells. Such a flower would exhibit more floral organs, particularly in whorls 1 and 2. Alternatively, a similar increase of the CZ size may also result in an increase of the PZ as more CZ cells may be partitioned into the PZ. In this case, a flower may not only show an increased number of whorl 4 and 3 organs (carpels and stamens) but also more petals and sepals. This raises the question of whether, and if so how, these different cell populations communicate to enable and coordinate meristem development.

4.3.1 Cell–cell communication, pattern formation in the meristem and meristem size

The analysis of chimeras has helped to establish a firm experimental basis for the existence of cell communication between layers essential for normal floral meristem development (Tilney-Bassett, 1986; Szymkowiak and Sussex, 1996). For example, a clonal analysis involving the *fasciated* (*f*) mutant of tomato provided early evidence of a size-relevant cross-talk between the clonal cell layers (Szymkowiak and Sussex, 1992). Using periclinal chimeras in which the individual L1, L2 and L3 layers differed in their genetic makeup, Szymkowiak and Sussex (1992) showed that the internal L3 layer plays a crucial role in the regulation of the size of the entire floral primordium. In doing so the L3 layer must influence the behaviour of the clonally distinct L1 and L2 layers. Petal initiation in tomato also depends on the interaction between the meristem cell layers. The *LATERAL SUPPRESSOR* (*LS*) locus encodes a likely transcription factor of the VHIID class and plants defective in *ls* fail to initiate axillary meristems and petals (Schumacher *et al.*, 1999). A periclinal chimera carrying a *ls* mutant L1 layer but wild-type L2 and L3 layers showed normal petal initiation (Szymkowiak and Sussex, 1993). Additional examples for cell layer communication with relevance to floral development include the activities of floral meristem identity genes (Carpenter and Coen, 1995; Hantke *et al.*, 1995; Sessions *et al.*, 2000) and floral homeotic genes (Perbal *et al.*, 1996; Sieburth *et al.*, 1998; Jenik and Irish, 2000). In addition, recent microsurgical and laser ablation studies have shown that the shoot apical meristem can reorganize a new CZ following ablation of the old CZ and that the L1 layer controls the cell division pattern of subtending layers and meristem maintenance (Reinhardt *et al.*, 2003a).

Arguably, the best understood cell–cell communication mechanism in the apical and floral meristem relates to the WUS–CLV feedback loop maintaining the stem cell population in the CZ (Weigel and Jürgens, 2002; Carles and Fletcher, 2003; Bäurle and Laux, 2003; Gross-Hardt and Laux, 2003). In Arabidopsis, a set of experiments indicates that the mRNA expression pattern of the *CLAVATA3* (*CLV3*) gene (see later) serves as a marker for the stem cells (Fletcher *et al.*, 1999; Brand *et al.*, 2000;

Schoof *et al.*, 2000). The *WUSCHEL* (*WUS*) gene, encoding a putative homeodomain transcription factor, is a central player in the maintenance of stem cells; in *wus* mutants the stem cells are substituted with partially differentiated cells (Laux *et al.*, 1996; Mayer *et al.*, 1998). The *WUS* gene is expressed in a small patch of central rib meristem cells located beneath the stem cells. Thus, either *WUS* acts in an indirect fashion on the overlying cells or the *WUS* protein moves to the neighbouring cells, as was shown in the case of the homeodomain protein KNOTTED1 (KN1) and LEAFY (Lucas *et al.*, 1995; Sessions *et al.*, 2000). The present evidence, however, favours the former scenario (Gross-Hardt *et al.*, 2002). The three *CLAVATA* genes (*CLV1-3*) are required for maintaining the proper number of undifferentiated stem cells within a meristem (Leyser and Furner, 1992; Clark *et al.*, 1993, 1995; Crone and Lord, 1993; Laux *et al.*, 1996; Kayes and Clark, 1998). Mutations in any of the three *CLV* genes result in an enlarged meristem and an increase in the number of lateral organs. The proteins encoded by the *CLV* loci are components of a receptor-mediated signal transduction mechanism (see later) that plays an important role in delimiting the upper and lateral borders of the *WUS* mRNA expression domain (Brand *et al.*, 2000; Schoof *et al.*, 2000). A feedback loop between *CLV3* and *WUS* appears to maintain the required number of stem cells (Brand *et al.*, 2000; Schoof *et al.*, 2000).

In recent years evidence has accumulated that the three CLV proteins function in a receptor complex. CLV1 is a leucine-rich repeat receptor-like kinase (LRR-RLK) (Clark *et al.*, 1997; Williams *et al.*, 1997; Stone *et al.*, 1998) and CLV2 is a LRR-receptor-like protein (LRR-RLP) with a short cytoplasmic tail lacking a kinase domain (Jeong *et al.*, 1999). CLV1 is present in a 185-kDa inactive complex and in an active 450-kDa complex which includes the 185-kDa complex but also contains a KINASE-ASSOCIATED PROTEIN PHOSPHATASE (KAPP) and a Rho GTPase-related protein (ROP) (Trotochaud *et al.*, 1999). It is possible that the 185-kDa complex consists of CLV1 disulfide-linked to CLV2. CLV2 is required for the protein stability of CLV1 and the calculated size of a CLV1/CLV2 dimer would fit the data (Jeong *et al.*, 1999, see later). The CLV3 gene encodes a small secreted protein of 96 residues that is considered to be the ligand of CLV1 (Fletcher *et al.*, 1999; Rojo *et al.*, 2002). Intercellular movement of CLV3 appears to be restricted in part by CLV1 (Lenhard and Laux, 2003). Furthermore, CLV3 is required for the assembly of the 450-kDa complex (Trotochaud *et al.*, 1999). KAPP was originally isolated in an interaction cloning experiment using the catalytic domain of the receptor kinase RLK5 (recently renamed HAESA; Jinn *et al.*, 2000) as bait (Stone *et al.*, 1994). KAPP is a promiscuous protein that is able to bind to several different RLKs (Braun *et al.*, 1997). These interactions occur through its kinase interaction domain (KI) that carries a forkhead associated (FHA) homologous region (Li *et al.*, 1999). KAPP is able to associate with the phosphorylated form of CLV1 to dephosphorylate CLV1 and to act as a negative regulator of CLV1 signalling (Williams *et al.*, 1997; Stone *et al.*, 1998). *POLTERGEIST* (*POL*) encodes a putative, phosphatase 2C and appears to be a novel regulator of CLV signalling (Yu *et al.*, 2000, 2003).

A second prominent regulator of stem cell development is *SHOOT MERISTEMLESS* (*STM*) (Barton and Poethig, 1993; Clark *et al.*, 1996; Endrizzi *et al.*, 1996), an Arabidopsis homologue of the maize gene *KNOTTED1* (Vollbrecht *et al.*, 1991, 2000; Long *et al.*, 1996; Kerstetter *et al.*, 1997). In strong *stm* mutants, no meristem can be detected. Moreover, *STM* appears to act independently of *WUS* and the *CLV* genes. A major role of *STM* appears to be the negative regulation of the *ASYMMETRIC LEAVES* genes *AS1* and *AS2* in the vegetative shoot apical meristem (Byrne *et al.*, 2000, 2002).

ULTRAPETALA (*ULT*) is another recently identified gene involved in regulating meristem development (Fletcher, 2001; Carles *et al.*, 2004). The predicted ULT protein is cysteine-rich, carries a domain resembling a B box and may thus be a transcription factor (Carles *et al.*, 2005). Plants carrying a *ult* mutation exhibit enlarged inflorescence and floral meristems, which produce more flowers and floral organs, respectively. This indicates that *ULT* limits the activity of shoot and floral meristems in wild-type plants. Genetic studies indicate that the *ULT* and *CLV* genes act in separate pathways which nevertheless overlap in their negative regulation of inflorescence and floral meristem size. Athough *ULT* acts through *WUS* in the control of floral determinacy (see later), ULT is not part of the *WUS* and *STM* pathways that promote meristem activity (Fletcher, 2001; Carles *et al.*, 2004).

4.3.2 Cellular factors regulating floral meristem development

It is reasonable to assume that the basic cellular machinery somehow affects floral meristem size and thus the number of floral organs. Defects in this machinery often result in smaller meristems and, therefore, fewer organs. Thus, the analysis of mutants exhibiting larger floral meristems should facilitate the identification of factors with a specific role in regulating floral meristem size. This rationale is exemplified by the research on the *CLV* genes. Alterations in some basic cellular mechanisms can also lead to increases in floral meristem size. As described in more detail later, the diameter of floral meristems in corresponding mutants is increased but the height of the meristem shows only minor alterations, if any at all. Such effects indicate that the PZ is predominantly, but not necessarily exclusively, affected. Furthermore, the organ number of the perianth is more variable than the number of stamens and carpels. This contrasts with the situation in *clv* mutants. For example, mutations in *CLV3* primarily cause an increase in the population of CZ cells and thus a very noticeable elevation in meristem height. The increase in the girth of the meristem, and thus the size of the PZ, is assumed to be a secondary consequence of the hyperproliferation of the CZ cells. In agreement with this interpretation, the number of carpels and stamens is more dramatically affected than the number of perianth organs in *clv* mutants. Apart from alterations in floral meristem size, mutants with defects in those basic cellular processes also show clear, sometimes quite dramatic, pleiotropic phenotypes. These include, e.g. disorganized meristems, alterations in cell proliferation, cell size and shape, malformed floral organs or stem fasciation. Similar considerations hold true for a second class of data dealing with floral organ initiation. Such pleiotropic phenotypes can pose difficulties in infering if or what aspect of the affected process has a direct role in regulating meristem size and/or

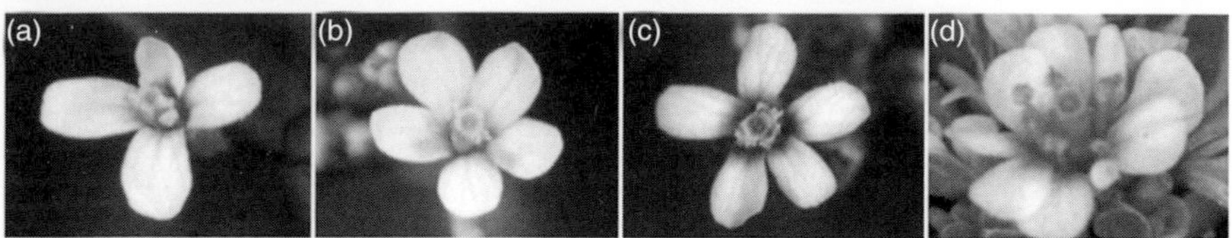

Figure 4.1 Flower morphology of wild-type Arabidopsis and mutants with increased floral organ numbers. (a) Wild-type flower. (b) *perianthia-1* flower. (c) *wiggum-1* flower. (d) *pluripetala-1* flower. Photos courtesy of Mark Running.

floral organ number, size and shape. Nevertheless, a gene defective in a mutant with a large meristem phenotype is very likely to play an important role in the regulation of meristem development.

The study of the *ENHANCED RESPONSE TO ABSISIC ACID 1* (*ERA1*)/*WIGGUM* (*WIG*) and *PLURIPETALA* (*PLP*) genes reveals that the prenylation of proteins affects meristem size and floral organ number (Figure 4.1). Prenylation describes the attachment of either the 15-carbon farnesyldiphosphate (FPP) or the 20-carbon geranylgeranyldiphosphate (GGPP) to cysteines close to the carboxy-terminus of a protein (Rodríguez-Concepción *et al.*, 1999). There is a single protein farnesyl transferase (PTF) and two protein geranylgeranyl transferases (PGGT-I, PGGT-II). PTF and PGGT-I are heterodimeric enzymes with a shared α subunit but distinct β subunits. The β subunits are important for substrate specificity (Rodríguez-Concepción *et al.*, 1999). Prenylation occurs in a large variety of proteins such as small GTP-binding proteins, protein kinases or transcription factors. Thus, prenylation is important for a diverse set of processes including signal transduction, membrane transport and transcriptional regulation. *PLP* encodes the α subunit shared between PFT and PGGT-I (Running *et al.*, 2004). *ERA1/WIG* encodes the β subunit of PFT (Cutler *et al.*, 1996; Ziegelhoffer *et al.*, 2000). Genetic evidence indicates that *ERA1/WIG* and *PLP* function independently from the *CLV* pathway (Running *et al.*, 1998, 2004). Given the nature of the ERA1/WIG and PLP proteins, it is not surprising that the two genes not only affect meristem size but many aspects of plant development and function. For example, *era1/wig* and *plp* mutants also show a slower overall growth rate, decreased internode length, aberrant meristem integrity and defects in the response to the hormone abscisic acid (ABA) (Cutler *et al.*, 1996; Pei *et al.*, 1998; Yalovsky *et al.*, 2000; Ziegelhoffer *et al.*, 2000; Running *et al.*, 2004). A future challenge will consist of identifying the prenylated proteins that influence meristem size.

The outcome of work on the *FASCIATA* (*FAS*) genes represents a second example for a basic cellular process regulating meristem size. Mutations in the two *FAS* genes lead to wider floral meristems bearing more sepals but fewer petals (Leyser and Furner, 1992). *FAS1* and *FAS2* encode components of the Arabidopsis chromatin assembly factor CAF-1 that is involved in chromatin assembly during replication and DNA repair (Kaya *et al.*, 2001). CAF-1 is thought to enable the stable propagation of epigenetic states.

The two *MGOUN* (*MGO*) genes appear to affect floral organ outgrowth in a more direct fashion. The *MGO* genes positively regulate organ initiation; in *mgo* mutants larger floral meristems with fewer floral organs are found (Laufs *et al.*, 1998a). The *MGO* genes probably function by allocating cells from the meristem to primordia. Organ initiation is also regulated by the nuclear kinase TOUSLED (TSL) (Roe *et al.*, 1993, 1997a,b). Interestingly, the activity of the mammalian TSL homologs is cell-cycle regulated, high during S phase, and coupled to the moving replication fork (Sillјé *et al.*, 1999). Furthermore, individual TSL proteins can phosphorylate histone H3 (Li *et al.*, 2001) and the human chromatin assembly factor AS1 (Sillјé and Nigg, 2001). Thus, the Arabidopsis TSL kinase may also function in the regulation of chromatin assembly during DNA replication. This hypothesis reasonably explains the pleiotropic phenotype of *tsl* mutants. Indeed, Arabidopsis TSL was recently found to behave in a comparable fashion to its mammalian homologs (Ehsan *et al.*, 2004). *TSO1* is required for cytokinesis and cell expansion (Liu *et al.*, 1997; Hauser *et al.*, 1998, 2000; Song *et al.*, 2000). As a result, primordium initiation is affected as well. *TSO1* encodes a novel cysteine-rich protein with homology to polycomb proteins. The analysis of CPP1, a close relative from soybean, indicated that CPP1 is a DNA-binding protein involved in negatively regulating the leghemoglobin c3 gene (Cvitanich *et al.*, 2000).

4.4 Early control of organogenesis in the flower

All mutants described earlier show alterations in organ number that are in part related to defects in meristem size or integrity. The corresponding genes are likely to regulate floral organ number in an indirect fashion. What then is known about mechanisms that specifically affect floral organ patterning aspects such as organ spacing and orientation? A set of genes is likely to be more directly involved in very early aspects of floral organ formation. For the most part, these genes function independently of floral homeotic gene activity.

The *PERIANTHIA* (*PAN*) gene is a salient example of a gene with an early role in specifying where floral organs initiate within a whorl (Running and Meyerowitz, 1996; Chuang *et al.*, 1999). *PAN* encodes a bZIP-like putative transcription factor. *PAN* appears to be required for rendering Arabidopsis flowers tetramerous; flowers of *pan* mutants are pentamerous, displaying five sepals, five petals, five stamens and two carpels (Figure 4.1). Plants mutant in *PAN* do not show any other obvious defects. Floral meristem size of *pan* mutants is apparently normal despite the increase in floral organ number in the outer two whorls. This suggests that *PAN* specifically functions in the regulation of floral organ spacing although the exact mechanism is not yet known. *PAN* seems to act independently from factors such as *CLV* and *ERA1/WIG*, but also from floral homeotic genes (Running and Meyerowitz, 1996). Furthermore, *PAN* expression does not depend on floral meristem identity genes even though its protein activity does depend on those genes. This indicates that *PAN* functions together with other floral factors. From an evolutionary standpoint, it is interesting to note that *pan* mutants exhibit pentamerous rather than tetramerous flowers, which are typical for members of the Brassicaceae. Pentamerous flowers are present in more ancestral families. Thus, it is tempting to speculate that *PAN* was a crucial factor in the evolution of tetramerous flowers in the Brassicaceae (Chuang *et al.*, 1999).

Another gene with a role in floral organ spacing is *ETTIN* (*ETT*) (Sessions *et al.*, 1997). *ETT* resembles *PAN* in its genetic interactions with the *CLV*, *ERA1/WIG* and the floral homeotic genes and *ETT* and *PAN* regulate organ spacing in a redundant fashion. However, the relationship between *ETT* and *PAN* remains unclear. The function of *ETT* is not restricted to floral organ spacing but is also required for the apical–basal patterning of the gynoecium. *ETT* is related to *AUXIN RESPONSE FACTOR 1*, and thus is likely to be a transcription factor involved in mediating downstream auxin responses (Sessions *et al.*, 1997). Auxin is known to influence the positioning of lateral organs around the circumference of the peripheral zone (Reinhardt *et al.*, 2000, 2003b; Stieger *et al.*, 2002; Vernoux *et al.*, 2000).

Lateral organ development in Arabidopsis flowers depends on *PRESSED FLOWER* (*PRS*) function (Matsumoto and Okada, 2001). *PRS* encodes a putative WUS-related homeodomain transcription factor. Flowers of *prs* mutants are characterized by absent or diminutive lateral sepals and stamens. Furthermore, the marginal cells of the medial (adaxial–abaxial) sepals are missing as are the stipules at the base of leaf petioles (Matsumoto and Okada, 2001; Nardmann *et al.*, 2004). *PRS* is expressed in two lateral foci flanking the anlagen of floral primordia. In addition, *PRS* expression can be detected in a few L1 cells that are part of the lateral domains of all lateral organs, including the cotyledons (Matsumoto and Okada, 2001; Nardmann *et al.*, 2004). Recent work has indicated the maize genes *NARROW SHEATH1* (*NS1*) and *NS2* to be close homologs of *PRS* (Nardmann *et al.*, 2004). The *NS* genes are important for the development of the two baso-lateral domains in maize leaves, possibly by helping to recruit lateral leaf founder cells (Scanlon *et al.*, 1996, 2000). It has been proposed that in maize and Arabidopsis the primary function of *NS* and *PRS* is to recruit meristem cells in a specific lateral domain to become lateral organ founder cells (Nardmann *et al.*, 2004). According to this view, the main function of *NS* and *PRS* resides in the specification of this particular lateral domain within the meristem and not in the elaboration of this domain during later steps of organogenesis. This hypothesis can explain the absence of entire lateral sepals, not just the lateral parts, combined with the defects observed in the lateral margins of the two medial sepals, and the absence of lateral stamens in *prs* mutants (Nardmann *et al.*, 2004).

Mutations in *PETAL LOSS* (*PTL*) usually lead to an absence of petals (Griffith *et al.*, 1999). Sometimes petals are produced but then often are misoriented. *PTL* encodes a putative transcription factor of the trihelix class, also known as GT factors (Brewer *et al.*, 2004). Curiously, *PTL* expression in very early floral development could not be detected in petal primordia but was observed in the four inter-sepal regions (Brewer *et al.*, 2004). It was only later, during the petal expansion phase, that *PTL* became detectable in petals. The early expression in the inter-sepal domains was, however, essential for wild-type development. In addition, *PTL* expression was found in marginal regions of floral primordia, sepals, petals and stamens as well as in young leaves. Indeed, *PTL* also regulates sepal separation (Brewer *et al.*, 2004). *PTL* appears to act in a whorl-specific and not organ-specific manner. For example, mutations in the floral homeotic gene *APETALA3* (*AP3*) lead to the formation of sepals in place of petals and carpels in place of stamens (Bowman *et al.*, 1989; Jack *et al.*, 1992). Flowers of *ap3 ptl* double mutants exhibit a reduction in the number of whorl 2 sepals (Griffith *et al.*, 1999). The main function of *PTL* appears to reside in growth suppression in the inter-sepal regions, thereby allowing full sepal separation. The effect on petal number and orientation seems to be an indirect consequence of *PTL* activity in the inter-sepal domains. It was speculated that overgrowth in inter-sepal zones of *ptl* mutants weakens a petal initiation signal (Griffith *et al.*, 1999). The role of *PTL* in margin development of a set of plant organs remains presently unclear. Given the growth-inhibitory function of *PTL* it is reasonable to suggest that *PTL* is required to restrict lateral expansion. It is possible that redundancy renders the analysis more difficult as *PTL* is a member of a multi-gene family (Brewer *et al.*, 2004).

Recently, *RABBIT EARS* (*RBE*) has been identified as a positive regulator of second whorl organ development (Takeda *et al.*, 2004). In *rbe* mutants, the petals fail to initiate or are malformed. *RBE* encodes a nuclear-localized putative transcription factor carrying one Cys2His2 zinc finger domain and its expression domain is restricted to the likely petal precursor cells and the early petal primordia (Takeda *et al.*, 2004). *RBE* function, however, is not specific to petals per se but rather to organs developing in the second whorl. *RBE* expression was still detected in the second whorl organ primordia of *ap3* mutants and the corresponding sepals were either reduced or filamentous in *ap3 rbe* double mutants (Takeda *et al.*, 2004). *PTL* is a positive regulator of *RBE*, as *RBE* expression could not be detected in *ptl* mutants (Takeda *et al.*, 2004). Genetic evidence indicates that *RBE* is not part of the *CLV1* pathway and that *RBE* also acts independently of *PAN* (Takeda *et al.*, 2004). *RBE* is a *SUPERMAN* (*SUP*)-like protein. However, in contrast to *RBE*, which is a positive regulator of cell division in petal precursor cells, *SUP* appears to be a negative regulator of cell proliferation at the boundary of the third and fourth whorls (Sakai *et al.*, 1995, 2000). *JAGGED* (*JAG*) represents another gene with homology to *RBE* and *SUP* (Dinneny *et al.*, 2004; Ohno *et al.*, 2004). Plants defective in *JAG* function show lateral organs, such as sepals and petals, with serrated margins. Apparently, *JAG* suppresses premature differentiation during the morphogenesis of lateral organs (Dinneny *et al.*, 2004, Ohno *et al.*, 2004).

Interestingly, *UNUSUAL FLORAL ORGANS* (*UFO*) affects petal primordium initiation and outgrowth (Durfee *et al.*, 2003; Laufs *et al.*, 2003). *UFO* differs from *RBE* or *PTL* as its activity appears to be required to counteract a long-range influence exhibited by the floral homeotic gene *AGAMOUS* (*AG*). *UFO* is the ortholog of the *Antirrhinum* gene *FIMBRIATA* (*FIM*) and encodes an F-box protein (Simon *et al.*, 1994; Ingram *et al.*, 1995, 1997; Bai *et al.*, 1996; Lee *et al.*, 1997; Samach *et al.*, 1999). Thus, UFO is likely to target proteins for degradation through the 26S proteasome pathway. Flowers of *ufo* mutants display a complex phenotype. The most well-known aspect of *UFO* and *FIM* function relates to their roles as intermediates between the meristem identity and floral organ identity genes. But the mutant phenotype also indicated that *UFO* regulates cell proliferation and organ growth. In accordance with its multiple roles, *UFO* has a dynamic expression profile during early floral development. Eventually, however, its expression becomes restricted to the developing petals (Ingram *et al.*, 1995; Lee *et al.*, 1997). The direct role of *UFO* in petal development was discovered by two different approaches. One approach consisted in an analysis of mutations located in the promoter of *UFO*. Mutations such as *ufo-11* or *ufo-12* lead to the absence of the *UFO* expression during petal development and result in a petal-less phenotype (Durfee *et al.*, 2003). In the second approach, *UFO* was transiently expressed in *ufo* mutants during early floral development using an ethanol-inducible system (Laufs *et al.*, 2003). This strategy bypassed the early requirements for *UFO* and allowed the investigation of the later aspects of *UFO* function. As in the case of *PTL* and *RBE*, it appears that *UFO* regulation of organ initiation in whorl 2 is independent of the type of organ produced in this whorl as the flowers of *pi ufo-11* double mutants did not generate the expected sepals in whorl 2. There is evidence that *UFO* antagonizes the activity of the floral homeotic gene *AGAMOUS* (*AG*) activity in whorl 2 (Durfee *et al.*, 2003). *AG* is normally expressed in whorls 3 and 4 and suppresses petal development when ectopically expressed in whorl 2 (Bowman *et al.*, 1991; Drews *et al.*, 1991; Mizukami and Ma, 1992; Jack *et al.*, 1997). Indeed, flowers of *ag-1 ufo-11* double mutants form petals in whorl 2 (Durfee *et al.*, 2003). Interestingly, however, ectopic expression of *AG* was not observed in *ufo-12* mutants suggesting that *AG* interferes with petal development in a non-cell autonomous fashion. One function of *UFO* consists in blocking this negative influence of *AG*. In the absence of *AG* function, *UFO* becomes dispensable for petal formation. These results are in accordance with the findings that *AG* regulates the relative contributions of the L2 and L3 layers to the petals in a non-cell autonomous manner (Jenik and Irish, 2000). How does *UFO* relate to *PTL* and *RBE*? Taking into account the genetic interactions of *UFO* and *PTL* with *AG*, it is likely that *UFO* and *PTL* act in different pathways. The *ag ptl* double mutants showed an additive phenotype and lacked whorl 2 organs (Griffith *et al.*, 1999). The relationship between *UFO* and *RBE* is not known at present.

4.5 Generating organ boundaries

As a rule, floral organs arise as distinct units on the floral meristem. Clonal analysis indicates that only a relatively small and fixed number of organ founder cells in the early meristem contribute to a given floral organ indicating that this separation may occur before morphological evidence is apparent (Bossinger and Smyth, 1996; Furner, 1996). Furthermore, alterations in organ separation have been of great evolutionary consequence. For example, sympetaly, the congenital fusion of petals, represents a key innovation in the evolution of flower morphology, particularly in the asterids (Endress, 2001b). Sympetaly is achieved by meristem fusion of the petal primordia and by the action of an intercalary meristem that produces the corolla tube. Sympetaly apparently allows the architecture of the flower to become more stable and provides the basis for the generation of the largest flowers on earth. In addition it allowed the evolution of very elaborate pollination modes (Endress, 2001b). These considerations raise the obvious question as to what separates developing floral primordia and floral organ primordia from the surrounding cells? It is believed that specialized lateral organ boundaries are necessary for this process to happen.

What characterizes floral meristem and floral organ boundaries at the cellular level and what are the molecular mechanisms governing the development of lateral organ boundaries? A detailed analysis of cell proliferation patterns during floral development in Arabidopsis, by assaying the incorporation of the thymidine analogue BrdU in the DNA of flower meristem cells, revealed clear evidence for the existence of two- to three-cell wide bands of non-dividing cells between floral organ primordia (Breuil-Broyer *et al.*, 2004). In addition, those non-proliferating cells also failed to express a set of genes encoding different components of the basic cell-cycle machinery (Breuil-Broyer *et al.*, 2004). Perhaps differences in cell expansion are also essential for boundary formation. It was noted that during the separation of floral meristems from inflorescence meristems, cells located in a narrow band corresponding to the floral boundary region exhibited a lack in cell expansion (Reddy *et al.*, 2004). Thus, potential organ boundaries in developing flowers are likely to be characterized, at least in part, by a narrow band of cells with reduced cell cycle and cell expansion activity.

Analysis of orthologous genes in three plant species has increased the knowledge regarding the mechanism required for the separation of lateral organs. *NO APICAL MERISTEM* (*NAM*) in petunia (Souer *et al.*, 1996), the three closely related and redundant *CUP-SHAPED COTYLEDON* (*CUC1-3*) genes in Arabidopsis (Aida *et al.*, 1997; Takada *et al.*, 2001; Vroemen *et al.*, 2003) and the *CUPULIFORMIS* (*CUP*) gene from Antirrhinum (Stubbe, 1966; Weir *et al.*, 2004) are involved in setting lateral organ boundaries. These genes are members of the plant-specific NAC gene family (Duval *et al.*, 2002) that encode proteins that can act as transcription factors (Taoka *et al.*, 2004). They have a very intriguing expression pattern as their expression is found in a narrow band of cells at the boundary of lateral organs and the meristem, e.g. in the case of cotyledons, leaves, flowers and floral organs (Souer *et al.*, 1996; Aida *et al.*, 1997, 1999; Ishida *et al.*, 2000; Vroemen *et al.*, 2003; Breuil-Broyer *et al.*, 2004; Weir *et al.*, 2004). In addition, some are expressed in the peri-organ boundaries within floral whorls. Formation of boundaries around cotyledons appears to be important for early SAM development, as in *nam*, *cuc1 cuc2* or *cup* seedlings, a SAM is not present (Souer *et al.*, 1996; Aida *et al.*, 1997, 1999; Ishida *et al.*, 2000; Vroemen *et al.*, 2003; Weir *et al.*, 2004). The study of later aspects of the function of these genes is possible since escape shoots often develop or shoots can be regenerated from calli. For example, flowers of *cuc1 cuc2* double mutants, or *nam* single mutants, display sepal fusion or stamen fusions (Souer *et al.*, 1996; Aida *et al.*, 1997). In *cup* mutants, sepals and petals are fused, as are ovules (Weir *et al.*, 2004). The phenotype suggests that those genes regulate the separation of organs by suppressing cell division at the organ boundaries. Interestingly, a single gene in Antirrhinum appears to regulate a process which in Arabidopsis is controlled by at least three redundant genes. There is evidence that CUP interacts with TIC (*TCP-Interacting with CUP*; Weir *et al.*, 2004) a member of the TCP-family of transcription factors (Cubas *et al.*, 1999b; Kosugi and Ohashi, 2002). Many of the *TCP*-family genes appear to be involved in the suppression of cell division and organ outgrowth (Luo *et al.*, 1996; Doebley *et al.*, 1997; Cubas *et al.*, 1999a; Nath *et al.*, 2003, Palatnik *et al.*, 2003) and TCP proteins, such as PCF1 and PCF2 in rice or AT-TCP20 in Arabidopsis, bind to the promoter of the proliferating cell nuclear antigen gene (*PCNA*) that plays a role in the regulation of DNA synthesis and the cell cycle (Kosugi and Ohashi, 1997; Trémousaygue *et al.*, 2003). Thus, Weir *et. al.* suggest a more specific model for lateral organ boundary formation in Antirrhinum; in their view, boundary formation is achieved through the interaction of the boundary-specific, NAC-domain factor CUP, with the more generally expressed TCP-domain factor, TIC. This interaction eventually results in the inactivity of genes positively regulating cell division (Weir *et al.*, 2004). Obviously, this scenario could easily be expanded to a formal description of the more complex situation in Arabidopsis. Recent evidence in Arabidopsis, however, indicates that the activity of *CUC* genes does not lead to noticeable global reduction in cell division during the early phase of sepal boundary formation (Laufs *et al.*, 2004). These data raise the possibility that *CUC* gene expression needs to be shut down quickly and precisely to avoid enlargement of the boundary. Such a hypothesis is in accordance with the recent finding that *CUC1* and *CUC2* are under control of a micro-RNA-based mechanism (Laufs *et al.*, 2004; Mallory *et al.*, 2004; Baker *et al.*, 2005).

The molecular basis of boundary formation is more complex than outlined earlier. At least in the Arabidopsis embryo there is differential expression of individual *CUC* genes in response to alterations in auxin transport. For example, in *pin-formed1* mutants *CUC1* expression is expanded while the *CUC2* expression domain is reduced (Aida *et al.*, 2002). In addition, another factor likely to be involved in lateral organ boundary formation is the Arabidopsis gene *LATERAL ORGAN BOUNDARIES* (*LOB*) (Shuai *et al.*, 2002). While a *lob* mutant analysis did not reveal a noticable phenotype, expression studies using an enhancer-trap line or promoter-GUS constructs, indicate that *LOB* is expressed in organ boundaries in a similar manner as the *CUP* or *CUC* genes.

4.6 Floral organ size

Apart from a function in the control of phyllotaxy and the outgrowth of floral and floral organ primordia there is evidence to indicate a specific role for auxin and polar auxin transport in the coordination of cell proliferation and growth at the floral organ level. For example, plants with a defect in *AUXIN-RESISTANT1* (*AXR1*) that encodes a protein with similarity to ubiquitin-activating enzyme E1 (Leyser *et al.*, 1993), show smaller lateral organs due to a reduction in cell proliferation (Lincoln *et al.*, 1990). *AXR1* represents an early component of the auxin-signalling process (Leyser, 2002; Dharmarsiri and Estelle, 2004). In addition, mutations in *REVOLUTA* (*REV*), also known as *INTERFASCICULAR FIBERLESS1* (*IFL1*), lead to pleiotropic phenotypes (Talbert *et al.*, 1995; Zhong *et al.*, 1997; Otsuga *et al.*, 2001). However, *rev/ifl1* mutants regularly produce enlarged leaves, albeit of abnormal morphology, and sometimes also fertile flowers with enlarged floral organs (Talbert *et al.*, 1995). The increase in leaf size appears to be due to supernumerary cell divisions, indicating that one function of *REF/IFL1* consists in limiting cell proliferation during organ formation (Talbert *et al.*, 1995). *REV/IFL1* encodes a putative class III HD-ZIP transcription factor (Zhong and Ye, 1999; Ratcliffe *et al.*, 2000; Otsuga *et al.*, 2001) and *rev/ifl1* mutants show a defect in polar auxin transport in the stem (Zhong and Ye, 2001).

The study of two genes, *AUXIN-REGULATED GENE INVOLVED IN ORGAN SIZE* (*ARGOS*) (Hu *et al.*, 2003) and *AINTEGUMENTA* (*ANT*) (Krizek, 1999; Mizukami and Fischer, 2000), generated very convincing evidence for a prominent role of auxin in the regulation of floral organ size. ARGOS represents a novel, plant-specific protein while ANT is an AP2-type transcription factor (Weigel, 1995; Elliott *et al.*, 1996; Klucher *et al.*, 1996; Vergani *et al.*, 1997; Nole-Wilson and Krizek, 2000; Krizek, 2003). *ARGOS* transcription was found to be upregulated upon exposure of roots of seven-day-old seedlings to the auxin naphthylacetic acid (NAA) (Hu *et al.*, 2003). This auxin-mediated transcriptional upregulation was dependent on *AXR1* but not on other auxin signalling factors such as *AXR2*, *TIR1* or *IAA28*. *ARGOS* appears to have a prominent role in regulating organ size. Plants with reduced levels of *ARGOS* showed smaller rosette leaves, thinner stems and flowers with smaller floral organs. In contrast, plants transgenic for a *35S::ARGOS* construct, constitutively expressing *ARGOS* at high levels, exhibited organ hyperplasia. The plants had enlarged leaves, thicker stems and flowers with the normal complement of organs but which were of larger size. Interestingly, the larger leaves and floral organs of *35S::ARGOS* plants displayed normal overall morphology. The variation in leaf and floral organ size was found to be due to differences in cell number. In particular, ectopic expression of *ARGOS* did not alter the rate of cell proliferation but led to a prolonged period of cell division during organogenesis, accompanied by an extended expression of *CYCLIN D3;1*. Thus, it was proposed that *ARGOS* links auxin signalling, cell proliferation and organ size control.

A role in coordinating cell proliferation and organ size was previously proposed for *ANT* as well (Mizukami and Fischer, 2000). Loss-of-function mutations in *ANT* affect all aboveground lateral organs and also result in a smaller floral meristem, a variable loss of floral organs, as well as smaller and narrower floral organs (Elliott *et al.*, 1996; Klucher *et al.*, 1996; Krizek, 1999). Constitutive expression of *ANT* does not affect floral meristem size and results in flowers with a normal complement of floral organs. However, in a manner similar to plants with constitutive *ARGOS* activity, *35S::ANT* plants carry floral organs which exhibit an increased size (Krizek, 1999; Mizukami and Fischer, 2000). Thus, modulating *ARGOS* or *ANT* activity leads to very similar effects on floral organ size. *ANT* appears to act downstream of *ARGOS* as *ANT* expression responds to *ARGOS* activity and the formation of enlarged floral organs in *35S::ARGOS* plants is dependent on *ANT* activity (Hu *et al.*, 2003). *ANT*, in turn, is likely to function upstream of *CYCLIN D3;1* (Dewitte *et al.*, 2003). There exists some discrepancy in the literature as to what is the cellular function of *ANT*. In one paper the authors observed a prolonged phase of cell proliferation eventually leading to enlarged organs ranging from leaves to carpels and seeds. Those results led the authors to postulate that *ANT* has a general function in cell proliferation control and meristematic competence (Mizukami and Fischer, 2000). Krizek (1999) showed that the increased organ size of *35S::ANT* plants was apparently restricted to floral organs. Furthermore, an elevated level of cell divisions was only observed for sepals while in petals and stamens mainly cell size appeared to be increased. On the basis of her data the author raised the hypothesis that *ANT* plays a primary role in the control of cell size (Krizek, 1999). While presently unknown, likely reasons for the observed discrepancies include differences in the genetic background or growth conditions. In the future it will be important to corroborate these studies as the authors propose rather contrary roles for *ANT* in organ size control. Whatever the exact function of *ANT*, the combined evidence discussed here earlier clearly suggests that auxin regulates floral organ size through a regulatory pathway including *AXR1*, *ARGOS*, *ANT* and *CYCD3;1*. The identification of additional components of this pathway in future studies should lead to interesting new insights in the coordination of cell proliferation and organ size control.

4.7 Flower shape and symmetry

The shape of a flower obviously depends on the type and number of organs present. Patterning processes that establish the determinacy and the dorsoventral polarity of the flower and the polarity within floral organs contribute essentially to flower shape (Coen and Carpenter, 1993; Weigel and Meyerowitz, 1994; Coen, 1996; Sessions and Yanofsky, 1999; Ng and Yanofsky, 2000; Endress, 2001a; Lohmann and Weigel, 2002).

4.7.1 Floral determinacy

Floral determinacy intrinsically impacts on floral shape. The inflorescence and floral meristems differ not only because they generate different types of lateral organs. As a rule the inflorescence meristem continues development and thus repeatedly initiates lateral organs until the plant eventually dies. Therefore, the inflorescence meristem is considered indeterminate. In contrast, floral meristems are determinate as they produce a defined small set of floral organs (Steeves and Sussex, 1989). The regulation of floral determinacy is best understood in Arabidopsis. The determined state of the floral meristem implies that the pool of stem cells in the central zone is only transiently maintained. How does this behaviour tie in with the negative *WUS–CLV* feedback loop that maintains the size of the stem cell population? In addition, the stem cells should stop development after the floral meristem has produced the innermost organs, the carpels. Therefore, it is conceivable that the regulation of floral determinacy somehow relates to specification of the carpels. Indeed, recent data indicate that another negative feedback loop involving *WUS*, the floral meristem gene *LEAFY* (*LFY*) (Schultz and Haughn, 1991; Huala and Sussex, 1992; Weigel *et al.*, 1992), and the floral homeotic gene *AG*, is important for the regulation of floral determinacy (Lenhard *et al.*, 2001; Lohmann *et al.*, 2001). *AG* regulates the identity of the third and fourth whorl, i.e. stamen and carpels, and encodes a MADS-domain transcription factor (Bowman *et al.*, 1989; Yanofsky *et al.*, 1990; Shiraishi *et al.*, 1993; Mizukami *et al.*, 1996; Riechmann *et al.*, 1996). In addition, *AG* is required for floral determinacy. For example, flowers of *ag* mutants show extended development, with whorl 4 being replaced by an essentially unlimited number of extra whorls carrying either sepals or petals (Bowman *et al.*, 1989; Mizukami and Ma, 1995, 1997; Sieburth *et al.*, 1995). In addition, ectopically expressing *AG* in the inflorescence meristem results in the formation of a central flower (Mizukami and Ma, 1997). *LFY* encodes a transcription factor that directly regulates the activity of *AG* (Parcy *et al.*, 1998; Busch *et al.*, 1999).

WUS is required for stem cell development in both the inflorescence and the floral meristems (Laux *et al.*, 1996). *WUS* is only transiently active in the floral meristem as indicated, e.g. by the fact that *WUS* expression becomes undetectable halfway through flower development (Mayer *et al.*, 1998). The down-regulation of *WUS* expression requires *AG* activity. In addition, *LFY* and *WUS* are needed to activate AG expression (Lenhard *et al.*, 2001; Lohmann *et al.*, 2001). LFY is known to act as a direct activator of *AG* (Parcy *et al.*, 1998; Busch *et al.*, 1999). Thus, would WUS also act directly on the *AG* promoter? The *in planta* manipulation of *WUS* activity and promoter analyses (Lenhard *et al.*, 2001; Lohmann *et al.*, 2001), combined with yeast transactivation tests and *in vitro* DNA-binding assays (Lohmann *et al.*, 2001), strongly indicate that WUS also functions as a direct activator of *AG* expression. The data show that LFY and WUS bind to adjacent regulatory elements in the *AG* promoter. On the basis of these studies, regulation of floral determinacy in Arabidopsis involves an autoregulatory loop in which the combined action of LFY and WUS activates the expression of *AG*, which in turn is required for the repression of *WUS*. It remains to be seen if AG directly regulates *WUS* expression. The negative role of *AG* on *WUS* expression is in agreement with the finding that *AG* needs to be active only in the L2 and L3 layers of the floral meristem to regulate floral determinacy (Sieburth *et al.*, 1998). In the young flower, and in contrast to the inflorescence meristem where *WUS* RNA is localized to a group of L3 and L4 cells, *WUS* is expressed in a small group of cells restricted to parts of the L2 and L3 cell layers (Mayer *et al.*, 1998). It is noteworthy to point out the major differences between the *WUS–CLV* and the *WUS–AG* loops. The first loop involves communication between cells while the second loop, in principle, could act in a cell autonomous fashion.

Another distinction is that the *WUS* and *CLV* genes act simultaneously while the activities of *WUS* and *AG* are separated in time. Both regulatory mechanisms seem to act in a largely additive manner (Clark *et al.*, 1993).

As a rule, floral determinacy is tight and thus it may not surprise us that the expression of the stem cell regulator *WUS* in the floral meristem is under complex negative control. Genetic evidence suggests that apart from *AG*, additional factors play a role in floral determinacy (Mizukami and Ma, 1997). Recently, work on *ULTRAPETALA* (*ULT*) uncovered another pathway with a role in floral determinacy and the regulation of *WUS* expression (Fletcher, 2001; Carles *et al.*, 2004). A role for *ULT* in floral determinacy is suggested by the variable development of supernumerary whorls of carpels, stamens or undifferentiated tissue within gynoecia of *ult* mutants (Fletcher, 2001). As outlined earlier, *ULT* acts separately from *CLV*, *WUS* and *STM* in the regulation of inflorescence and floral meristem size. As far as floral determinacy is concerned, however, *ULT* and *WUS* function in the same pathway and act in an antagonistic fashion (Carles *et al.*, 2004). In a similar fashion to *AG*, the *ULT* gene is required to terminate *WUS* transcription around halfway through floral development (Carles *et al.*, 2004). Perhaps *ULT* represses *WUS* through *AG* as *AG* expression is delayed in the centre of *ult* flowers (Fletcher, 2001). Although experimental support is still lacking, *ULT* could act together with *LFY* and *WUS* to activate *AG*.

4.8 Dorsoventral symmetry

Flowers of certain species display a very prominent asymmetry (Endress, 1999). Antirrhinum flowers, e.g. display a distinct dorsoventral asymmetry and exhibit a single plane of symmetry. They are thus variably referred to as monosymmetric, asymmetric, irregular, bilaterally symmetrical or zygomorphic. At the morphological level, this dorsoventral asymmetry is particularly evident in whorls 2 and 3. The flower carries two dorsal petals, two lateral petals and a single ventral petal. While the petals are fused at the base, thereby forming a tube, the unfused parts still form distinct petal lobes. The individual petal types differ in size and shape. In addition, the dorsal and lateral petals exhibit a dorsoventral asymmetry at the organ level itself (see later). The ventral petal, however, shows bilateral symmetry. The dorsal stamen primordium only develops into a staminode, a stamen arrested at an early stage, and the two lateral stamens are shorter than the ventral stamens. In contrast, no such clear dorsoventral distinctions characterize mature Arabidopsis flowers, which are thus called symmetric, regular, poly- or radial symmetric, or actinomorphic (Figure 4.2). A combination of genetic and molecular work has identified a set of genes that are required for the elaboration of dorsal–ventral asymmetry in Antirrhinum flowers (Coen, 1996; Sessions and Yanofsky, 1999). It appears that the corresponding mechanism is only at work in flowers. Aberrations in plants carrying mutations in dorsoventral patterning genes are restricted to the flowers while other organs exhibiting a pronounced dorsoventral (adaxial–abaxial) polarity, such as leaves, do

Figure 4.2 Flowers with variant petal and corolla contortion patterns. (a) Arabidopsis Landsberg *erecta* flower. The two Arabidopsis mutants, (b) *lefty-1* and (c) *spiral2-1*, present petal contortions in the left direction and right direction, respectively. (d) Enantiomorphy in *Hibiscus tiliaceus* (Malvaceae). Two flowers, located on the same plant, depict naturally occurring, alternative forms of corolla contortion. The flower on the left shows a sinistrorse (left direction) arrangement; the flower on the right, a dextrorse (right direction) arrangement. Photos (a–c) courtesy Takashi Hashimoto; (d) courtesy of Peter Endress.

not seem to be altered. Two related genes, *CYCLOIDEA* (*CYC*) and *DICHOTOMA* (*DICH*), are required for proper dorsal development and encode putative TCP-class DNA-binding proteins (Stubbe, 1966; Carpenter and Coen, 1990; Luo *et al.*, 1996, 1999; Almeida *et al.*, 1997; Cubas *et al.*, 1999a). In contrast, *DIVARICATA* (*DIV*) encodes a putative MYB-type transcription factor and is involved in the development of ventral characteristics (Almeida *et al.*, 1997; Galego and Almeida, 2002).

The *CYC* and *DICH* genes play partially overlapping roles. Mutations in *CYC* result in a partial loss of the dorsoventral asymmetry in the flower and mainly affect the dorsal half of the flower (Luo *et al.*, 1996). Flowers from *cyc* mutants are also termed semipeloric and are characterized by an altered size and shape of petals and by the elevated number of sepals, petals and stamens. The dorsal petals show a mixture of dorsal and lateral characteristics while the lateral petals are ventralized. In addition, up to six sepals, petals and stamens (but no staminode) develop. Organ number is not affected in plants defective for *DICH* activity as they carry flowers with the normal complement of organs (Luo *et al.*, 1999). However, the dorsal parts of the two dorsal petals develop in a more symmetrical fashion in *dich* mutants. Interestingly, *cyc dich* double mutants display fully radialized peloric flowers carrying six sepals, six fully ventralized petals and six stamens. The spatial and temporal mRNA expression domains of *CYC* and *DICH* are not identical and their early expression does not coincide with the boundaries of particular floral organs. Still, the two expression patterns mostly overlap and are restricted to a dorsal region in the floral meristem and later to a variable degree in the developing dorsal petals and the staminode. During later stages of petal morphogenesis, *DICH* expression is easier to detect in the developing dorsal half of the dorsal petal lobe, while *CYC* expression is found more equally spread through the dorsal petal lobe. Careful morphological analysis of *cyc* mutants indicates that *CYC* normally retards growth and restricts organ number at an early stage irrespective of whorl identity (Luo *et al.*, 1996). At later stages, however, *CYC* seems to promote growth of dorsal petal lobes but to arrest development of the dorsal stamen. Thus, late *CYC* function may depend on the interaction with other, perhaps whorl-specific, factors. Indeed late *CYC* expression in whorl 2 depends on the B-function gene *DEFICIENS* (*DEF*)

(Clark and Coen, 2002). All in all, the results are compatible with the idea that *CYC* and *DICH* overlap in function during early floral meristem and dorsal organ development. Somewhat later in organogenesis *DICH* plays a particular role in elaborating the dorsoventral asymmetry of the dorsal petals themselves. Irrespective of this distinction both genes also appear to regulate dorsoventral asymmetry within individual organs (Luo *et al.*, 1996, 1999). Interestingly, epigenetic regulation of the *CYC* locus may be responsible for the occurrence of at least some variants with altered asymmetry present in natural populations. For example, there exists a naturally occurring peloric mutant of *Linaria vulgaris* (toadflax), initially described by Linnaeus, which displays polysymmetry rather than monosymmetry. It was found that the mutant phenotype depends on the methylation status of the *Linaria CYC* locus (Cubas *et al.*, 1999a).

Mutations in *CYC* and *DIV* are recessive. Therefore, the ventralized phenotype of flowers of *cyc* or *cyc dich* mutants indicates that the wild-type genes restrict an activity, which promotes the manifestation of ventral characteristics, to the ventral regions of the flower. The *DIV* gene is part of such a mechanism (Almeida *et al.*, 1997; Galego and Almeida, 2002). Flowers of *div* mutants display a lateralized phenotype as the ventral petal and the ventral parts of the lateral petals assume characteristics normally restricted to more lateral petals. Curiously, the mRNA expression pattern of *DIV* does not appear to be restricted ventrally as it was detectable throughout the floral meristem and in all young floral primordia. At later stages, however, *DIV* expression appears to be restricted to the corolla, but is also expressed in dorsal petals, although its expression is asymmetric within the ventral and lateral petals. The expression domains of *CYC*, *DICH* and *DIV* overlap in both time and space. In addition, all three genes encode putative transcription factors. Thus, it seems that the negative regulation of *DIV* by *CYC* and *DICH*, clearly suggested by the genetic analysis, does not involve a direct regulation at the transcriptional level but some other, as yet unknown, mechanism.

4.8.1 Petal asymmetry and contort aestivation

As described above for *Antirrhinum*, asymmetry is not a feature restricted to the flower as a unit, but is also evident at the level of individual floral organs. Little is known of how asymmetry is regulated at the floral organ level in general. In *Antirrhinum*, mutant analysis indicates a role for *DICH* in dorsal petal lobe asymmetry, but the detailed mechanistic basis of *DICH* function is not understood. Interestingly, the asymmetric shape of the dorsal petal lobe appears to depend on the overall direction of anisotropic growth, rather than regional differences in growth rates within the tissue (Rolland-Lagan *et al.*, 2003). In addition, a long-range signal seems to orient the growth direction in the petal lobe relative to the proximal–distal axis of the entire petal (Rolland-Lagan *et al.*, 2003).

Contort aestivation is a particularly curious type of floral asymmetry in which each petal is asymmetric but the whole corolla has a rotational symmetry (Endress, 1999). Direction can either be right or left. Some species display fixed flowers with only one-sided handedness, others show individuals carrying left- and right-handed contortions (Figure 4.2). While an intriguing feature of particularly representatives of the asterids and some members of the rosids, and of obvious aesthetic value to the human eye, the biological relevance of contort aestivation remains elusive (Endress, 1999). In any case, perhaps the cytoskeleton and its effects on the direction of cell expansion are responsible for this asymmetry, at least in some of the cases. Work in Arabidopsis has provided evidence that alterations in the arrangement of cortical microtubules during the expansion phase of cell growth may result in petals displaying left- or right-handed helical growth (Figure 4.2; Furutani *et al.*, 2000; Thitamadee *et al.*, 2002; Buschmann *et al.*, 2004; Shoji *et al.*, 2004). The *lefty* mutations result in left-handed helical growth of petals (Thitamadee *et al.*, 2002). Further analysis revealed that *lefty* mutations result in single amino acid exchanges in two different α-tubulin proteins at the contact site with β-tubulin. The mutant α-tubulins are incorporated into the cortical microtubules altering their orientation and somehow affecting the directionality of cell expansion. In addition, the *SKU6/SPIRAL1* (*SPR1*) *TORTIFOLIA1* (*TOR1*)/*SPIRAL2* (*SPR2*) genes encode plant-specific microtubule-associated proteins (MAP, Buschmann *et al.*, 2004; Sedbrock *et al.*, 2004; Shoji *et al.*, 2004). Mutations in *TOR1/SPR2* lead to a shift in cortical microtubule orientation, alterations in the cell expansion pattern and a switch from straight to normal organ growth (Furutani *et al.*, 2000; Buschmann *et al.*, 2004). The effects are restricted to the cell-expansion phase of organ growth; however, it remains to be seen how an alteration of cortical microtubule dynamics is linked to helical growth. It will also be interesting to investigate if *lefty* or *tor1/spr2*-type mutations relate to examples of contort aestivation found in nature.

4.9 Outlook: to boldly go where no one has gone before . . .

Tremendous progress has been made in recent years regarding the mechanisms governing the size and shape of a flower. It has become obvious, however, that important gaps still exist in our knowledge. What will the future hold? Of course opinions will vary. Nevertheless we shortly highlight just a few developments that we think are important for future progress.

Arguably, the dynamics of growth is one of the most important issues that we need to better incorporate in our hypotheses. Often, genetics excels at the identification of key factors involved in a given process. Genetic models, however, tend to provide a rather static view of a mechanism. Future assessment of the dynamics of floral meristem and floral organ development will rely on the advancement of modern imaging techniques which allow the recording of events over time, both at the cellular and at the molecular level (Laufs *et al.*, 1998b; Grandjean *et al.*, 2004; Reddy *et al.*, 2004). This, of course, must integrate the precise definition of growth parameters and techniques to measure them reliably over time (Nath *et al.*, 2003; Rolland-Lagan *et al.*, 2003; Coen *et al.*, 2004). Furthermore, the regulation of organ size and shape is difficult to envision by the human mind and may even be counter-intuitive. Thus, the development of computer models will be instrumental in better understanding such complex processes, both in space and time (Rolland-Lagan *et al.*, 2003; Shapiro *et al.*, 2003; Prusinkiewicz, 2004). Flowers of different species are characterized by a bewildering diversity in their size, form and shape, providing an enormous pool of variability based on genetic variation. Comparative morphological investigations, combined with a molecular study of the evolutionary basis of floral development, will therefore continue to have an enormous impact and a number of groups are already discussing the best strategies for future research (Irish, 2000; Theissen *et al.*, 2000; Baum *et al.*, 2002; Soltis *et al.*, 2002). Last but not least we need to connect the organ-level regulation with the basic cell-cycle machinery (Fleming, 2002; Beemster *et al.*, 2003). Finally, and stating the obvious, we would like to emphasize that substantial progress will of course be enormously facilitated by bringing together the different approaches under one roof.

Acknowledgements

We thank Peter Endress, Takashi Hashimoto and Mark Running for providing figures. We apologize to colleagues whose work we could not cite due to space restrictions.

References

Aida, M., Ishida, T., Fukaki, H., Fujisawa, H. and Tasaka, M. (1997) *The Plant Cell*, **9**, 841–857.
Aida, M., Ishida, T. and Tasaka, M. (1999) *Development*, **126**, 1563–1570.
Aida, M., Vernoux, T., Furutani, M., Traas, J. and Tasaka, M. (2002) *Development*, **129**, 3965–3974.
Almeida, J., Rocheta, M. and Galego, L. (1997) *Development*, **124**, 1387–1392.
Bai, C., Sen, P., Hofmann, K. *et al.* (1996) *Cell*, **86**, 263–274.
Baker, C. C., Sieber, P., Wellmer, F. and Meyerowitz, E. M. (2005) *Current Biology*, **15**, 303–315.
Barton, M. K. and Poethig, R. S. (1993) *Development*, **119**, 823–831.
Baum, D. A., Doebley, J., Irish, V. F. and Kramer, E. M. (2002) *Trends in Plant Science*, **7**, 31–34.
Bäurle, I. and Laux, T. (2003) *BioEssays*, **25**, 961–970.
Beemster, G. T. S., Fiorani, F. and Inzé, D. (2003) *Trends in Plant Science*, **8**, 154–158.
Benjamins, R., Quint, A., Weijers, D., Hooykaas, P. and Offringa, R. (2001) *Development*, **128**, 4057–4067.
Benková, E., Michniewicz, M., Sauer, M. *et al.* (2003) *Cell*, **115**, 591–602.
Bennett, S. R. M., Alvarez, J., Bossinger, G. and Smyth, D. R. (1995) *The Plant Journal*, **8**, 505–520.
Bossinger, G. and Smyth, D. R. (1996) *Development*, **122**, 1093–1102.
Bowman, J. L., Smyth, D. R. and Meyerowitz, E. M. (1989) *The Plant Cell*, **1**, 37–52.
Bowman, J. L., Smyth, D. R. and Meyerowitz, E. M. (1991) *Development*, **112**, 1–20.
Brand, U., Fletcher, J. C., Hobe, M., Meyerowitz, E. M. and Simon, R. (2000) *Science*, **289**, 617–619.
Braun, D. M., Stone, J. M. and Walker, J. C. (1997) *The Plant Journal*, **12**, 83–95.
Breuil-Broyer, S., Morel, P., de Almeida-Engler, J., Coustham, V., Negrutiu, I. and Trehin, C. (2004) *The Plant Journal*, **38**, 182–192.
Brewer, P. B., Howles, P. A., Dorian, K. *et al.* (2004) *Development*, **131**, 4035–4045.
Busch, M., Bomblies, K. and Weigel, D. (1999) *Science*, **285**, 585–587.
Buschmann, H., Fabri, C. O., Hauptmann, M. *et al.* (2004) *Current Biology*, **14**, 1515–1521.
Byrne, M., Barley, R., Curtis, M. *et al.* (2000) *Nature*, **408**, 967–971.

Byrne, M. E., Simorowski, J. and Martienssen, R. A. (2002) *Development*, **129**, 1957–1965.
Carles, C. C. and Fletcher, J. C. (2003) *Trends in Plant Science*, **8**, 394–401.
Carles, C. C., Choffnes-Inada, D., Reville, K., Lertpiriyapong, K. and Fletcher, J. C. (2005) *Development* **132**, 897–911.
Carles, C. C., Lertpiriyapong, K., Reville, K. and Fletcher, J. C. (2004) *Genetics*, **167**, 1893–1903.
Carpenter, R. and Coen, E. S. (1990) *Genes and Development*, **4**, 1483–1493.
Carpenter, R. and Coen, E. S. (1995) *Development*, **121**, 19–26.
Christensen, S. K., Dagenais, N., Chory, J. and Weigel, D. (2000) *Cell*, **100**, 469–478.
Chuang, C.-F., Running, M. P., Williams, R. W. and Meyerowitz, E. M. (1999) *Genes and Development*, **13**, 334–344.
Clark, J. I. and Coen, E. S. (2002) *The Plant Journal*, **30**, 639–648.
Clark, S. E., Jacobsen, S. E., Levin, J. Z. and Meyerowitz, E. M. (1996) *Development*, **122**, 1567–1575.
Clark, S. E., Running, M. P. and Meyerowitz, E. M. (1993) *Development*, **119**, 397–418.
Clark, S. E., Running, M. P. and Meyerowitz, E. M. (1995) *Development*, **121**, 2057–2067.
Clark, S. E., Williams, R. W. and Meyerowitz, E. M. (1997) *Cell*, **89**, 575–585.
Coen, E. S. (1996) *EMBO Journal*, **15**, 6777–6787.
Coen, E. S. and Carpenter, R. (1993) *The Plant Cell*, **5**, 1175–1181.
Coen, E. S. and Meyerowitz, E. M. (1991) *Nature*, **353**, 31–37.
Coen, E., Rolland-Lagan, A. G., Matthews, M., Bangham, J. A. and Prusinkiewicz, P. (2004) *Proceedings of the National Academy of Sciences of the United States of America*, **101**, 4728–4735.
Conlon, I. and Raff, M. (1999) *Cell*, **96**, 235–244.
Crone, W. and Lord, E. (1994) *Canadian Journal of Botony*, **72**, 384–401.
Crone, W. and Lord, E. M. (1993) *American Journal of Botony*, **80**, 1419–1426.
Cubas, P., Lauter, N., Doebley, J. and Coen, E. (1999a) *The Plant Journal*, **18**, 215–222.
Cubas, P., Vincent, C. and Coen, E. (1999b) *Nature*, **401**, 157–161.
Cutler, S., Ghassemian, M., Bonetta, D., Cooney, S. and McCourt, P. (1996) *Science*, **273**, 1239–1241.
Cvitanich, C., Pallisgaard, N., Nielsen *et al.* (2000) *Proceedings of the National Academy of Sciences of the United States of America*, **97**, 8163–8168.
Dewitte, W., Riou-Khamlichi, C., Scofield, S. *et al.* (2003) *The Plant Cell*, **15**, 79–92.
Dharmarsiri, N. and Estelle, M. (2004) *Trends in Plant Science*, **9**, 302–308.
Dinneny, J. R., Yadegari, R., Fischer, R. L., Yanofsky, M. F. and Weigel, D. (2004) *Development*, **131**, 1101–1110.
Ditta, G., Pinyopich, A., Robles, P., Pelaz, S. and Yanofsky, M. F. (2004) *Current Biology*, **14**, 1935–1940.
Doebley, J., Stec, A. and Hubbard, L. (1997) *Nature*, **386**, 485–488.
Drews, G. N., Bowman, J. L. and Meyerowitz, E. M. (1991) *Cell*, **65**, 991–1002.
Durfee, T., Roe, J. L., Sessions, R. A. *et al.* (2003) *Proceedings of the National Academy of Sciences of the United States of America*, **100**, 8571–8576.
Duval, M., Hsieh, T. F., Kim, S. Y. and Thomas, T. L. (2002) *Plant Molecular Biology*, **50**, 237–248.
Ehsan, H., Reichheld, J. P., Durfee, T. and Roe, J. L. (2004) *Plant Physiology*, **134**, 1488–1499.
Elliott, R. C., Betzner, A. S., Huttner, E. *et al.* (1996) *The Plant Cell*, **8**, 155–168.
Endress, P. K. (1999) *International Journal of Plant Sciences*, **160**, S3–S23.
Endress, P. K. (2001a) *Current Opinion in Plant Biology*, **4**, 86–91.
Endress, P. K. (2001b) *Journal of Experimental Zoology*, **291**, 105–115.
Endrizzi, K., Moussian, B., Haecker, A., Levin, J. Z. and Laux, T. (1996) *The Plant Journal*, **10**, 967–979.
Estelle, M. (1998) *The Plant Cell*, **10**, 1775–1778.
Fleming, A. J. (2002) *Planta*, **216**, 17–22.
Fletcher, J. C. (2001) *Development*, **128**, 1323–1333.
Fletcher, J. C., Brand, U., Running, M. P., Simon, R. and Meyerowitz, E. M. (1999) *Science*, **283**, 1911–1914.
Friml, J. (2003) *Current Opinion in Plant Biology*, **6**, 7–12.
Friml, J. and Palme, K. (2002) *Plant Molecular Biology*, **49**, 273–284.
Friml, J., Yang, X., Michniewicz, M. *et al.* (2004) *Science*, **306**, 862–865.

Furner, I. J. (1996) *The Plant Journal*, **10**, 645–654.
Furutani, I., Watanabe, Y., Prieto, R. *et al.* (2000) *Development*, **127**, 4443–4453.
Galego, L. and Almeida, J. (2002) *Genes and Development*, **16**, 880–891.
Gälweiler, L., Guan, C., Müller, A. *et al.* (1998) *Science*, **282**, 2226–2230.
Geldner, N., Anders, N., Wolters, H. *et al.* (2003) *Cell*, **112**, 219–230.
Geldner, N., Frimi, J., Stierhof, Y.-D., Jürgens, G. and Palme, K. (2001) *Nature*, **413**, 425–428.
Goto, N., Starke, M. and Kranz, A. R. (1987) *Arabidopsis Information Service*, **23**, 66–71.
Grandjean, O., Vernoux, T., Laufs, P., Belcram, K., Mizukami, Y. and Traas, J. (2004) *The Plant Cell*, **16**, 74–87.
Griffith, M. E., da Silva Conceicao, A. and Smyth, D. R. (1999) *Development*, **126**, 5635–5644.
Gross-Hardt, R. and Laux, T. (2003) *Journal of Cell Sciences*, **116**, 1659–1666.
Gross-Hardt, R., Lenhard, M. and Laux, T. (2002) *Genes and Development*, **16**, 1129–1138.
Hafen, E. and Stocker, H. (2003) *PLoS Biology*, **1**, E86.
Hantke, S. S., Carpenter, R. and Coen, E. S. (1995) *Development*, **121**, 27–35.
Hauser, B. A., He, J. Q., Park, S. O. and Gasser, C. S. (2000) *Development*, **127**, 2219–2226.
Hauser, B. A., Villanueva, J. M. and Gasser, C. S. (1998) *Genetics*, **150**, 411–423.
Hill, J. P. and Lord, E. M. (1989) *Canadian Journal of Botony*, **67**, 2922–2936.
Honma, T. and Goto, K. (2001) *Nature*, **409**, 525–529.
Hu, Y., Xie, Q. and Chua, N. H. (2003) *The Plant Cell*, **15**, 1951–1961.
Huala, E. and Sussex, I. M. (1992) *The Plant Cell*, **4**, 901–913.
Ingram, G. C., Doyle, S., Carpenter, R., Schultz, E. A., Simon, R. and Coen, E. S. (1997) *EMBO Journal*, **16**, 6521–6534.
Ingram, G. C., Goodrich, J., Wilkinson, M. D., Simon, R., Haughn, G. W. and Coen, E. S. (1995) *The Plant Cell*, **7**, 1501–1510.
Irish, V. F. (2000) *Genome Biology*, **1**, reviews:1015.1011–1015.1014.
Ishida, T., Aida, M., Takada, S. and Tasaka, M. (2000) *The Plant Cell Physiology*, **41**, 60–67.
Ito, T., Wellmer, F., Yu, H. *et al.* (2004) *Nature*, **430**, 356–360.
Jack, T., Brockman, L. L. and Meyerowitz, E. M. (1992) *Cell*, **68**, 683–697.
Jack, T., Sieburth, L. and Meyerowitz, E. (1997) *The Plant Journal*, **11**, 825–839.
Jenik, P. D. and Irish, V. F. (2000) *Development*, **127**, 1267–1276.
Jeong, S., Trotochaud, A. E. and Clark, S. E. (1999) *The Plant Cell*, **11**, 1925–1934.
Jinn, T.-L., Stone, J. M. and Walker, J. C. (2000) *Genes and Development*, **14**, 108–117.
Kaya, H., Shibahara, K.-I., Taoka, K.-I., Iwabuchi, M., Stillman, B. and Araki, T. (2001) *Cell*, **104**, 131–142.
Kayes, J. M. and Clark, S. E. (1998) *Development*, **125**, 3843–3851.
Kerstetter, R. A., Laudencia-Chingcuanco, D., Smith, L. G. and Hake, S. (1997) *Development*, **124**, 3045–3054.
Klucher, K. M., Chow, H., Reiser, L. and Fischer, R. L. (1996) *The Plant Cell*, **8**, 137–153.
Kosugi, S. and Ohashi, Y. (1997) *The Plant Cell*, **9**, 1607–1619.
Kosugi, S. and Ohashi, Y. (2002) *The Plant Journal*, **30**, 337–348.
Krizek, B. A. (1999) *Developmental Genetics*, **25**, 224–236.
Krizek, B. A. (2003) *Nucleic Acids Research*, **31**, 1859–1868.
Laufs, P., Coen, E., Kronenberger, J., Traas, J. and Doonan, J. (2003) *Development*, **130**, 785–796.
Laufs, P., Dockx, J., Kronenberger, J. and Traas, J. (1998a) *Development*, **125**, 1253–1260.
Laufs, P., Grandjean, O., Jonak, C., Kiêu, K. and Traas, J. (1998b) *The Plant Cell*, **10**, 1375–1389.
Laufs, P., Peaucelle, A., Morin, H. and Traas, J. (2004) *Development*, **131**, 4311–4322.
Laux, T., Mayer, K. F., Berger, J. and Jürgens, G. (1996) *Development*, **122**, 87–96.
Lee, I., Wolfe, D. S., Nilsson, O. and Weigel, D. (1997) *Current Biology*, **7**, 95–104.
Lenhard, M. and Laux, T. (2003) *Development*, **130**, 3163–3173.
Lenhard, M., Bohnert, A., Jürgens, G. and Laux, T. (2001) *Cell*, **105**, 805–814.
Leyser, H. M., Lincoln, C. A., Timpte, C., Lammer, D., Turner, J. and Estelle, M. (1993) *Nature*, **364**, 161–164.

Leyser, H. M. O. and Furner, I. J. (1992) *Development*, **116**, 397–403.
Leyser, O. (2002) *Annual Review of Plant Biology*, **53**, 377–398.
Li, J., Smith, G. and Walker, J. (1999) *Proceedings of the National Academy of Sciences of the United States of America*, **96**, 7821–7826.
Li, Y., DeFatta, R., Anthony, C., Sunavala, G. and De Benedetti, A. (2001) *Oncogene*, **20**, 726–738.
Lincoln, C., Britton, J. H. and Estelle, M. (1990) *The Plant Cell*, **2**, 1071–1080.
Liu, Z., Running, M. and Meyerowitz, E. M. (1997) *Development*, **124**, 665–672.
Lohmann, J. U. and Weigel, D. (2002) *Developmental Cell*, **2**, 135–142.
Lohmann, J. U., Hong, R. L., Hobe, M. *et al.* (2001) *Cell*, **105**, 793–803.
Long, J. A., Moan, E. I., Medford, J. I. and Barton, M. K. (1996) *Nature*, **379**, 66–69.
Lucas, W. J., Bouché-Pillon, S., Jackson, D. P. *et al.* (1995) *Science*, **270**, 1980–1983.
Luo, D., Carpenter, R., Copsey, L., Vincent, C., Clark, J. and Coen, E. (1999) *Cell*, **99**, 367–376.
Luo, D., Carpenter, R., Vincent, C., Copsey, L. and Coen, E. (1996) *Nature*, **383**, 794–799.
Luschnig, C., Gaxiola, R. A., Grisafi, P. and Fink, G. R. (1998) *Genes and Development*, **12**, 2175–2187.
Mallory, A. C., Dugas, D. V., Bartel, D. P. and Bartel, B. (2004) *Current Biology*, **14**, 1035–1046.
Matsumoto, N. and Okada, K. (2001) *Genes and Development*, **15**, 3355–3364.
Mayer, F. X., Schoof, H., Haecker, A., Lenhard, M., Jürgens, G. and Laux, T. (1998) *Cell*, **95**, 805–815.
Meijer, M. and Murray, J. A. (2001) *Current Opinion in Plant Biology*, **4**, 44–49.
Menand, B. and Robaglia, C. (2004) In *Cell Growth: Control of Cell Size* (eds M. N. Hall, M. Raff and G. Thomas), Cold Spring Harbor Laboratory Press, New York, pp. 625–637.
Meyerowitz, E. M. (1997) *Cell*, **88**, 299–308.
Mizukami, Y. (2001) *Current Opinion in Plant Biology*, **4**, 533–539.
Mizukami, Y. and Fischer, R. L. (2000) *Proceedings of the National Academy of Sciences of the United States of America*, **97**, 942–947.
Mizukami, Y. and Ma, H. (1992) *Cell*, **71**, 119–131.
Mizukami, Y. and Ma, H. (1995) *Plant Molecular Biology*, **28**, 767–784.
Mizukami, Y. and Ma, H. (1997) *The Plant Cell*, **9**, 393–408.
Mizukami, Y., Huang, H., Tudor, M., Hu, Y. and Ma, H. (1996) *The Plant Cell*, **8**, 831–845.
Mouradov, A., Cremer, F. and Coupland, G. (2002) *The Plant Cell*, **14**, S111–S130.
Müller, A., Guan, C., Galweiler, L. *et al.* (1998) *EMBO Journal*, **17**, 6903–6911.
Nardmann, J., Ji, J., Werr, W. and Scanlon, M. J. (2004) *Development*, **131**, 2827–2839.
Nath, U., Crawford, B. C., Carpenter, R. and Coen, E. (2003) *Science*, **299**, 1404–1407.
Ng, M. and Yanofsky, M. F. (2000) *Current Opinion in Plant Biology*, **3**, 47–52.
Nole-Wilson, S. and Krizek, B. A. (2000) *Nucleis Acids Research*, **28**, 4076–4082.
Ohno, C. K., Reddy, G. V., Heisler, M. G. and Meyerowitz, E. M. (2004) *Development*, **131**, 1111–1122.
Okada, K., Ueda, J., Masako, K. K., Bell, C. J. and Shimura, Y. (1991) *The Plant Cell*, **3**, 677–684.
Otsuga, D., DeGuzman, B., Prigge, M. J., Drews, G. N. and Clark, S. E. (2001) *The Plant Journal*, **25**, 223–236.
Palatnik, J. F., Allen, E., Wu, X. *et al.* (2003) *Nature*, **425**, 257–263.
Palme, K. and Gälweiler, L. (1999) *Current Opinion in Plant Biology*, **2**, 375–381.
Parcy, F., Nilsson, O., Busch, M. A., Lee, I. and Weigel, D. (1998) *Nature*, **395**, 561–566.
Parenicova, L., de Folter, S., Kieffer, M. *et al.* (2003) *The Plant Cell*, **15**, 1538–1551.
Pei, Z. M., Ghassemian, M., Kwak, C. M., McCourt, P. and Schroeder, J. I. (1998) *Science*, **282**, 287–290.
Pelaz, S., Ditta, G. S., Baumann, E., Wisman, E. and Yanofsky, M. F. (2000) *Nature*, **405**, 200–203.
Pelaz, S., Tapia-Lopez, R., Alvarez-Buylla, E. R. and Yanofsky, M. F. (2001) *Current Biology*, **11**, 182–184.
Perbal, M.-C., Haughn, G., Saedler, H. and Schwarz-Sommer, Z. (1996) *Development*, **122**, 3433–3441.
Potter, C. J. and Xu, T. (2001) *Current Opinion in Genetics and Development*, **11**, 279–286.
Prusinkiewicz, P. (2004) *Current Opinion in Plant Biology*, **7**, 79–83.
Ratcliffe, O. J., Riechmann, J. L. and Zhang, J. Z. (2000) *The Plant Cell*, **12**, 315–317.
Reddy, G. V., Heisler, M. G., Ehrhardt, D. W. and Meyerowitz, E. M. (2004) *Development*, **131**, 4225–4237.

Reinhardt, D., Frenz, M., Mandel, T. and Kuhlemeier, C. (2003a) *Development*, **130**, 4073–4083.
Reinhardt, D., Mandel, T. and Kuhlemeier, C. (2000) *The Plant Cell*, **12**, 507–518.
Reinhardt, D., Pesce, E. R., Stieger, P. *et al.* (2003b) *Nature*, **426**, 255–260.
Riechmann, J. L., Krizek, B. A. and Meyerowitz, E. M. (1996) *Proceedings of the National Academy of Sciences of the United States of America*, **93**, 4793–4798.
Rodríguez-Concepción, M., Yalovsky, S. and Gruissem, W. (1999) *Plant Molecular Biology*, **39**, 865–870.
Roe, J. L., Durfee, T., Zupan, J. R., Repetti, P., McLean, B. G. and Zambryski, P. C. (1997a) *Journal of Biological Chemistry*, **272**, 5838–5845.
Roe, J. L., Nemhauser, J. L. and Zambryski, P. C. (1997b) *The Plant Cell*, **9**, 335–353.
Roe, J. L., Sessions, R. A., Feldmann, K. A. and Zambryski, P. C. (1993) *Cell*, **75**, 939–950.
Rojo, E., Sharma, V. K., Kovaleva, V., Raikhel, N. V. and Fletcher, J. C. (2002) *The Plant Cell*, **14**, 969–977.
Rolland-Lagan, A.-G., Bangham, J. A. and Coen, E. (2003) *Nature*, **422**, 161–163.
Running, M. P. and Meyerowitz, E. M. (1996) *Development*, **122**, 1261–1269.
Running, M. P., Fletcher, J. C. and Meyerowitz, E. M. (1998) *Development*, **125**, 2545–2553.
Running, M. P., Lavy, M., Sternberg, H. *et al.* (2004) *Proceedings of the National Academy of Sciences of the United States of America*, **101**, 7815–7820.
Sablowski, R. W. M. and Meyerowitz, E. M. (1998) *Cell*, **92**, 93–103.
Sakai, H., Krizek, B. A., Jacobsen, S. E. and Meyerowitz, E. M. (2000) *The Plant Cell*, **12**, 1607–1618.
Sakai, H., Medrano, L. J. and Meyerowitz, E. M. (1995) *Nature*, **378**, 199–203.
Samach, A., Klenz, J. E., Kohalmi, S. E., Risseeuw, E., Haughn, G. W. and Crosby, W. L. (1999) *The Plant Journal*, **20**, 433–445.
Satina, S. and Blakeslee, A. F. (1941) *American Journal of Botony*, **28**, 862–871.
Scanlon, M., Schneeberger, R. and Freeling, M. (1996) *Development*, **122**, 1683–1691.
Scanlon, M. J. (2000) *Development*, **127**, 4573–4585.
Schoof, H., Lenhard, M., Haecker, A., Mayer, K. F. X., Jürgens, G. and Laux, T. (2000) *Cell*, **100**, 635–644.
Schultz, E. A. and Haughn, G. W. (1991) *The Plant Cell*, **3**, 771–781.
Schumacher, K., Schmitt, T., Rossberg, M., Schmitz, G. and Theres, K. (1999) *Proceedings of the National Academy of Sciences of the United States of America*, **96**, 290–295.
Schwarz-Sommer, Z., Huijser, P., Nacken, W., Saedler, H. and Sommer, H. (1990) *Science*, **250**, 931–936.
Sedbrock, J. C., Ehrhardt, D., Fisher, S. E., Scheible, W. R. and Somerville, C. R. (2004) *The Plant Cell* **16**, 1506–1520.
Sessions, A. and Yanofsky, M. F. (1999) *Genes and Development*, **13**, 1051–1054.
Sessions, A., Nemhauser, J. L., McColl, A., Roe, J. L., Feldmann, K. A. and Zambryski, P. C. (1997) *Development*, **124**, 4481–4491.
Sessions, A., Yanofsky, M. F. and Weigel, D. (2000) *Science*, **289**, 779–781.
Shapiro, B. E., Levchenko, A., Meyerowitz, E. M., Wold, B. J. and Mjolsness, E. D. (2003) *Bioinformatics*, **19**, 677–678.
Shiraishi, H., Okada, K. and Shimura, Y. (1993) *The Plant Journal*, **4**, 385–398.
Shoji, T., Narita, N. N., Hayashi, K. *et al.* (2004) *Plant Physiology* **136**, 3933–44.
Shuai, B., Reynaga-Peña, C. G. and Springer, P. S. (2002) *Plant Physiology*, **129**, 747–761.
Sieburth, L. E., Drews, G. N. and Meyerowitz, E. M. (1998) *Development*, **125**, 4303–4312.
Sieburth, L. E., Running, M. P. and Meyerowitz, E. M. (1995) *The Plant Cell*, **7**, 1249–1258.
Silljé, H. H. and Nigg, E. A. (2001) *Current Biology*, **11**, 1068–1073.
Sillјé, H. H., Takahashi, K., Tanaka, K., Van Houwe, G. and Nigg, E. A. (1999) *EMBO Journal*, **18**, 5691–5702.
Simon, R., Carpenter, R., Doyle, S. and Coen, E. (1994) *Cell*, **78**, 99–107.
Simpson, G. G. and Dean, C. (2002) *Science*, **296**, 285–289.
Soltis, D. E., Soltis, P. S., Albert, V. A. *et al.* (2002) *Trends in Plant Science*, **7**, 22–31.
Song, J.-Y., Leung, T., Ehler, L. K., Wang, C. and Liu, Z. (2000) *Development*, **127**, 2207–2217.

Souer, E., van Houwelingen, A., Kloos, D., Mol, J. and Koes, R. (1996) *Cell*, **85**, 159–170.

Steeves, T. A. and Sussex, I. M. (1989) *Patterns in Plant Development,* Cambridge University Press, Cambridge.

Stieger, P. A., Reinhardt, D. and Kuhlemeier, C. (2002) *The Plant Journal*, **32**, 509–517.

Stone, J. M., Collinge, M. A., Smith, R. D., Horn, M. A. and Walker, J. C. (1994) *Science*, **266**, 793–795.

Stone, J. M., Trotochaud, A. E., Walker, J. C. and Clark, S. E. (1998) *Plant Physiology*, **117**, 1217–1225.

Stubbe, H. (1966) *Genetik und Zytologie von Antirrhinum L. sect. Antirrhinum*, VEB Gustav Fischer Verlag, Jena.

Szymkowiak, E. J. and Sussex, I. M. (1992) *The Plant Cell*, **4**, 1089–1100.

Szymkowiak, E. J. and Sussex, I. M. (1993) *The Plant Journal*, **4**, 1–7.

Szymkowiak, E. J. and Sussex, I. M. (1996) *Annual Review of Plant Physiology and Plant Molecular Biology*, **47**, 351–376.

Takada, S., Hibara, K.-i., Ishida, T. and Tasaka, M. (2001) *Development*, **128**, 1127–1135.

Takeda, S., Matsumoto, N. and Okada, K. (2004) *Development*, **131**, 425–434.

Talbert, P. B., Adler, H. T., Parks, D. W. and Comai, L. (1995) *Development*, **121**, 2723–2735.

Taoka, K., Yanagimoto, Y., Daimon, Y., Hibara, K., Aida, M. and Tasaka, M. (2004) *The Plant Journal*, **40**, 462–473.

Theissen, G., Becker, A., Di Rosa, A. *et al.* (2000) *Plant Molecular Biology*, **42**, 115–149.

Thitamadee, S., Tuchihara, K. and Hashimoto, T. (2002) *Nature*, **417**, 193–196.

Tilney-Bassett, R. A. E. (1986) *Plant Chimeras*, E. Arnold, London.

Trémousaygue, D., Garnier, L., Bardet, C., Dabos, P., Herve, C. and Lescure, B. (2003) *The Plant Journal*, **33**, 957–966.

Trotochaud, A. E., Hao, T., Wu, G., Yang, Z. and Clark, S. E. (1999) *The Plant Cell*, **11**, 393–406.

Vergani, P., Morandini, P. and Soave, C. (1997) *FEBS Letters*, **400**, 243–246.

Vernoux, T., Kronenberger, J., Grandjean, O., Laufs, P. and Traas, J. (2000) *Development*, **127**, 5157–5165.

Vollbrecht, E., Reiser, L. and Hake, S. (2000) *Development*, **127**, 3161–3172.

Vollbrecht, E., Veit, B., Sinha, N. and Hake, S. (1991) *Nature*, **350**, 241–243.

Vroemen, C. W., Mordhorst, A. P., Albrecht, C., Kwaaitaal, M. A. and de Vries, S. C. (2003) *The Plant Cell*, **15**, 1563–1577.

Weigel, D. (1995) *The Plant Cell*, **7**, 388–389.

Weigel, D. and Jürgens, G. (2002) *Nature*, **415**, 751–754.

Weigel, D. and Meyerowitz, E. M. (1994) *Cell*, **78**, 203–209.

Weigel, D., Alvarez, J., Smyth, D. R., Yanofsky, M. F. and Meyerowitz, E. M. (1992) *Cell*, **69**, 843–859.

Weir, I., Lu, J., Cook, H., Causier, B., Schwarz-Sommer, Z. and Davies, B. (2004) *Development*, **131**, 915–922.

Williams, R. W., Wilson, J. M. and Meyerowitz, E. M. (1997) *Proceedings of the National Academy of Sciences of the United States of America*, **94**, 10467–10472.

Yalovsky, S., Kulukian, A., Rodriguez-Concepcion, M., Young, C. A. and Gruissem, W. (2000) *The Plant Cell*, **12**, 1267–1278.

Yanofsky, M. F., Ma, H., Bowman, J. L., Drews, G. N., Feldmann, K. A. and Meyerowitz, E. M. (1990) *Nature*, **346**, 35–39.

Yu, L. P., Miller, A. K. and Clark, S. E. (2003) *Current Biology*, **13**, 179–188.

Yu, L. P., Simon, E. J., Trotochaud, A. E. and Clark, S. E. (2000) *Development*, **127**, 1661–1670.

Zhong, R. and Ye, Z. H. (1999) *The Plant Cell*, **11**, 2139–2152.

Zhong, R. and Ye, Z. H. (2001) *Plant Physiology*, **126**, 549–563.

Zhong, R., Taylor, J. J. and Ye, Z. H. (1997) *The Plant Cell*, **9**, 2159–2170.

Ziegelhoffer, E. C., Medrano, L. J. and Meyerowitz, E. M. (2000) *Proceedings of the National Academy of Sciences of the United States of America*, **97**, 7633–7638.

5 Inflorescence architecture – Moving beyond description to development, genes and evolution

Susan R. Singer

5.1 Overview

The tremendous diversity in angiosperm inflorescence architecture, the shoot system that gives rise to flowers, has long intrigued biologists. Environmental and genetic regulation of flowering time, branch complexity (e.g. number of iterations of branching prior to flowering), number of flowers per node and the extent of terminal meristem growth in the reproductive phase contribute to the overall pattern of an inflorescence. Over the past decade, research on plant architecture has focused on the model systems *Arabidopsis thaliana* and *Antirrhinum majus*, along with a few model crops. A better understanding of the molecular–genetic regulation of plant form, especially branching and determinacy, will enable the improvement of crop plants (Reinhardt and Kuhlmeir, 2002). Specifically, modification of inflorescence architecture could enhance fruit and seed yield or quality. This chapter provides an overview of inflorescence typologies, inflorescence development, evolution of inflorescence architecture and future directions for research with an emphasis on the potential of modeling inflorescence development.

5.2 Inflorescence typologies

Inflorescences range from an un-branched main axis terminating with the production of a single flower to incredibly complex branching systems that produce numerous flowers over an extended period of time. For almost two centuries these branching patterns have intrigued botanists and led to extensive typologies and nomenclatures for inflorescences (Roeper, 1826; Rickett, 1944; Weberling, 1989). More recently, genetic, molecular and cell biological advances have moved us toward mechanistic understandings of branching architecture.

The iterative nature of plants makes it possible to decompose these patterns into simpler, repetitive units (Figure 5.1). The focus of this section is the structure of inflorescences. Morphological patterns are the endpoints of developmental (proximate) and evolutionary (ultimate) processes that are explored later in this chapter. Investigations of inflorescence architecture offer insight into the evolution of an astounding array of reproductive shoot systems in the angiosperms, as well as the potential to genetically manipulate these branching patterns to improve crop yield and enhance the aesthetics of horticultural species.

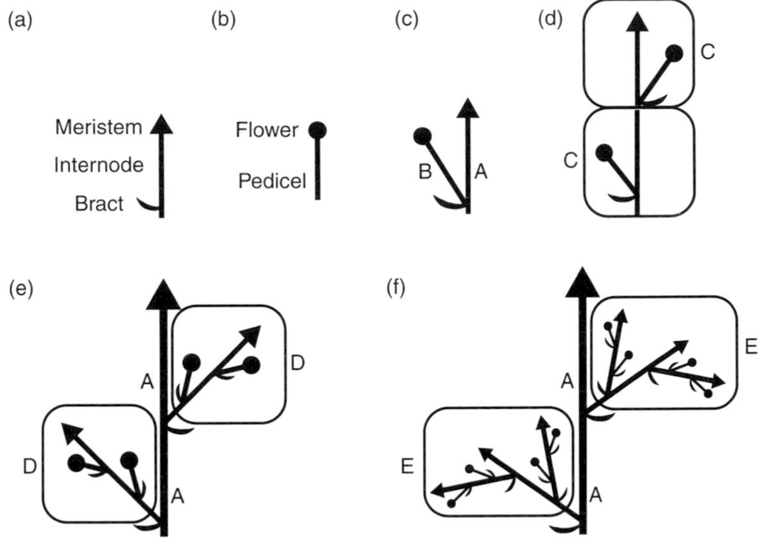

Figure 5.1 Complex inflorescence patterns can be decomposed into simpler, iterative units. Letters are used to identify iterative units that are found in simple (D) and compound (E and F) inflorescences.

Most of the variation among inflorescences can be attributed to four factors:

(1) extent of growth in each of three dimensions of stem and stem-like structures;
(2) determinacy or indeterminacy of meristems within the shoot system;
(3) specification of meristem identity; and
(4) relative positions of lateral shoots and/or flowers (phyllotaxy).

Morphological diversity of the angiosperms increases exponentially when one considers the range of floral forms (see Chapter 4). The aim of this chapter is to elucidate the shoot systems from which these varied flowers form.

Internode length is a simple, but contributory factor to diverse inflorescence typologies. The transition to flowering in rosette plants is accompanied by internode elongation called bolting (Figure 5.2). The marked difference in internode length between the vegetative and reproductive stems of rosette plants distinguishes them morphologically from relatives with more uniform internode length, affecting the height at which flowers are presented to pollinators. It is, however, relative internode and pedicel lengths within the reproductive portion of the plant, rather than between the vegetative and reproductive portions of the plant, that traditionally distinguish between inflorescence typologies (Figure 5.3). A raceme is characterized by lateral flowers with visible pedicels forming in sequential axils that are separated by visibly identifiable internodes. Severe reduction of pedicel length leads to sessile flowers and an inflorescence called a spike. A radial increase in stem growth converts a spike to a spadix. Returning to the raceme and reducing internode length of the stem

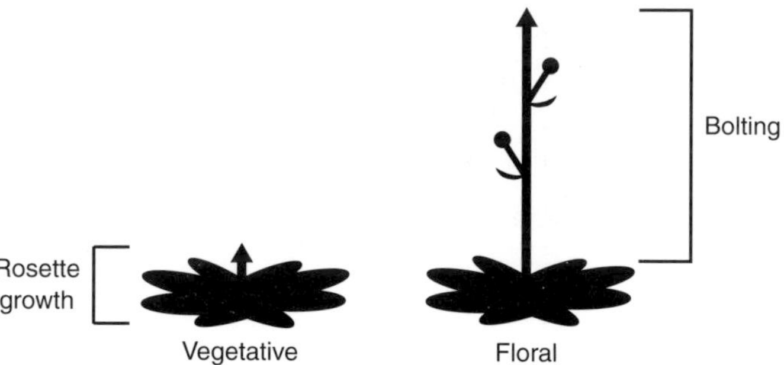

Figure 5.2 Bolting and inflorescence patterns. In some species, including *Arabidopsis*, bolting accompanies the transition to reproductive development having a significant morphological effect on the inflorescence.

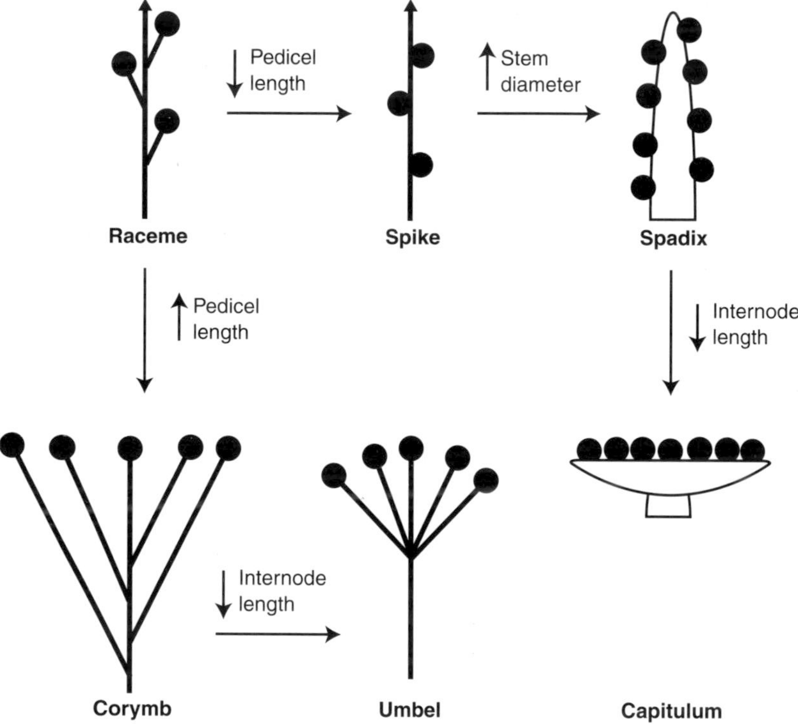

Figure 5.3 The length and diameter of stems and pedicels determine inflorescence typology.

results in an umbel. Individual flowers of an umbel are separated by pedicels, but the pedicels appear to originate from a common point of origin because of minimal internode growth between sequential nodes. In the case of an umbel, a radial increase in stem growth in the inflorescence results in a capitulum.

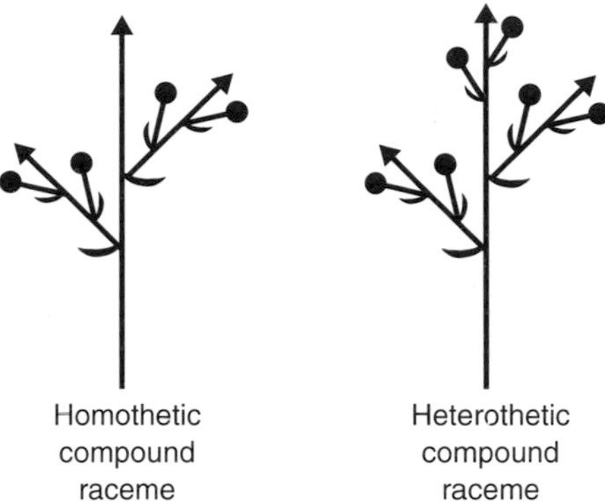

Figure 5.4 Role of determinacy in raceme architecture. Homothetic compound racemes have an indeterminate first order meristem. Heterothetic compound racemes also have an indeterminate first order meristem, but that meristem initiates a terminal raceme.

Determinacy of shoot meristems determines both the complexity of inflorescence branching and the number of flowers initiated on a floral branch. Stebbins (1974) proposed that the main axis of ancestral flowering plants terminated with a flower. Terminal flowers are determinate structures, as are tendrils that form in the inflorescences of grapevine (Calonje *et al.*, 2004). Grapevine tendrils are not only determinate, but express some floral meristem identity genes. Returning to the raceme, the main axis has indeterminate growth and the second order meristems have determinate growth. In this case, the second order meristem has a floral identity. Compound racemes (Figure 5.1) have indeterminate first and second order growth, resulting in lateral (axillary) racemes. Third order meristems have a floral identity. Weberling (1989) cites and confirms Troll's further categorization of compound racemes into homothetic and heterothetic compound racemes (Figure 5.4). Homothetic racemes initiate second order racemes as previously described. The first order meristem of heterothetic racemes also maintains its indeterminacy, but shifts to producing second order flowers after a number of lateral racemes have formed. The result is multiple lateral racemes and a terminal raceme.

Unlike the raceme, all shoot meristems in the panicle inflorescence eventually become specified for floral meristem identity and terminate with the production of a flower (Figure 5.5). In addition, the terminal flowers are dominant. When nutrients are limited, the terminal flowers may be the only flowers to develop fully (Weberling, 1989). How many third order inflorescences and/or flowers a second order panicle produces is a function of position. Basal inflorescence axils have more extensive branching than apical axils.

Some plants, including tomato, have lateral cymes with either a determinate or an indeterminate main axis. Cymes exhibit sympodial growth which can result in a zigzag stem pattern (Figure 5.6). The second order meristem is consumed in the production of a flower. A subtending third order meristem goes on to produce

Figure 5.5 Unlike racemes, all shoot meristems within the panicle inflorescence eventually terminate with flowers.

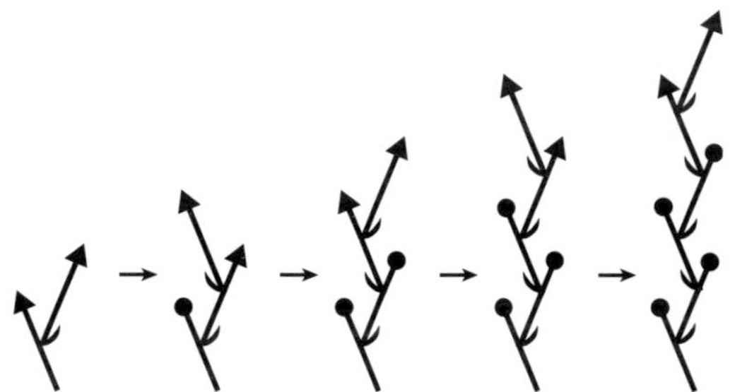

Figure 5.6 Sympodial growth produces lateral cymes. The terminal meristem produces an axillary meristem before terminating as a flower. The axillary meristem then initiates another axillary meristem before it terminates as a flower. This pattern reiterates indefinitely in indeterminate inflorescences. All shoot meristems are eventually consumed in the production of a flower in a determinate inflorescences.

a flower, while its subtending fourth order meristem also is specified for floral development.

Within inflorescence typologies, additional patterns can be generated through phyllotactic changes. Floral shoots or flowers that form in axils with alternate, decussate or spiral phyllotaxy contribute to inflorescences with distinctive morphologies. Further variation occurs among spiral patterns that correlate with the relative rates of shoot apex growth and primordial initiation, yielding patterns corresponding to different sequential Fibonacci numbers (Richards, 1951; Jean, 1988). Consider the correspondence between changes in apex growth affecting whether an inflorescence is a raceme or a capitulum and parallel changes in the complexity of the spiral phyllotactic pattern (Figure 5.7).

While most work on phyllotaxy has concentrated on leaf arrangements, Kirchoff (2003) has shown that Hofmeister's (1868) rule for phyllotaxy applies equally to organ placement in inflorescences. Hofmeister's rule is foundational to most phyllotactic models. The rule states that the distance from the last initiated primoridium

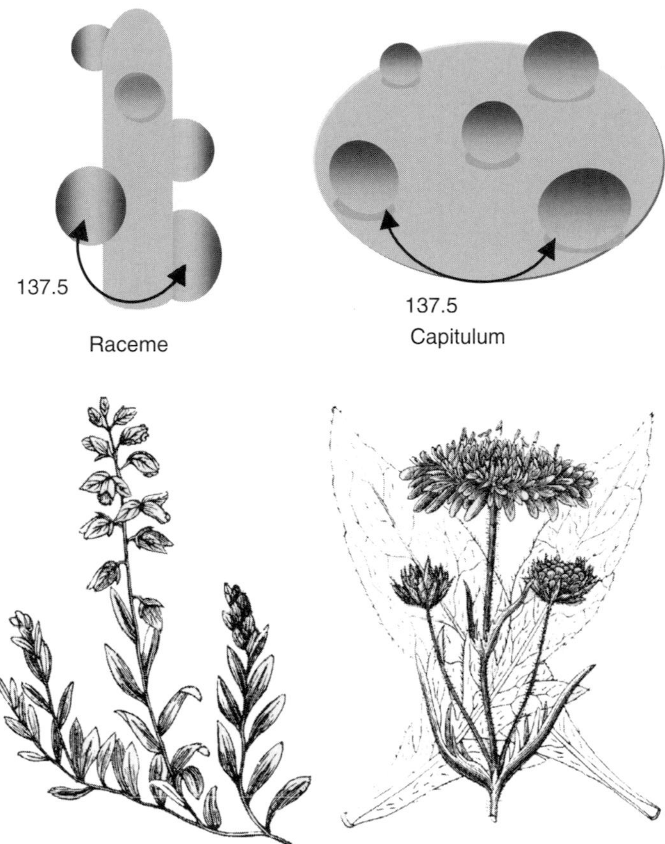

Figure 5.7 Inflorescence architecture is affected by phyllotaxy of axillary flowers. In these examples, both inflorescences have spiral phyllotaxy, but the overall phyllotactic pattern increases in complexity as the apex diameter increases relative to internode length. That is, wider and flatter meristems have more complex phyllotactic patterns than more narrow, elongate meristems.

and next primordium, as well as the largest available space on the apex, limits organ positioning. The predictive value of the rule has been experimentally verified numerous times for leaf positioning (Snow and Snow, 1933). Kirchoff investigated the positioning of *Phenakospermum guyannense* (South American travelers' palm) sepals relative to their subtending bracts. These plants have three sepals and bilaterally symmetric flowers. Bract position reliably predicts sepal position and thus the orientation of the flower in the lateral cyme (Figure 5.8). As a result of the organ primorida interactions, successive flowers on a branch are mirror images of each other.

As illustrated with the travelers' palm, orientation of bracts and asymmetric flowers within the shoot system of an inflorescence are significant components of inflorescence architecture. Weberling (1989) emphasizes that branching patterns are more definitive in typologies than the distinctions between vegetative and

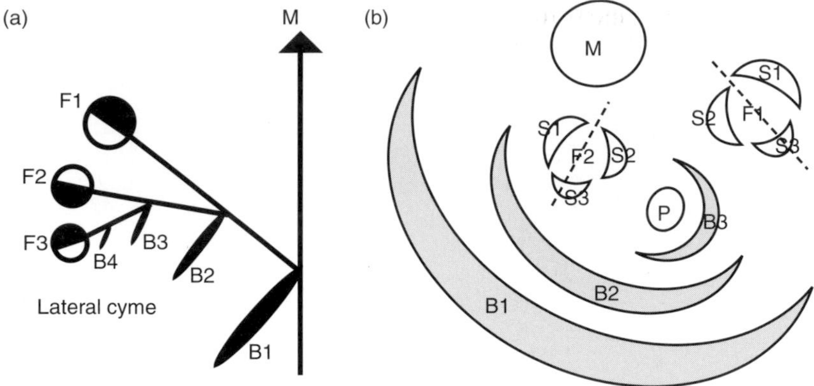

Figure 5.8 Hofmeister's rule predicts floral orientation in *P. guyannense* cymes. The first sepal will form in the first available space on the apex, as far away as possible from the subtending bract. The first sepal placement in turn determines the orientation of the flower to its inflorescence branch. (a) Schematic of a lateral cyme of *P. guyannens.* (b) Cross section of the lateral cyme shown in (a), but earlier in development. The plane of floral symmetry in these bilaterally symmetric flowers is indicated by a dashed line. B = bract, F = flower, M = terminal inflorescence meristem, S = sepal. Organs are numbered based on their order of initiation. Figure is adapted from Kirchoff (2003).

inflorescence leaves/bracts. However, suppression of organs, particularly leaf-like organs, can have phylogenetic significance. In the case of legumes, bracteoles appear to be ancestral with suppression of bracteoles associated with the Papilionoideae (Prenner, 2004). Typologies provide an organizing frame within which to probe developmental and evolutionary questions.

5.3 Inflorescence development

5.3.1 *Inflorescence meristem*

Defining the morphology of an inflorescence is quite different from defining an inflorescence meristem. Functionally, an inflorescence meristem produces structures that contribute to the inflorescence. Inflorescence meristems can be part of primary, secondary or higher branch order shoot apices. What is less clear is when, developmentally, a shoot apical meristem becomes an inflorescence meristem. As discussed in Chapter 2, phase change from vegetative to adult development involves a series of changes that are necessary but not sufficient for reproductive development. Morphological changes in meristems associated with phase change can be independent of inflorescence status (Marc and Hackett, 1992). An adult meristem may be neither an inflorescence nor a floral meristem. As our understanding of gene expression and long distance signaling in the transition to flowering deepens, it is increasingly clear that the meristem(s) of an inflorescence are moving through a complex series of changes and that there is not a single configuration of cell and molecular traits that constitute the inflorescence meristem state.

Operational definitions for developmental events can provide a temporal and spatial reference point to look for distinguishing characteristics of developmental states. In *Pisum sativum* (garden pea), determination for inflorescence development is a stable state that is separable from determination for flower development (Ferguson *et al.*, 1991). Pea has a double or compound raceme. Eight nodes before the terminal meristem initiates axillary inflorescences, both cultured shoot tips and isolated meristems are committed to producing an inflorescence. These shoot tips and meristems produce the same number of vegetative nodes prior to inflorescence initiation in situ and in culture.

In *Zea mays*, however, similar culture experiments showed that determination for tassel development is a multistep process and does not occur until near the time of floral morphogenesis (Irish and Nelson, 1991). Unlike pea with its indeterminate compound raceme, *Nicotiana tabacum* has second order cymes and a terminal flower. Here the meristem is determined to produce a terminal flower about four nodes before floral organ development begins (Singer and McDaniel, 1986). The formation of a terminal flower makes it more difficult to separate cleanly commitment to inflorescence development and commitment to flower development. *Helianthus annuus* forms a capitulum and, like pea, has extensive temporal and spatial separation between inflorescence determination and initiation. Cultured apices of 5 to 8-day-old seedlings initiate an inflorescence after producing just a few nodes (Paterson, 1984). Expression of inflorescence development appears to be inhibited by root signals in *Helianthus* (Habermann, 1962). These studies and others (Larkin et al., 1990; McDaniel et al., 1991) show no consistent pattern between inflorescence typology and the timing of commitment for inflorescence development in monocots or eudicots.

5.3.2 Regulating determinacy

Inflorescence meristems may have limited growth patterns that fall into two categories. Flower and/or fruit development may limit continued growth of an inflorescence meristem. In this case, the removal of flowers or fruits often results in renewed growth. These meristems are indeterminate. Botanically, determinate meristems are consumed in the production of an organ such as a flower or tendril. In this section we will consider what is known about the genetic regulation of determinacy.

Indeterminacy of inflorescence meristems of *Antirrhinum* and *Arabidopsis* is maintained by the homologous genes *CENTRORADIALUS* and *TERMINAL FLOWER1*, respectively (Shannon and Meeks-Wagner, 1991; Bradley *et al.*, 1996, 1997). The two sequences have a high level of sequence similarity and loss-of-function mutants both terminate with the production of a terminal flower. The CEN/TFL1 gene family is highly conserved (Pnueli *et al.*, 1998; Amaya *et al.*, 1999; Jensen *et al.*, 2001, Foucher *et al.*, 2003).

Tomato has a sympodial shoot system. *SELF-PRUNING* (*SP*) is the tomato ortholog of *TFL1* and is necessary for the maintenance of shoot indeterminacy (Pnueli *et al.*, 1998). Vegetative and reproductive phases alternate regularly during sympodial growth in tomato with three vegetative nodes separating inflorescences. *SP* appears to play a role in regulating the switch between vegetative and inflorescence development that characterizes a sympodial inflorescence. Thus the *TFL1*-like genes regulate inflorescence architecture in eudicot sympodial and monopodial inflorescences by suppressing the vegetative to reproductive transition. This conservation of function extends to the monocots. Overexpression of the *Lolium perenne* (perennial ryegrass) homolog, *LpTFL1*, rescued *Arabidopsis tfl1-14* mutants eudicots (Jensen *et al.*, 2001).

While both *TFL1* and *CEN* suppress development of a terminal flower, *TFL1* also regulates flowering time and *LpTFL1* levels are affected by vernalization. In pea, five *CEN/TFL1* family genes have been cloned (Foucher *et al.*, 2003). One corresponds to the *DETERMINATE* (*DET*) gene that maintains indeterminacy in the first order inflorescence meristem of pea (Singer *et al.*, 1990). A second is *LATE FLOWERING* (*LF*), a vernalization-sensitive gene that regulates the time of flowering (Murfet, 1978). The findings of Foucher and co-workers indicate that regulation of determinacy and flowering time are separable in pea, although a single gene serves both functions in *Arabidopsis*.

TFL1, CEN and *DET* all suppress determinate growth in the inflorescence and provide an opportunity to compare the role of these genes in simple (*Arabidopsis* and *Antirrhinum*) and compound (pea) racemes. In the simple raceme systems, the mutants form true terminal flowers. However, *det* plants terminate with a hairy stub, the determinate structure produced by second order inflorescence meristems. Loss of function of a second gene *VEG1* is required for the first order inflorescence meristem of pea to produce a true terminal flower, i.e. *det veg1* plants form a terminal flower, as do *cen* and *tfl1* plants (Singer *et al.*, 1999). Curiously, *veg1* plants never initiate flowers (Reid and Murfet, 1984).

TFL1 and *TFL1*-like genes have been classified as shoot meristem identity genes (Jack, 2004), but what this means mechanistically is still an open question. TFL and FLOWERING LOCUS T (FT) have opposing functions in terms of suppressing versus activating floral meristem development, respectively (Hanzawa *et al.*, 2005). The two proteins are less than 60% similar, but a single amino acid change converts TFL1 to FT function. The amino acid change results in a swap in the interacting protein partners of TFL1 and FT. Relative to plant architecture, the TFL1 to FT conversion is a striking example of how a single base pair change could dramatically alter inflorescence architecture.

The conserved role of *TFL1*-like genes has been studied primarily through a candidate gene approach. Within the monocots, quantative trait loci (QTL) approaches to identifying other inflorescence architecture genes have been fruitful. Green millet is the presumed progenitor of the cereal grain foxtail millet and a comparison of the two species offers insight into the role of determinacy in inflorescence architecture. While green millet is a noxious weed, foxtail millet is an important crop species in China. The primary differences between the two species is a more complex branching pattern in foxtail millet, accompanied by a reduction in vegetative axillary branching (Doust *et al.*, 2004). The difference between five orders of lateral branching in the green millet tassel and more than eight orders of branching in the foxtail millet branching substantially increases grain yield. While maize *TEOSINTE BRANCHED1* (*TB1*) plays a major role in inflorescence architecture in maize, its role in foxtail millet inflorescence architecture is minimal. Fourteen QTLs have now been identified that are associated with the difference in inflorescence traits, some of which correspond to transcription factors and hormone pathway genes that are putative inflorescence branching regulators (Doust *et al.*, 2005). Doust and colleagues hypothesize that these QTLs have been affected by natural selection in the over 300 members of the *Setaria-Pennisetum* clade within the monocots.

Regulation of determinacy of lateral meristems also has implications for inflorescence architecture. While sweet peas have indeterminate second order inflorescence meristems, garden peas produce one or two flowers on laterals before the inflorescence meristems differentiate into a hairy stub. The garden pea *nep-1* mutant, however, can produce up to five flowers on an inflorescence branch before terminating. The potential to manipulate the number of flowers and resultant fruits on an inflorescence branch has agronomic potential. Additional flowers could potentially increase yield in the case of some legumes, while, in species like apple, the ability to limit the number of developing fruits would lead to increased fruit size.

Co-ordinated manipulation of terminal and lateral inflorescence meristem development could result in synchronous or asynchronous fruit development. Consider the case of garden pea which has asynchronous fruit development. With large-scale production and mechanical harvesting, only a limited number of mature pods can be obtained from each plant. Synchronous flower development is possible through a combination of limited terminal growth and increased flowers per inflorescence branch. Limited terminal growth results from the loss of *DET* function. When *det* is coupled with a mutation such as *neptune* (*nep-1*) that increases the number of flowers and pods on each inflorescence branch, synchronous flowering is observed (Singer *et al.*, Figure 5.9).

Figure 5.9 Comparison of the *det nep-1* garden pea inflorescence (left) that flowers synchronously and the wild-type pea (right) with asynchronous reproductive development.

5.3.3 Regulating meristem identity

Genetic regulation of determinacy is necessary for inflorescence development, but there is insufficient evidence to make a case for inflorescence meristem identity genes. Inflorescence architecture cannot be viewed as a cassette isolated from the highly integrated developmental continuum of flowering. Flowering time can be modified through changes in the juvenile, adult vegetative and adult reproductive phases, which are separable (Wiltshire *et al.*, 1994; Chen *et al.*,1997; Diggle, 1999; Vega *et al.*, 2002; Hunter *et al.*, 2003). Ultimately, flowering time genes function to control the activity of two subclasses of meristem identity genes: shoot meristem identity genes and floral meristem identity genes which in turn regulate the floral organ identity genes (Jack, 2004). Environmental and autonomous floral signaling pathways are integrated by *LEAFY(LFY)* and *APETALA1 (AP1)* which then regulate expression of floral meristem identity genes (Blazquez and Weigel, 2000). Examples of the involvement of floral meristem identity genes affecting inflorescence architecture are known. Down-regulation of *RFL*, the *LEAFY* (*LFY)* homolog of rice, is accompanied by branching of the panicle (Kyozuka *et al.*, 1998).

The case of *RFL* in rice is not anomalous. Phenotypic profiling of 80 early flowering mutants in *Arabidopsis* revealed numerous pleiotropic effects, including altered inflorescence architecture (Pouteau *et al.*, 2004). The mutants were selected only for altered flowering time, emphasizing the intricate interactions between the many actors in the developmental pathway leading to inflorescence and floral development. Given the highly conserved nature of genes involved in the floral transition, perhaps the central question is of how the same actors interact to give rise to such diverse inflorescence typologies (Beveridge *et al.*, 2003; Ferrario et al., 2004; Hecht *et al.*, 2005)? This evolution of development question is a particularly exciting area of emergent research.

5.4 Evolution of inflorescence architecture

As seen in the foxtail millet example earlier, comparative genetic approaches to inflorescence architecture can provide a starting point for addressing evolutionary questions. To illustrate the possibilities, three inflorescence architecture cases are developed here: the origin of maize from teosinte, the genetic basis of broccoli and cauliflower inflorescences in the Brassicaceae and role of phenotypic plasticity in inflorescence architecture from an evolutionary perspective (for a more general review of plant architecture, see Sussex and Kerk, 2001a,b).

5.4.1 Teosinte

A single domestication event is the most probable explanation for the evolution of maize from wild teosinte (Matsuoka *et al.*, 2002). This finding is striking because of the tremendous morphological diversity of maize landraces. The dramatic alteration in inflorescence architecture is a common thread. While teosinte has long inflorescence branches with axillary female inflorescences and terminal male inflorescences, the inflorescence branches of maize are short and highly feminized (Iltis, 2000). Expression of the *tb1* gene in tissues in maize and teosinte corresponds to regions of suppressed growth. Thus *tb1* affects inflorescence architecture through the regulation of determinacy. Comparatively, the different expression patterns in the two species correlates with developmental changes that alter the architecture of the plant. These data and prior results from the Doebley laboratory suggest that changes in *tb1* expression were critical to the evolution of maize (Hubbard *et al.*, 2002).

5.4.2 Cauliflower and Broccoli

Brassica oleracea is a perennial, herbaceous herb found in Europe and Mediterranean areas. The amazing range of phenotypes, including cauliflower, broccoli, kale and wild cabbage, among the subspecies of *B. oleracea* provide extraordinary examples of morphological change during domestication. Wild cabbage is a perennial plant found in southern Britain, western France, northern Spain and along rocky cliffs overlooking the Mediterranean. Domestication was most likely to have occurred in this region as people selected variants of the wild species. The result, over thousands of years, is *B. oleracea* plants with such distinct morphological forms. Despite the morphological differences, all these plants are considered to be members of the same species by systematists. To distinguish between the forms, subspecies (ssp.) names have been assigned. The wild *B. oleracea* is named *B. oleracea* ssp. *oleracea*; cauliflower is *B. oleracea* ssp. *botrytis*; broccoli is *B. oleracea* ssp. *italica*; while kale is *B. oleracea* ssp. *acephala*.

It is likely that a nonsense mutation in the *CAULIFLOWER* (*CAL*) gene, first cloned in *Arabidopsis* (Kempin *et al.*, 1995), contributes to the extensive branching and arrest of prefloral and floral development in cauliflower and broccoli, respectively (Purugganan et al., 2000). In the Brassicacea, *AP1*, a floral integrator and floral meristem identity gene, duplicated and diverged. The *B. oleracea BoCAL* gene is orthologous to *CAL*, and paralogous to *AP1*. Orthologs of *CAL* have been sequenced from 37 different accessions of *B. oleracea* of various subspecies. A base pair change in exon 5 of *BoCAL* that results in a premature stop codon corresponds to the altered inflorescence of cauliflower and broccoli (Purugganan *et al.*, 2000).

This nonsense polymorphism was also found in some *B. oleracea* ssp. *acephala* (kale) and *B. oleracea* ssp. *oleracea* (wild cabbage). Purugganan's research group concluded that specific alleles of *BoCAL* were selected by early farmers for modified inflorescence structures in *B. oleracea.* The moderate levels of BoCAL in other subspecies of *B. oleracea* indicate that the *BoCAL* premature stop codon is not sufficient to explain the cauliflower and broccoli phenotypes. However, this mutation and the resultant selection by farmers is certainly a critical aspect of this inflorescence architecture modification.

While Purugganan's group has demonstrated that the cauliflower phenotype arose within the Brassicacea, our laboratory has found a similar phenotype in pea. In this example of convergent evolution, the *AP1* ortholog (*PROLIFERATING INFLORESCENCE MERISTEM, PIM*) has a loss-of-function phenotype that resembles *ap1* and has additional orders of branching prior to floral initiation. A second mutation, *broccoli (broc)*, has a wild-type phenotype in a wild-type background. However, the *pim broc* double mutant resembles the *ap1 cal* cauliflower/broccoli phenotype (Singer *et al.*, 1999). It is not known whether this more complex inflorescence phenotype is also attributable to a duplication and divergence of the *AP1* ortholog, *PIM*.

5.4.3 Phenotypic plasticity

Within a species, inflorescence architecture is substantially affected by the environment. Phenotypic plasticity and inflorescence architecture have generally been investigated in an ecological context and outcomes are often driven by resource allocation (Diggle, 1997, 1999). For example, *Euphorbia nicaeensis* has three variations on its inflorescence architecture, dependent upon the environment (Al Samman *et al.*, 2001).

Studies of *Pancratium maritimum* fruit and seed set showed that the earliest opening flowers on an inflorescence are more likely to set fruit and produce more seeds than later opening flowers. Numerous manipulations, including supplementary pollination do not substantially alter the outcome. The within-inflorescence patterns of fruit and seed production in *P. maritimum* are mainly attributable to competition for resources (Medrano *et al.*, 2000).

More recently, the probability that phenotypic plasticity is being studied indirectly in breeding programs is being discussed. Genotype-environment interactions are being assessed through the use of wild accessions and recombinant inbreds (Pouteau *et al.*, 2004). In this study, reduced floral repression was associated with an increase in phenotypic plasticity. Pouteau and colleagues make a compelling case that mutations (early flowering mutants in their study) distort the wild-type network systems, affecting the expression of numerous unrelated genes and thus increasing phenotypic plasticity. They build on Greenspan's (2001) claim that developmental network systems require robustness and flexibility for developmental stability. Mutations, even single mutations, diminish the overall robustness of the system and confer more extensive phenotypic plasticity. The perspective offered by Pouteau et al. and Greenspan is a cautionary note about separating out direct, single gene effects on the range of inflorescence architectures within a species and the possibility that a complex, redundant network has been destabilized, affording increased plasticity. One could certainly make a case that increased flexibility in the absence of a crucial gene could potentially increase fitness.

The ability to adapt to environmental constraints is critical for the survival of a species. While phenotypic plasticity provides flexibility, there are also genetic differences among species that appear to be optimal in specific environments. For example, wind interacts with inflorescence architecture in a complex manner to facilitate pollination. An intriguing hypothesis is that the extensive diversity of inflorescence architecture within the Poaceae represents evolutionary solutions to the physical constraints of wind pollination (Friedman and Harder, 2004).

5.5 Future directions: modeling inflorescence architecture

It is increasingly clear that a range of factors need to be considered in analysing inflorescence architecture. The complexity of the problem begs for a modeling approach that is predictive and both drives and virtually tests hypotheses regarding gene interactions and morphological outcomes. Such models could integrate genetic and hormonal components of the developmental processes leading the stunning array of architectural patterns orchestrated by a shared set of orthologues and paralogues.

A number of approaches to predictive and hypothesis driven modeling are underway. L-systems and L-studio software offer a way to visualise models of plant architecture (Hanan *et al.*, 1988; Davidson *et al.*, 1994; Prusinkiewicz, 2004). Numerous other software approaches are being considered (see Meyerowitz's Computable Plant website: www.computableplant.org)

Computational approaches have been called for to understand the relationship between vegetative and reproductive shoot structure in pea (Beveridge *et al.*, 2003). Computational modeling has been instructive in elucidating the genetic and hormonal regulation of vegetative axillary branching in pea (Foo *et al.*, 2005). Extending this approach to inflorescence branching appears promising. Computational models would allow for the integration of the floral induction, floral integration and inflorescence/flower developmental pathways. Hypotheses such as the one by Poteau and colleagues, who propose a destabilization of networks when a gene is mutated, could be tested or used to design experiments by introducing information about various mutations into the model.

While models offer inroads for prediction and hypothesis testing, the potential for agronomic applications should not be overlooked. A synthetic model that accounts for environmental variables, gene response to sensory input, evidence based gene networks, gene expression levels from a variety of sources (including relative quantification PCR), genomic data (including microarry profiles) and physiological data could result in a robust tool to further our understanding of inflorescence architecture. Modeling is an effective way to attack complex biological problems at multiple levels (Katagiri, 2003). To provide valuable analytical tools, these models must not be solely descriptive, but predictive.

References

Al Samman, N., Martin, A. and Puech, S. (2001) *Botanical Journal of the Linnean Society*, **136**, 99–105.

Amaya, I., Ratcliffe, O. and Bradley, D. (1999) *The Plant Cell*, **11**, 1405–1417.

Beveridge, C. A., Weller, J. L., Singer, S. R. and Hofer, J. M. I. (2003) *Plant Physiology*, **131**, 927–934.

Blazquez, M. A. and Weigel, D. (2000) *Nature*, **404**, 889-892.

Bradley, D., Carpenter, R., Copsey, L., Vincent, C., Rothstein, S. and Coen, E. (1996) *Nature*, **379**, 791–797.

Bradley, D., Ratcliffe, O., Vincent, C., Carpenter, R. and Coen, E. (1997) *Science*, **275**, 80–83.
Calonje, M., Cubas, P., Martinez-Zapater, J. M. and Carmona, M. J. (2004) *Plant Physiology*, **135**, 1491–1501.
Chen, L., Cheng, C., Castle, L. and Sung, Z. R. (1997) *The Plant Cell*, **9**, 2011–2024.
Davidson, C. G., Hammell, M. S., Prusinkiewicz, P. W. and Remphrey, W. R. (1994) *Canadian Journal of Botany-Revue Canadienne De Botanique*, **72**, 701–714.
Diggle, P. K. (1997) *International Journal of Plant Sciences*, **158**, S99–S107.
Diggle, P. K. (1999) *International Journal of Plant Sciences*, **160**, S123–S134.
Doust, A. N., Devos, K. M., Gadberry, M. D., Gale, M. D. and Kellogg, E. A. (2004) *Proceedings of the National Academy of Sciences of the United States of America*, **101**, 9045–9050.
Doust, A. N., Devos, K. M., Gadberry, M. D., Gale, M. D. and Kellogg, E. A. (2005) *Genetics*, **169**, 1659–1672.
Ferguson, C. J., Huber, S. C., Hong, P. H. and Singer, S. R. (1991) *Planta*, **185**, 518–522.
Ferrario, S., Immink, R. G. and Angenent, G. C. (2004) *Current Opinion in Plant Biology*, **7**, 84–91.
Foo, E., Bullierb, E., Goussotb, M., Foucherb, F., Rameau, C. and Beveridge, C. A. (2005) *The Plant Cell*, **17**, 464–474.
Foucher, F., Morin, J., Courtiade, J. *et al.* (2003) *The Plant Cell*, **15**, 2742–2754.
Friedman, J. and Harder, L. D. (2004) *Functional Ecology*, **18**, 851–860.
Greenspan, R. J. (2001) *Nature Reviews Genetics*, **2**, 383–387.
Habermann, H. M. (1962) In *XVIth International Horticultural Congress, Brussels, Belgium*, pp. 243–251.
Hanan, J., Lindenmayer, A. and Prusinkiewicz, P. (1988) *Computer Graphics*, **22**, 141–150.
Hanzawa, Y., Money, T. and Bradley, D. (2005) *Proceedings of the National Academy of Sciences of the United States of America*, **102**, 7748–7753.
Hecht, V., Foucher, F., Ferrandiz, C. *et al.* (2005) *Plant Physiology*, **137**, 1420–1434.
Hofmeister, W. (1868) *Allgemeine morphologie der Gewachse*, Engelmann, Leipzig.
Hubbard, L., McSteen, P., Doebley, J. and Hake, S. (2002) *Genetics*, **162**, 1927–1935.
Hunter, C., Sun, H. and Poethig, R. S. (2003) *Science*, **301**, 334–336.
Iltis, H. H. (2000) *Economic Botany*, **54**, 7–42.
Irish, E. E. and Nelson, T. M. (1991) *Development*, **112**, 891–898.
Jack, T. (2004) *The Plant Cell*, **16**, S1–S17.
Jean, R. V. (1988) *Annals of Botany*, **61**, 293–303.
Jensen, C. S., Salchert, K. and Nielsen, K. K. (2001) *Plant Physiology*, **125**, 1517–1528.
Katagiri, F. (2003) *Plant Physiology*, **132**, 417–419.
Kempin, S. A., Savidge, B. and Yanofsky, M. F. (1995) *Science*, **267**, 522–525.
Kirchoff, B. K. (2003) *International Journal of Plant Sciences*, **164**, 505–517.
Kyozuka, J., Konishis, S., Nemoto, K., Izawa, T. and Shimamoto, K. (1998) *Proceedings of the National Academy of Sciences of the United States of America*, **95**, 1979–1982.
Larkin, J. C., Felsheim, R. and Das, A. (1990) *Developmental Biology*, **137**, 434–443.
Marc, J. and Hackett, W. (1992) *Planta*, **186**, 503–510.
Matsuoka, Y., Vigouroux, Y., Goodman, M. M., Sanchez, G. J., Buckler, E. and Doebley, J. (2002) *Proceedings of the National Academy of Sciences of the United States of America*, **99**, 6080–6084.
McDaniel, C. N., King, R. W. and Evans, L. T. (1991) *Planta*, **185**, 9–16.
Medrano, M., Guitian, P. and Guitian, J. (2000) *American Journal of Botany*, **87**, 493–501.
Murfet, I. C. (1978) *Pisum Newsletter*, **10**, 48–52.
Paterson, K. E. (1984) *American Journal of Botany*, **71**, 925–931.
Pnueli, L., Carmel-Goren, L., Hareven, D. *et al.* (1998) *Development*, **125**, 1979–1989.
Pouteau, S., Ferret, V., Gaudin, V. *et al.* (2004) *Plant Physiology*, **135**, 201–211.
Prenner, G. (2004) *Annals of Botany*, **93**, 537–545.
Prusinkiewicz, P. (2004) *Acta Hort*, **630**, 15–28.
Purugganan, M. D., Boyles, A. L. and Suddith, J. I. (2000) *Genetics*, **155**, 855–862.
Reid, J. B. and Murfet, I. C. (1984) *Annals of Botany*, **53**, 369–382.

Reinhardt, D. and Kuhlmeir, C. (2002) *EMBO Reports*, **3**, 846–851.
Richards, F. J. (1951) *Philosophical Transactions of the Royal Society of London Series B-Biological Sciences*, **235**, 509–564.
Rickett, H. W. (1944) *Botanical Review*,**10**, 187–231.
Roeper, J. A. C. (1826) *Linnaea*, **1**, 433–466.
Shannon, S. and Meeks-Wagner, D. R. (1991) *The Plant Cell*, **3**, 877–892.
Singer, S. and McDaniel, C. (1986) *Developmental Biology*, **118**, 587–592.
Singer, S., Sollinger, J., Maki, S. *et al.* (1999) *Botanical Review*, **65**, 385–410.
Singer, S. R., Hsiung, L. P. and Huber, S. C. (1990) *American Journal of Botany*, **77**, 1330–1335.
Snow, M. and Snow, R. (1933) *Philosophical Transactions of Royal Society of London B*, **222**, 353–400.
Stebbins, L. G. (1974) In *Flowering Plants*, The Belknap Press of Harvard University Press, Cambridge, pp. 261–282.
Sussex, I. M. and Kerk, N. M. (2001a) *Current Opinion in Plant Biology*, **4**, 33–37.
Sussex, I. M. and Kerk, N. M. (2001b) *Biofutur*, **2001**, 52–56.
Vega, S. H., Sauer, M., Orkwiszewksi, J. A. J. and Poethig, R. S. (2002) *The Plant Cell*, **14**, 133–147.
Weberling, F. (1989) *Morphology of Flowers and Inflorescences*, Cambridge University Press, New York.
Wiltshire, R. J. E., Murfet, I. C. and Reid, J. B. (1994) *Journal of Evolutionary Biology*, **7**, 447–465.

Part II Specialised components of development

6 Close, yet separate: patterns of male and female floral development in monoecious species

Rafael Perl-Treves and Prem Anand Rajagopalan

6.1 Introduction

Much of our knowledge on flower development comes from hermaphroditic model species such as *Arabidopsis*, rice, petunia and snapdragon, whose perfect (bisexual) flowers bear both stamens and pistils. Indeed, most higher-plants are hermaphrodites, producing bisexual flowers in each individual. However, many plant taxa have evolved mechanisms to separate male and female reproductive functions, either within the same plant (monoecious plants) or on separate male and female individuals (dioecious plants). Dioecious plants have evolved genetic mechanisms for sex determination at the individual plant level. Here we turn our attention to monoecious plants, where sex determination mechanisms operate, in a temporally and spatially regulated pattern, within the same plant. We shall approach this developmentally fascinating phenomenon from two main angles. From an ecological/evolutionary viewpoint, we can try to evaluate the costs and benefits of such reproductive strategy. Several species, mostly wild plants, have been the focus of this type of study and have provided exciting insights on sex determination in plants. I will not attempt to comprehensively review such studies, but rather will give a few selected examples. Typically, they raise questions such as: 'Why should some plants be monoecious?' 'How did monoecy evolve?' 'What are the consequences for plant fitness?' The second approach deals with the production of sex organs in the flowers; monoecious plants pose several intriguing developmental questions that can be answered using genetic and molecular tools. Here, the underlying question is 'How is a unisexual flower formed', and typical genetic/developmental terminology has been used. In this chapter I will review two species – maize and cucumber – that were studied more intensively. The focus on two cultivated species probably reflects the agricultural interest in genes that control separation of flower sexes, due to their economic importance to hybrid seed production.

6.2 Ecological and evolutionary aspects of monoecious plant development

6.2.1 Evolution of diverse reproductive strategies in land plants

Plants have evolved an impressive diversity of reproductive strategies, all aimed at assuring reproductive success in the short-term scale, as well as evolutionary success in the longer run. Much of the diversity in flower form and color relates to pollination mechanisms. On the other hand, variation in sex expression and breeding systems probably reflects the need to optimize two contrasting urges: to ensure fertilization of a non-motile organism, whose mating partners and pollinators could be limited, against the drive to generate genetic variation. Inbreeding depression due to lack of cross-fertilization is considered an important selective force that drives plant reproductive evolution. Since selective forces probably vary, different plant taxa vary greatly in the extent of self-fertilization that they allow, with dioecious plants representing one extreme in a continuum of breeding systems. Hermaphrodites, too, vary in the degree by which they facilitate or prevent self-pollination. Some hermaphrodite species encourage self-pollination in closed flowers (cleistogamy), whereas others have different rates of stamen and pistil maturation (dichogamy), floral architectures that minimize selfing (heterostyly), or self-incompatibility mechanisms that completely prevent self-pollination (reviews by Dellaporta and Calderon-Urrea, 1993; Barrett, 2002).

In this context, the monoecious situation probably represents a compromise between allogamy and autogamy: it combines separate male and female flowers on the same individual. Other less frequent variants include sex types such as andromonoecious (staminate plus perfect flowers), and gynomonoecious (pistillate plus perfect flowers). Together with the monoecious type, they can all be classified as 'monomorphic' species whose individuals show variable ratios of male, female or bisexual flowers. This is opposed to gender dimorphism manifested as a bimodal segregation of individuals between two genders. Monoecious plants represent about 7% of the species, dioecious plants 4%, while hermaphrodites comprise about 72% of the species; the remainder species have populations of hermaphrodite, male and female individuals in various combinations (Dellaporta and Calderon-Urrea, 1993; Barrett 2002).

Flower unisexuality found in monoecious and dioecious plants was probably derived from hermaphroditism, which is regarded as the ancestral state in land plants (Tanurdzic and Banks, 2004). Unisexual flowers have appeared in divergent lineages of the plant phylogeny, implying multiple mechanisms for achieving monoecy or dioecy. In fact, genetic mechanisms for sex determination range from single-gene control through oligogenic inheritance to sex chromosomes with different degrees of cytological diversification. Sex determination mechanisms in different taxa are differentially affected by specific plant hormones. The same hormone could have opposite effects in different plants and in different organs within the same plant, e.g. gibberellin causes maleness in cucumber and femaleness in maize. They also differ by the stages at which the organs of the 'other sex' are inhibited or aborted, following an initially bisexual stage of the floral bud (Dellaporta and Calderon-Urrea, 1993; Grant *et al.*, 1994). In spinach (*Spinacia oleracea*) and mercury (*Mercurialis annua*), the inappropriate organs do not emerge at all from the floral meristem; in cucumber and maize, they are arrested at early developmental stages, whereas in *Asparagus officinalis* they are arrested later, just before meiosis. Theoretical models on the evolution of monoecy from hermaphroditism, and of dioecy from either monoecy or gynodioecy, were reviewed by Dellaporta and Calderon-Urrea (1993); Grant *et al.* (1994); Charlesworth and Guttman (1999); Barrett (2002).

6.2.2 *Selective forces that favor the evolution of unisexual flowers*

Studies that adopt phylogenetic, ecological and marker-assisted population genetics methods are beginning to shed light on the selective forces that generated the monoecious and dioecious mating systems. These studies try to assess the cost of self-fertilization ('inbreeding depression') in different species. Under what conditions does it become too high, and favor the selection of unisexual flower types (or other mechanisms that encourage allogamy)? When does allogamy become an advantage? What are the metabolic costs of producing male versus female flowers and gametes, and how do these factors vary?

In the Solanaceae phylogenetic tree, the evolution of unisexual flowers appears to be associated with polyploidy. Polyploidy could have broken down self-incompatibility mechanisms present in the original diploid species, creating the need of alternate ways to enhance cross-fertilization (Miller and Venable, 2000). In very large plants – or in plants that reproduce vegetatively forming large clonal populations – it has been argued the rates of fertilization among flowers of same individual or clone (geitonogamy) become very large, exerting selection pressure toward dioecy or monoecy (Reusch, 2001).

6.2.3 *Flexibility in sex ratios in monoecious plants*

Compared to dioecious plants, monoecious species have two advantages: the possibility to self-pollinate when mates and/or pollinators are limited, and flexibility in sex ratios. The idea that monoecious plants benefit from plastic sex expression programs has motivated several studies that tried to document changes in 'gender ratio' in response to natural or experimentally induced changes in the environment. It implies that under different conditions, the gender ratio that is optimal for reproductive success will vary, and that plants can adapt and modify their sex ratios.

The 'functional gender' of a monoecious plant is the product of sex phases that occur during development; not only are the male and female flowers spatially separated, but we usually observe developmental changes that result in some temporal separation between the two sexes. The gender ratio could thus vary along the season, both within the plant and at the population level, resulting in different rates of cross-versus self-pollination. This level of complexity should be considered when we try to assess the implications of the monoecious reproductive strategy (Barrett, 2002).

Flexible resource allocation could be an important advantage of monoecious systems. Because female reproductive costs are usually higher, we expect sex ratios to tend toward femaleness with increased plant size. Sexual plasticity was observed in the aquatic plant *Sagittaria sagittifolia* (Alismataceae), where larger plants tend to have more female flowers. Sex ratios differed according to plant age and also according to nutrient supply and water depth (Dorken and Barrett, 2003). In *Sagittaria latifolia*, monoecious and dioecious populations occur in the same geographic range but tend to occupy different ecological niches; monoecious populations colonize disturbed, ephemeral water bodies, consistent with the reproductive assurance they enjoy due to the self-pollination option; dioecious populations flourish in stable but more competitive environment with deeper water (Dorken and Barrett, 2004). Femaleness correlated in both gynoecious and monoecious plants with nutrient-rich environments. Nevertheless, the authors used fertilizer treatments and demonstrated that the two sex types differed in plasticity of their responses. Monoecious plants were more 'plastic', and this agreed with their adaptation to disturbed environments and supported the view of ecological adaptivity as an advantage of monoecy. Genetic variation in sex ratio among monoecious individuals was measured as well, supporting the hypothesis that dioecy could evolve from monoecy by gradual selection of genotypes that display quantitative variation in sex ratios.

In cucurbits, reproductive development responds to a variety of factors, such as photosynthetic capacity, nitrogen supply and plant age. Reduced photosynthesis or presence of a developing fruit could prevent fruit set, induce young fruit abortion and even modulate sex expression by reducing pistillate flower production (Krupnick *et al.*, 2000; Hirosaka and Sugiyama 2004). Inter-organ communication would be required to integrate photosynthetic status and sink demand, and respond by altering sex development. Interestingly, Krupnick *et al.* (1999, 2000) observed that endogenous ethylene levels in *Cucurbita texana* were lowest when branches carried two or more fruits; injection of ethylene into the stem cavity increased pistillate bud numbers. Ethylene, a major sex hormone in cucurbits (see Section 6.3.3.4), probably plays a role in expressing such plasticity in response to environmental/developmental cues.

The wild cucurbit *Ecballium elaterium* is known for its peculiar seed dispersal mechanism: sudden squirting of the seeds from the dehiscent fruit. The monoecious, gynoecious and androecious sex types found in *Ecballium* are controlled by a single gene with three alleles (Galan, 1964). Its two subspecies, one monoecious and the other dioecious, were studied at the flower morphological level, in an attempt to estimate the selective forces that shape monoecious versus dioecious breeding systems (Costich and Meagher, 2001). Male flowers were larger and more numerous than female flowers in both subspecies, suggesting differential reproductive costs of the two sexes. Male and female flowers were larger in dioecious compared to monoecious plants, supporting a model that predicts a need for 'reproductive compensation' in unisexual individuals, where pollination services are more limited.

Monoecious populations were smaller and displayed lower isozyme and phenotypic variation compared to dioecious populations. Such population structures probably reflect the genetic gain in variation offered by increased allogamy in dioecious systems, as well as the ability of monoecious plants to assure reproduction by self-pollination and colonize novel niches, where they can found and maintain small populations.

6.2.4 *Relationship between monoecy and pollination*

Pollinators, too, can influence the evolution of a plant lineage from a hermaphrodite sex habit into a monoecious or dioecious habit. In some insect-pollinated monoecious plants, female flowers that have aborted their stamens face the problem of offering an adequate pollinator reward. In those species where pollen is the main reward, female flowers may be pollinated by deceit; insects are first 'trained' by rewarding male flowers and then pollinate the female flowers by occasional visits. In such species, female flowers must look as similar as possible (from an insect viewpoint) to male flowers. In wild *Begonia* (Begoniaceae), two monoecious species were investigated with the aim of understanding this somewhat problematic strategy (Le Corff *et al.*, 1998). Pollinator visits were recorded and the degree of discrimination displayed by different bee species was measured. As female flowers became more numerous during the season, pollination success diminished, but the strategy of having a predominant-female phase later in the season probably offers an evolutionary advantage.

6.2.5 *The andromonoecious option: more compromises?*

Huang (2003) analysed the male and female resource allocation and reproductive success of *Sagittaria guayanensis*. Most *Sagittaria* species are monoecious, but *S. guayanensis* is andromonoecious, each plant bearing male plus perfect flowers. According to one hypothesis, andromonoecy could have evolved as an intermediate step between hermaphroditism and monoecy. The andromonoecious situation could represent an attempt to optimize resource allocation: since female functions require more investment, the plant turned surplus flowers, which are unlikely to develop fruits due to limiting resources, into male flowers. The latter increase pollination probability by providing more pollen and attracting pollinator visits to the vicinity of the perfect flowers; pollen is also less likely to be wasted on self-pollination. However, these advantages are common to monoecious plants also; why are stamens retained in perfect flowers of andromonoecious species? In andromonoecious *S. guayanensis*, flowers received very few visits by insects, compared to monoecious sister species, and experiments involving emasculation and bagging of flowers demonstrated that cross-pollination in this species is quite rare; most sexual reproduction came from selfing. This, and the prevalence of monoecious species in the genus, indicates that andromonoecy in *Sagittaria* was secondarily derived from monoecy by selecting for stamen re-appearance in the female flowers, as a form of reproductive assurance where pollinators are so scarce. Interestingly, the more common monoecious species of *Sagittaria* can reproduce clonally, and extensive geitonogamy (self pollination between flowers of same individual/clone) could prevail if plants were not monoecious or dioecious, whereas *S.guayanensis* only reproduces sexually. Nevertheless, selective pressures that maintain some degree of allogamy must still exist, since the author shows larger allocation to male flowers (larger petals, larger pollen grains and four-fold more pollen per flower), compared to the stamens found within the perfect flowers. The fact that this species has not reverted to the hermaphrodite breeding system but remained andromonoecious may point in the same direction – of andromonoecy representing a balanced evolutionary compromise. In addition, retaining the stamens in the perfect flowers of andromonoecious plants solves the problem of pollinator reward.

A somewhat different situation was observed in another monoecious plant, *Belseria triflora*, that is pollinated by hummingbirds. Here, male flowers serve to increase the 'floral display' and were shown to attract the birds, and indirectly influence female success. Such flowers are 'low-cost structures' that produce six times less pollen than the hermaphrodite flowers (Podolsky, 1992), but help to direct the pollen to the perfect flowers more effectively, compared to additional perfect flowers that were artificially presented to the birds.

6.3 How do unisexual flowers develop?

6.3.1 Structural differences between male and female flowers

In this section we will turn our attention to genetic developmental programs that control the formation of unisexual flowers. Such programs were shown, in many species, to operate on an initially bisexual flower bud, and cause the arrest of development and in some cases the abortion of stamen primordia in female flowers, and of carpel primordia in male flowers. Emerging studies are focusing on the mechanism that sets off the sex determination pathway – how the unisexual flower is patterned, and how the 'wrong organs' are actually arrested. In many species, physiological modulation and even complete reversal of the sex expression programs are possible, and provide the researcher valuable handles to examine the system. Mutants and genetic variants that are affected in sex organ development are another important tool.

Sex determination could influence traits other than the sex organs themselves. In many species, male and female flowers differ also in inflorescence structure and perianth organs. A familiar example is maize, where the male and female inflorescences differ strikingly in traits such as branch length and spikelet-abortion pattern (see Section 6.3.2.1). The sessile ear (female inflorescence) has sessile spikelets supported by a strengthened stem segment, whereas the apical tassel (male inflorescence) has long ramifications, reflecting structural adaptation to wind pollination. Should we consider the differential architecture of male/female flowers as 'secondary sex traits'? Are they controlled by the same sex determining mechanism, e.g. the sex determining hormones?

In *Atriplex halimus* (Chenopodiaceae), male and female flowers have distinct architectures: typical male flowers have five tepals, stamens and an inhibited carpel. Typical female flowers are subtended by two bracts and have only one carpel; apart from the completely different structure, the latter have no primordial stamens even at the earliest stage, so the ontogeny of sex organs and the mode of elimination of the opposite sex must differ between the two flower types. Interestingly, the plant forms additional flower phenotypes, including bisexual flowers and unisexual ones that have the opposite sex architecture (bracteate male flowers and pentamerous female flowers). This raises the possibility that in *Atriplex*, floral architecture can be genetically uncoupled from sex organ development (Talamali *et al.*, 2003).

The 'ABC model' for flower development provides a conceptual framework to evaluate unisexual flowers against the hermaphrodite models. The 'building blocks' of such model are MADS-box genes that encode transcription factors and control flower organ identity. They have been implicated in the diversification of floral structures in plant phylogeny (Becker *et al.*, 2000). It appears that radiation events in the evolution of flowering plants coincide with the diversification of MADS gene lineages, and major floral architecture plans could have evolved by modification of the MADS proteins – their expression domains, or their ability to interact with other MADS protein partners (Winter *et al.*, 2002). In this context, unisexual flowers could be generated via modification in MADS proteins required for stamen and carpel specification (B, C classes). Changes in such genes could also affect organ identity in outer whorls and inflorescence characteristics (referred to as 'secondary traits'). However, so far there is no evidence for homeotic transformation as a direct mechanism underlying unisexual flower formation.

In the last two sections, I will examine in detail two species that can be regarded as models for the study of sex expression: a wind-pollinated monocot, maize, and an insect-pollinated dicot, cucumber. In both species, genetic and physiological studies initiated half-a-century ago have been recently expanded and re-interpreted at the molecular level.

6.3.2 *Unisexual flower development in maize*

6.3.2.1 *Development of the male and female florets*

Most species of the Poaceae (the grass family) are hermaphroditic, whereas maize and its relatives (tribe Andropogonae) are monoecious (Le Roux and Kellogg, 1999). The branched, indeterminate male inflorescence, the tassel, is borne on the top of the plant, while the unbranched female inflorescence, the ear, develops at a lower, axillary position. Pollen is carried by wind and captured by the long pistils ('silk'). Such neat spatial separation facilitates the emasculation of parental plants for F1 hybrid seed production; hybrid varieties that revolutionized plant breeding were first developed in maize in 1926. The developmental sequence leading to *Zea mays* unisexual flowers from bisexual primordia was described by Cheng *et al.* (1983) using scanning and transmission electron microscopy. The genetics of sex determination was examined and reviewed by Irish *et al.* (1994) and Irish (1999).

The basic unit of the maize inflorescence, either male or female, is the spikelet, with two subtending glumes enclosing two flowers (named florets in grasses). Florets have no typical sepals or petals; they develop a single lemma, a palea, three lodicules, three stamens and a pistil that bears a single ovule. After reaching a bisexual stage, the fate of the floret is determined. In the tassel, pistil primordia stop growing and eventually disintegrate in both florets of each spikelet, while the stamens fully develop, resulting in two functional staminate flowers. Gynoecium abortion, a programmed cell death (PCD) process, is manifested by vacuolation and loss of ribosomes and organelles in a small group of cells. Calderon-Urrea and Dellaporta (1999) observed localized DNA degradation and nuclear loss in the abortive tissue. However, a full description of the PCD process in maize is still lacking and we do not know whether it involves a classical apoptotic phenotype (e.g. DNA fragmentation, expression of specialized proteases, apoptotic cellular morphology; Buckner *et al.*, 2000).

In the ear, rows of spikelet pairs form, each with two florets: the second, lower one lags behind the first and is eventually aborted, exhibiting cellular disintegration as described earlier. The upper floret reaches the bisexual stage, then stamen primordia are eliminated by a similar abortive process. The gynoecium develops a single ovule and an elongated style, the silk. Mutants that are affected in the abortion of the opposite sex organs were isolated and characterized.

6.3.2.2 Mutations that affect stamen development: role of gibberellin

Several mutations, in which the stamens of the ear floret develop, yielding perfect ear-flowers, while the tassel remains unaffected, impart an andromonoecious sex phenotype (Figure 6.1a–c). They include several recessive mutants, *dwarf-1, -2, -3, -5* and *antherear*, as well as dominant ones, *Dwarf-8* and *Dwarf-9*. Biochemical and physiological analyses indicate that all the recessive mutants involve gibberellin synthesis: plants are dwarfed, GA synthesis is blocked and the phenotype can be cured by exogenous GA application. This indicates that gibberellin plays a major role in suppressing stamen development in the female flower. Moreover, Rood *et al.* (1980) reported 100-fold more GA in wild-type ears compared to the tassels, which could explain why stamens are inhibited in the ear but not in the tassel. In 1995, a transposon-tagged *antherear* allele was cloned (Bensen *et al.*, 1995), and shown to encode ent-kaurene-synthase A, an enzyme that catalyzes the first committed step in gibberellin synthesis. Winkler and Helentjaris (1995) cloned the maize *dwarf-3* gene and showed that it encoded a cytochrome P450-type enzyme, also involved in GA synthesis.

Does GA also act on the carpel by promoting its development or preventing its abortion? In the tassel, exogenous GA application reversed tassel sex, showing that high GA both repressed stamens and promoted pistil development. However, in GA-less mutants, stamens are released from suppression in the ear, but the ear's pistils are not affected, showing that they do not require high GA levels to develop, contrasting the situation in the tassel. The additive phenotypes of *dwarf*

Figure 6.1 Sex determination mutants in maize. (a) Schematic representation of main mutant phenotypes. (b) Ear spikelet of *anthererar1* plant at grain-filling stage. Anthers developed in both the left floret (where the pistil aborted) and the right one, where the pistil developed and was fertilized. (c) Mature ear of *anthererar1* mutant (b,c were reproduced from Bensen *et al.*, 1995, with permission and copyright of the American Society of Plant Biologists). (d,e) Feminized tassels borne on the top of tasselseed2 mutant plants. In (e), note the sessile ear-like florets and (circled) a single revertant spikelet resulting from transposon excision (reproduced from De Long *et al.*, 1993, with permission from Elsevier). (f) Expression of *TS2* mRNA in wild-type florets of the tassel: the in-situ hybridization signal localizes to the pistil primordia in both florets of the spikelet. (g) Propidium iodide staining reveals nuclear loss in group tassel-pistil cells that previously expressed *TS2* (f, g were reproduced from Calderon-Urrea and Dellaporta, 1999).

and *tasselseed* double mutants (*tasselseed* are deficient in pistil abortion, see later) probably indicate that the two pathways, leading to pistil abortion and to stamen abortion, are distinct (Irish *et al*., 1994).

The dominant mutations *Dwarf-8* and *Dwarf-9* impart a similar phenotype as is shown by the recessive mutants, but are not rescued by GA treatment. They have not been cloned yet, but could possibly encode homologues of GAI, a repressor of gibberellin responses that is negatively regulated by gibberellin. Mutated, it could remain constitutively activated and repress the gibberellin response.

6.3.2.3 Mutations that affect pistil abortion

Tasselseed (ts) mutations prevent carpel abortion in the tassel (Figure 6.1d,e). The recessive mutations *ts1* and *ts2* render the plant gynoecious: the ear is normal, but the tassel has pistillate instead of staminate flowers. These two mutations also prevent abortion of the secondary ear floret, resulting in the development of a pair of grains in each spikelet (Irish and Nelson 1993). 'Secondary sex-traits' of the flower were also affected – the mutant tassels had shorter glumes and sessile flowers instead of the long pedicels and glumes that characterize normal male flowers. Other *tasselseed* mutations have somewhat different phenotypes, e.g. varying patterns of feminized flowers in the tassel or changes in spikelet determinacy, in addition to changes in flower sex (Irish, 1999).

De-Long *et al*. (1993) studied a transposon tagged *ts2* mutant and cloned the *TS2* gene, whose product is responsible, in the wild type, for gynoecium abortion both in the tassel and in the secondary ear floret. The timing and location of wild-type *TS2* expression, in the L2 layer of gynoecium primordia just before their elimination, correlated well with its function (Figure 6.1f). Revertant tassel sectors that result from excision of the transposable element could be as small as a single spikelet, indicating a late timing of *TS2* expression, after spikelet and floret plans were laid down. In a later study (Calderon-Urrea and Dellaporta, 1999), *TS2* expression in the young pistils of both female florets was observed. Expression in the first floret, which is not destined to degenerate, was puzzling; it indicated that *TS2* is required but insufficient for organ abortion; other regulators/gene products are required.

It turned out that TS2 is homologous to short-chain alcohol dehydrogenases (De-Long *et al*., 1993), whose substrate is presently unknown. It could be a gibberellin, that is involved in maize sex development, or a brassinosteroid. Brassinosteroids and GA are important for male-organ development: GA- and brassinosteroid-deficient mutants of *Arabidopsis* were male-sterile (Sun *et al*., 1992; Li *et al*., 1996). We do not know yet whether TS2 removes a substrate required for cell viability, or generates a product that initiates cell death.

The cellular changes associated with pistil elimination were examined by Calderon-Urrea and Dellaporta (1999). Reduction in 4,6-diamine-2-phenylindole dihydrochloride (DAPI) or propidium-iodide fluorescence (DNA-binding stains), due to nuclear DNA loss, occurred in the same layer of the gynoecium where the

TS2 gene product is expressed (Figure 6.1g). Nuclei in adjacent cells outside the abortive gynoecium were brightly stained. In *ts2* mutants the nuclei appeared intact, demonstrating that *TS2* gene is required for the cell death process. A similar loss was apparent in the secondary floret gynoecium in the ear of wild-type plants, but not in the *ts2* mutant.

The *ts1* mutation has a similar phenotype as *ts2*. *TS2* expression was abolished in *ts1* mutants (Calderon-Urrea and Dellaporta, 1999), proving that TS1 regulates TS2 expression. Figure 6.2a presents a model for pistil fate during maize sex determination.

The *silkless1* (*sk1*) mutant is impaired in gynoecium development; the ear fails to develop female organs, but the tassel is unaffected (Figure 6.1a). Cell-death patterns in the pistils of *sk1* resemble those of the innapropriate organs mediated by the *TS2* gene. The *ts2* mutation is epistatic to *sk1*, double mutants showing the *ts2* phenotype: no inhibition of pistil in the ear and the appearance of pistillate flowers in the tassel. This suggests that the SK1 protein interacts with TS2 in the same organ–abortion pathway. The authors proposed that TS-2 causes PCD only when SK1 is not present, the latter playing a cell-protective role (Figure 6.2a). In the pistil of the ear's first floret, TS2 is present but abortion does not occur, because TS2 is ineffective where SK1 is expressed. In animals, anti-apoptotic proteins that inactivate cell-death factors by direct binding have been characterized (Goldstein, 1997). Molecular cloning and analysis of *sk1* has not been reported and so its role is still hypothetical.

6.3.2.4 Cytokinin counteracts pistil abortion

Young *et al.* (2004) engineered maize with isopentenyl transferase (IPT), a cytokinin-producing enzyme, under the control of an *Arabidopsis* senescence-inducible promoter. The promoter was activated in an age-dependant manner, and its expression in developing ears rescued the second-floret from abortion. This resulted in two developing pistils in the same spikelet. The two fertile ovules that developed formed a fused seed with two separate embryos (Figure 6.2b). Thus, cytokinin can prevent the entry of the second ear-florets to the PCD pathway. It did not, however, protect the gynoecium of male flowers, nor the stamens in female flowers. The reason could be lack of expression of the engineered promoter in these organs, or because cytokinin is not important there. Senescence, which is typically delayed by cytokinin, is distinct from PCD, but the two processes may have common regulatory features. For example, both are positively regulated by ethylene. Does cytokinin act by modulating *TS2* or *SK1* expression or by altering GA levels? The answer has yet to come. In another study involving transgenic maize, plants were engineered to over-express in their anthers cytokinin oxidase, a cytokinin-degrading enzyme (Huang *et al.*, 2003). This resulted in male sterility, suggesting that cytokinins are important also for anther and pollen development. Female fertility, too, was impaired in many plants, but it was unclear whether this was due to the transgene.

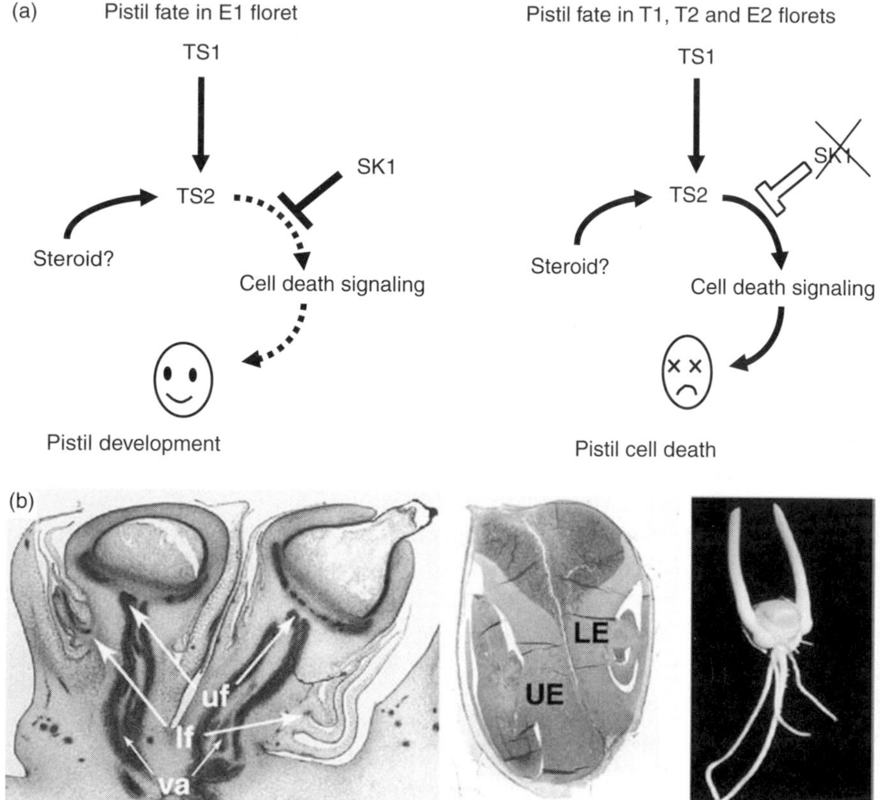

Figure 6.2 (a) Model for pistil determination in maize. T1, T2 are the first and second florets in the tassel; E1 and E2 are the first and second florets in the ear. In T1, T2 and E2, abortion is carried out through the activity of *TS2*, under the positive control of *TS1*. In the functional floret of the ear (E1), *TS2* is expressed, but *SK1* prevents cellular death (reproduced from Calderon-Urrea and Dellaporta, 1999). (b) Ectopic cytokinin production in transgenic plants that over-express a cytokinin biosynthetic enzyme, rescues the pistil of secondary florets in the ear's spikelets. Left: two ear-spikelets of SAG12:IPT transgenic plants, each with a fast developing upper floret (uf) and a lagging but lower floret (lf) that did not abort and will eventually catch up. Va, developing vascular tissue. Middle, right: the two florets are fertilized, and later fuse and form kernels with two functional, genetically distinct embryos (LE, UE) (reproduced from Young *et al.*, 2004, with permission from Blackwell Publishing).

6.3.2.5 Conservation of the maize sex determination pathway in other species
Tasselseed homologs were cloned from *Arabidopsis* (Brassicaceae) and *Silene latifolia* (Caryophyllaceae; Lebel-Hardenack *et al.*, 1997), and had approximately 50% amino-acid identity with *TS2*. *Silene* is a dioecious plant and it was asked whether pistil inhibition in male plants could involve the *TS2* homolog. This did not seem to be the case, because in both *Silene* and *Arabidopsis*, the homologous gene was expressed exclusively in the tapetal tissue of mature anthers, and was not linked to the Y chromosome of *S. latifolia*. These genes could play a role in cell death

of the tapetum. Maize could have recruited the mechanism for its pistil abortion program, whereas in *Silene*, where the rudimentary male-flower gynoecium shows no evidence of cell death, the mechanism for selective development of sex organs might be different.

On the other hand, comparative studies of sex expression within the grasses did reveal homology between maize and its wild monoecious relatives of the tribe Andropogonae. Le Roux and Kellogg (1999) made detailed microscopical observations of spikelet development in four disparate species of this group and noted the striking similarity in the pattern of floral development and organ abortion, suggesting that a common mechanism operates across the tribe. Li *et al.* (1997) studied *tasselseed*-like mutations in *Tripsacum*, a maize relative, and provided direct genetic and molecular evidence for the involvement of a *ts*-homologous gene in sex determination: *TS2* molecular probes were genetically linked to the *Tripsacum* mutation, and an inter-specific hybrid between *ts2*-maize and the *Tripsacum-tasselseed* mutant did not show complementation, suggesting that the two genes are 'allelic'.

6.3.3 Unisexual flower development in cucumber

6.3.3.1 Cucumber as an experimental model for sex expression

The cucurbit family is remarkable for its diversity of sex types. Between the 1950s and the 1970s, the genetic basis of sex determination in *Cucumis sativus*, the cucumber, was unraveled, and hormonal and environmental influences on sex expression were studied in detail (reviewed by Frankel and Galun, 1977; Perl-Treves 1999). The practical use of these discoveries by breeders has become routine, but for a long period little new was learned about the mechanisms underlying sex expression. In the past decade, there has been renewed interest in the topic, utilizing new advances in plant molecular biology and genome analysis to study the mode of action of sex genes, and their interactions with sex-modifying hormones.

6.3.3.2 Sex expression in the cucumber bud and along the shoot

Cucumber plants may produce male, female and, less commonly, hermaphrodite (bisexual) flowers; the typical varieties are monoecious. Unisexual male flowers develop from the initially bisexual buds as a result of inhibition of the carpel primordia, while female flowers form when stamen primordia are arrested (Figure 6.3). Different approaches were taken to determine the precise time when sexual differentiation of the bud begins, and the point at which such differentiation becomes irreversible by hormonal treatments (Atsmon and Galun 1960; Galun, 1961a; Galun *et al.*, 1963; Goffinet, 1990).

The distribution of different types of flowers along the cucumber shoot can give rise to various sex types. In monoecious genotypes, clusters of male flowers, or single female flowers, form in the leaf axils. Along the shoot, sex expression changes, and three phases may be recognized: a male phase with only staminate flowers, a mixed phase and a continuous-female phase. Thus, sex tendency gradually changes from

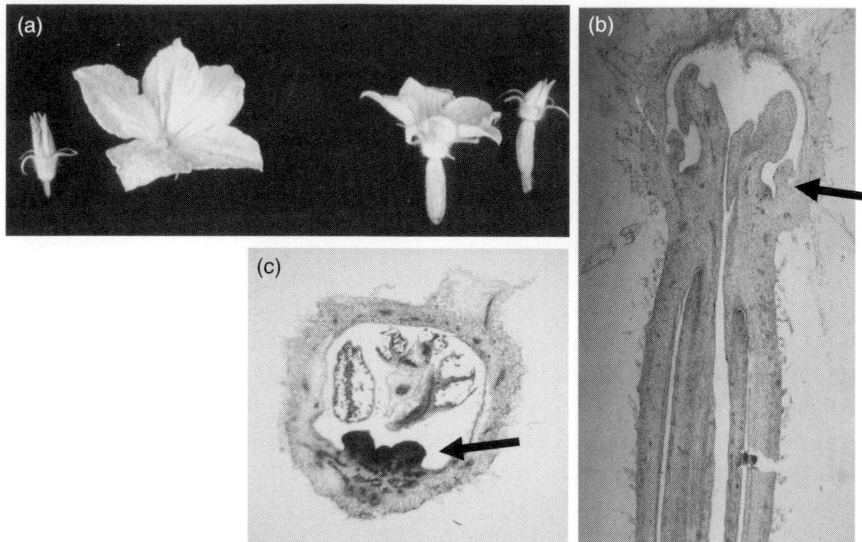

Figure 6.3 Male and female flowers of cucumber. (a) Male (left) and female (right) flowers at anthesis and one day before anthesis. (b) Section through intermediate stage (4 mm) female bud. (c) Section through intermediate stage (5 mm) male bud. Arrows indicate inhibited stamens and inhibited carpels, respectively (from Perl-Treves, 1999).

male to female. The mechanism responsible for such a 'sex-gradient' is not fully understood, but probably involves local changes in the concentration of endogenous hormones (see later).

6.3.3.3 Inheritance of sex in cucumber

Several genes affect gene expression by changing the fate of specific buds, or by affecting the succession of sex phases along the shoot. These genes can be combined to obtain sex types other than monoecious. Thus, gynoecious plants bear only pistillate flowers, androecious plants bear only staminate flowers, hermaphrodites have only bisexual flowers and andromonoecious plants have both male and bisexual flowers. Sex inheritance in cucumber has been fully elucidated (Galun 1961b; Kubicki, 1969a,b). Three major genes account for most sex phenotypes:

(1) F/f is a partially dominant gene that controls femaleness. The F allele causes the female phase to start much earlier; FF plants are gynoecious.
(2) A/a increases maleness. A is epistatic to F, *aaff* plants are androecious, and will never attain the mixed, or female phases.
(3) M/m is different from the F and A genes. It appears to act locally on the individual buds – those that are destined to develop an ovary. The dominant allele will only allow the formation of stamen-less female flowers as well as of male flowers, while in *mm* plants, bisexual flowers form (in addition

to male flowers). We may interpret its role as a stamen suppressor that acts only in buds determined to develop a carpel.

Additional genes that affect sex expression were described, e.g. a recessive femaleness gene (Przybecki *et al.*, 2003). Quantitative variation attributable to minor genes exists as well (Galun, 1961b) and has been mapped using quantitative trait loci (QTL) analysis (J.E. Staub, personal communication).

6.3.3.4 Environmental and hormonal regulation of sex expression

Several studies correlated changes in sex expression with environmental factors such as day length and temperatures. Short days and low night temperatures enhanced femaleness in a monoecious cultivar, while long day, warm night conditions promoted maleness (review by Perl-Treves, 1999). A variety of exogenously applied chemical compounds alter sex expression in cucumber. Gibberellic acid (GA) enhanced male tendency of cucumber (Galun 1959a) and postponed the continuous-female phase. GA apparently had a dual effect – it favored male flower initiation, and also inhibited existing pistillate buds from full development. Compounds that inhibit GA synthesis promoted femaleness (e.g. Mitchell and Wittwer, 1962). Auxins were reported to promote female flower formation (Galun, 1959b). In a recent study, application of brassinosteroid was shown to promote femaleness (Papadopoulou and Grumet, 2005).

Ethylene exerts a strong feminizing effect on cucumbers and is considered the main sex hormone; it also affects other cucurbit species such as pumpkin (*Cucurbita* spp.) and melon (*Cucumis melo*; Rudich *et al.*, 1969; Byers *et al.*, 1972). Ethylene and ethrel (an ethylene-releasing compound) increased the number of pistillate flowers in monoecious cucumbers, and decreased the number of male flowers (e.g. Rudich *et al.*, 1969); they also caused the first female flowers to form much earlier. Inhibitors of ethylene synthesis and action promoted maleness (e.g. Atsmon and Tabbak, 1979). Copper, which is a co-factor of ethylene receptors required for ethylene binding (Hirayama and Alonso, 2000), can also modify sex expression and increase femaleness (Mibus *et al.*, 2000; Saraf-Levy and Perl-Treves, unpublished results), presumably by increasing the proportion of active ethylene receptor that interacts with available ethylene molecules.

Since hormones can exogenously modify sex expression in cucumber, several studies have compared endogenous hormone levels among different sex genotypes, or in a single genotype undergoing a change in sex expression. For example, endogenous gibberellin levels were compared between monoecious and gynoecious isogenic lines (Atsmon *et al.*, 1968; Rudich *et al.*, 1972b). Monoecious samples had ten times more GA activity than gynoecious ones. A two-fold higher GA content was measured in gynoecious leaves under long day conditions that promote maleness, as compared to short day conditions (Friedlander *et al.*, 1977a). Apices and buds of gynoecious lines produced two–four times more ethylene than monoecious and androecious lines. Ethylene evolution was doubled under short days, as compare to long day conditions (Rudich *et al.*, 1972a, 1976; Makus *et al.*, 1975; Trebitsch *et al.*, 1987; Yamasaki *et al.*, 2001, 2003b).

What are the possible interactions between different hormones? Is the level, or action of one hormone, controlled by a second one? It has been suggested (Shannon and de La Guardia, 1969; Trebitsh *et al.*, 1987) that the feminizing effect of auxin is, in fact, mediated by ethylene, since auxin caused an increase in 1-aminocyclopropane carboxylate (ACC) and ethylene evolution that preceded the sex-reversing effect. The possible interaction between ethylene and gibberellin has not been elucidated yet and physiological data on the mutual effects of these two hormones are somewhat contradictory (Rudich *et al.*, 1972b; Atsmon and Tabbak, 1979; Yin and Quinn, 1995).

6.3.3.5 A model for cucumber sex expression

Yin and Quinn (1995) proposed a theoretical model for sex expression in cucumber, and for the possible relationships between hormones and sex determining genes. They treated plants either with GA, ethylene or their inhibitors, and concluded that each hormone not only promoted one sex, but also inhibited the formation of flowers of the other sex. The precise regions in which male, female or perfect flowers form are determined by the level of an endogenous hormone, and the sensitivity threshold of its receptors in different regions of the plant. The model made predictions regarding the mode of action of the two sex determining genes, *F* and *M*; the *F* gene could determine the range of sex hormone (i.e. ethylene) concentrations along the plant's shoot (Figure 6.4). A monoecious plant would have an increasing concentration of ethylene; lowering the ethylene level at given nodes will delay the female phase, while increasing it will promote femaleness. Hormone receptors provide the second element of the model, required to predict which organ type will form at a specific ethylene concentration. The female-organ receptor would perceive ethylene above a specific threshold concentration, and transmit a promoting signal to carpel primordia. A distinct male-organ receptor will transmit an inhibitory signal when stimulated above its own ethylene concentration-threshold. The *M* gene product specifies the sensitivity of male organs to the inhibitory effect of ethylene. Its dominant allele presumably encodes a sensitive variant of the receptor, preventing stamen development at an ethylene range that promotes pistil development, while the *m* allele, encoding a less sensitive isoform, will cause an 'overlapping' range of ethylene that is both female-promoting and male-permissive, and allow the formation of perfect flowers in *mm* genotypes. The model also predicts that some sterile flowers would form at the boundary between the male and mixed-female phases, where ethylene is sufficient to inhibit stamens but insufficient for carpel development. Interestingly, such flowers were observed by Mibus *et al.* (2000).

The model thus includes a 'systemic' component that can generate a gradient of hormone concentrations along branches: it can accommodate phenomena such as the rootstock influencing the sex of the scion, observed in grafting experiments, where gynoecious and monoecious cucumbers were grafted reciprocally

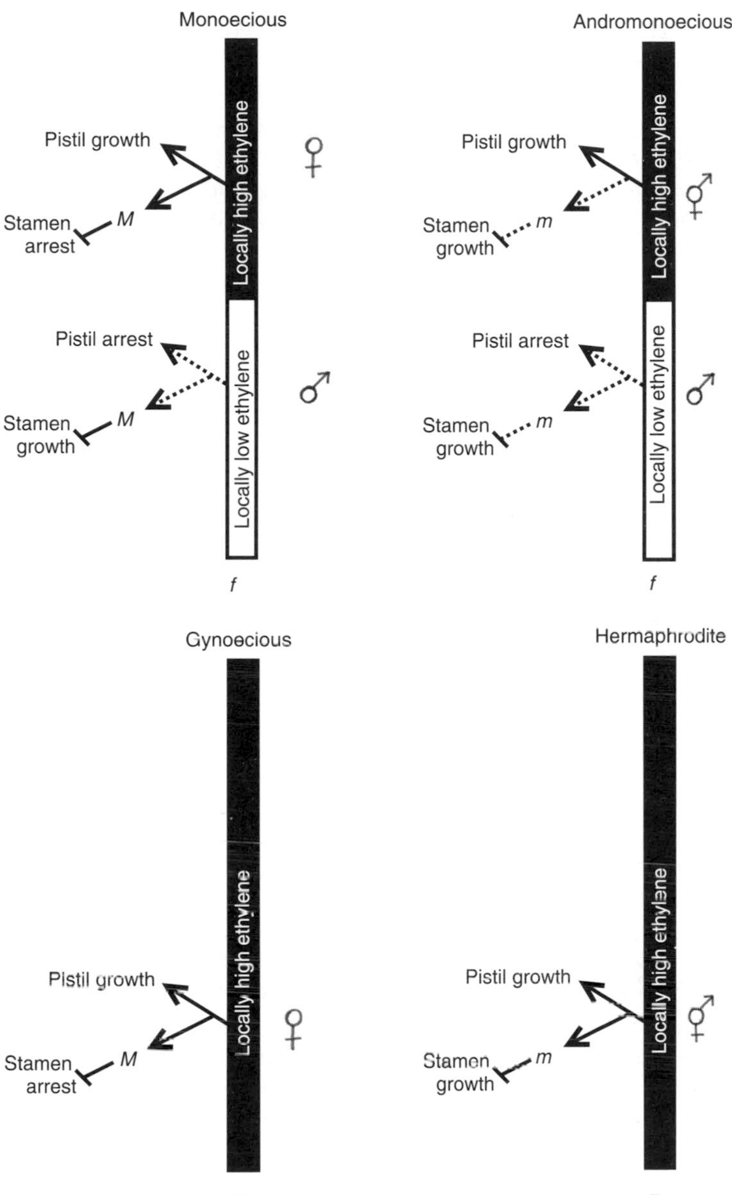

Figure 6.4 Theoretical model explaining the interaction of ethylene concentration, ethylene sensitivity and the two cucumber sex genes *F* and *M* based on Yin and Quinn (1995) and Yamasaki *et al.* (2001). Local ethylene concentration is genetically determined by *F* and can change along the shoot, causing sex-phase changes. At high-ethylene zones, ethylene induces the male-organ receptors, causing inhibition of the stamens in pistillate flowers; in low-ethylene zones, stamens will not be arrested and pistils will develop. A separate female-organ receptor responds to ethylene and induces carpel growth. Insensitive alleles of the *M* gene cause bisexual flowers to form.

(Friedlander *et al.*, 1977b; Takahashi *et al.*, 1982; Mibus *et al.*, 2000); ACC, the ethylene precursor, could be the diffusive agent. On the other hand, the effect of the *M* gene was not graft-diffusible (Mibus *et al.*, 2000); this is in agreement with a model predicting a receptor-like action, which should be localized in specific organ primordia within the flower.

A few features of sex expression in cucumber are not explained by the present model, such as the intercalation of male and female flowers in the mixed-phase of monoecious shoots – it is possible that the ethylene gradient is a fluctuating one. And why are the stamens inhibited only in carpel-producing flowers and not in male flowers? Could stamen inhibition be mediated by the developing pistil? Where do other hormones fit in? Important support for the model came in 1997, when Trebitsh *et al.* (1997) mapped an ACC synthase (ACS) gene to the *F* locus, suggesting that *F* could indeed determine ethylene levels. Recently, Yamasaki *et al.* (2000, 2001) reported that ethylene receptors in cucumber apices are induced by ethylene. They proposed that *M*, the putative stamen-primordia receptor (or other signaling molecule) that affects only stamens and not pistil development is located downstream of *F* (Figure 6.4). Therefore, *M* gene expression and subsequent stamen inhibition will only occur in buds that produce a sufficient amount of ethylene (i.e. female buds); in *mm* plants stamens are non-inhibited regardless of *F*.

6.3.3.6 Molecular studies on ethylene-synthesis and perception genes in Cucumis

Since the importance of ethylene in this system has been fully recognized, several studies have addressed the regulation of sex expression by ethylene synthesis/perception at the molecular level.

Ethylene synthesis. Ethylene is produced from *S*-adenyosyl methionine via a two-step process mediated by the enzymes ACS and ACC oxidase (ACO; Fluhr and Matto, 1996). ACS and ACO are encoded by multigene families. Trebitsch *et al.* (1997) discovered that gynoecious cucumber lines possess a second copy of the cucumber *ACS1* gene (designated *ACS-1G*) that is absent in near-isogenic monoecious lines. Two mapping populations were used to map *ACS-1G* with respect to *F*. A 100% linkage was found, strongly suggesting that the putative extra-copy of *ACS-1* is the *F* gene itself, whose function is to regulate ethylene production. Treatment of excised apices with auxin caused a strong, transient elevation of *ACS-1* transcript level, similar in both genotypes. Using Northern analysis, differences between untreated monoecious and gynoecious lines could not be detected. However, Kamachi *et al.* (2000) used more sensitive RT-PCR (reverse transcription polymerase chain reaction) assays and observed expression of *ACS-1* transcript in apices of gynoecious, but not isogenic monoecious, cucumbers. Additional members of the cucumber *ACS* gene family were also isolated. Expression of *ACS-2* and *ACS-4* was shown to increase with the transition to the female phase (Kamachi *et al.*, 1997; Yamasaki *et al.*, 2001, 2003a,b), and in response to ethrel. Thus, the ethylene signal that is probably initiated by the *F*-locus *ACS* may be amplified by other genes that are not necessarily linked to a sex-determining locus, but respond to ethylene.

Development of gene-specific probes and isolation of the two respective promoters may shed light on the function of the duplicated *ACS-1* genes, whereas a conclusive proof identifying *F* as *ACS-1* should come from transgenic complementation experiments. Recently, cloning of genomic sequences from the *ACS-1G* locus was reported (Mibus and Tatlioglu, 2004), providing a more direct evidence of the molecular duplication at the *F* locus: two identical *ACS* coding regions were isolated from gynoecious lines, but the respective promoter regions differed markedly; one could be amplified from both monoecious (*ff*) and gynoecious (*FF*) lines, the second was gynoecious-specific.

ACO catalyzes the second and final step in ethylene formation. Kahana *et al.* (1999) examined the expression of three different *ACO* genes in cucumber plants of various sex genotypes. All three were expressed in leaves and shoot apices of cucumber. In the leaf, ACO transcripts could be correlated with femaleness and ethrel treatment, whereas in the shoot apex, the relationships were more complex. Exogenous ethrel treatments acted differently on distinct *ACO* family members in different sex genotypes, showing that sex genotypes differentially regulate ethylene production. Local patterns of expression *ACO*, *ACS-2*, as well as ethylene receptors *ERS*, *ETR-1* and *ETR-2* transcripts were examined by in situ mRNA hybridization by Kahana *et al.* (1999) and Yamasaki *et al.* (2003a). The specific, localized patterns that were observed in these studies suggest that ethylene plays diverse roles in both early and late stages of flower development, other than sex determination. It appears that both ethylene synthesis and ethylene perception are regulated in spatially restricted patterns that differ between sex genotypes.

Sex expression in melon (*Cucumis melo*) is also influenced by ethylene (Rudich *et al.*, 1969). Grumet and co-workers produced transgenic melons of the andromonoecious cultivar Hale's Best Jumbo. They reported changes in sex expression of plants that over-expressed an ACS cDNA from petunia under the control of the 35S promoter (Papadopoulou *et al.*, 2005). The transgenic melons produced two–three-fold more ethylene, and displayed increased femaleness: hermaphrodite buds reached anthesis approximately ten nodes earlier than the control, resulting in larger numbers of bisexual flowers and higher fruit-set. In a parallel experiment, the critical role of ethylene for melon femaleness was further demonstrated by blocking ethylene perception. Melons that constitutively over-expressed Arabidopsis *etr1-1*, a defective *ETR1* allele, had nearly no bisexual flowers, whereas male flowers formed normally; the plants also displayed additional ethylene-insensitivity symptoms such as delayed flower abscission (Papadopoulou *et al.*, 2002).

Ethylene perception genes – roles in pistil development. Apart from ethylene synthesis, ethylene perception and signal transduction are important in determining cucumber sex. The ethylene transduction pathway has been elucidated through the study of ethylene insensitive mutants in *Arabidopsis* (Bleecker and Kende, 2000).

Levels of transcripts related to ethylene perception were correlated with sex expression in a number of studies. Yamasaki *et al.* (2000) and Saraf-Levy and Perl-Treves (unpublished results) studied the expression of *ETR-1, ETR-2* and *ERS-1* cucumber homologs in gynoecious, monoecious and androecious lines. The two latter genes were more strongly expressed in the female genotype, and induced by ethrel application, showing that ethylene sensing is important for femaleness. The 'autocatalytic effect' of ethylene (known from several ethylene-regulated pathways) is manifested during sex expression by increasing its own synthesis (see earlier) and also by enhancing its own perception. Consistent with the proposed role of ethylene receptors in female-organ development, Rajagopalan *et al.* (2004) produced transgenic monoecious cucumbers that over-expressed *ERS-1* under the control of the constitutive 35S promoter. In two out of seven independent plants, segregation of progeny that displayed enhanced femaleness was observed (Figure 6.5). Such plants formed their first female flower at least four nodes earlier than the control plants and the proportion of female flowers was several folds higher. This was correlated with higher levels of *ERS* transcripts, as well as with accelerated yellowing of detached leaves, linking the phenotype to the expression of the transgene. Ethylene receptors act as negative switches in *Arabidopsis*: the unoccupied receptor actively represses the ethylene response (Wang *et al.*, 2002), while ethylene binding relieves such

Figure 6.5 Enhanced femaleness in plants that over-express a 35S:ERS1 transgene. A monoecious cultivar, Poinsett76, was used. Left: some transgenic progeny attained an early female phase (many consecutive pistillate flowers) that is never observed in control plants. The feminized plants also exhibit accelerated senescence of detached leaves (right).

repression. Such a complicated mode of action made it impossible to foresee the effect of over-expressing *ERS*: depending on the physiologically limiting factor, one can rationalize either an increased ethylene response or an inhibited response, depending on whether or not the extra receptor protein is activated by copper ions and stimulated by endogenous ethylene. The strong feminization observed in the two families argues for ethylene receptors being important in female sex determination, and for receptor protein levels being limiting.

Ethylene perception: role in stamen development and relationship to the M *gene.* The model discussed in Section 6.3.3.5 (Figure 6.4) suggests that stamens of *mm* plants have an altered ethylene response compared to *M*-genotype stamens, whereas their pistils respond normally. Yamasaki *et al.* (2001) compared several ethylene responses between monoecious, gynoecious and andromonoecious cucumbers and concluded that in agreement with the model, *mm* plants seem to be defective in ethylene responses. Monoecious and andromonoecious plants had similar ethylene production, and similar ACS transcript levels, lower than gynoecious line, in agreement with the implication of *F* in ethylene synthesis. However, elongation of andromonoecious seedlings (*ffmm* genotype) was less inhibited by ethylene, compared to the gynoecious (*FFMM*) and monoecious (*ffMM*) lines, which have an *M* allele. The latter two lines showed a normal ethylene response, inducing the *ACS-2*, *ERS* and *ETR-2* transcripts in response to ethrel, whereas the andromonoecious line had a much reduced response (Yamasaki *et al.*, 2001; Figure 6.6). In another

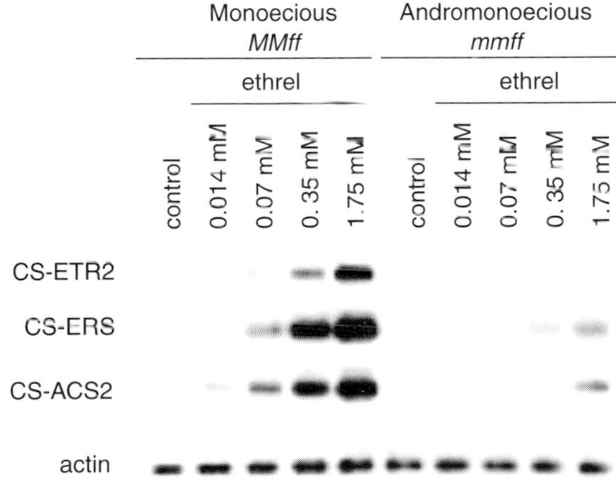

Figure 6.6 Ethrel treatments strongly induce the ETR2 and ERS1 ethylene receptor transcripts in the shoot apex of a monoecious cultivar that carries the *M* allele. Induction of the same transcripts in an andromonoecious genotype (homozygous for the *m* allele) is very limited, showing that *M* is required for the ethylene response. Adapted from Yamasaki *et al.* (2001), by permission of Oxford University Press.

study, Yamasaki *et al.* (2003b) showed that the increase in femaleness induced by short days was greater in a monoecious line as compared to an andromonoecious line. Shortening the photoperiod increased the mid-day peak of ethylene evolution; such an increase was pronounced in the monoecious line, but nearly absent in the andromonoecious one. A parallel increase in *ACS-2* and *ERS* transcripts in response to short days was recorded in the monoecious line, but not in the *mm* line. These studies provide a first evidence that the *M* gene is required for ethylene perception, but still do not identify which component of the ethylene-signaling cascade it actually encodes. Saraf-Levy *et al.* (2000) mapped *ERS-1*, taking advantage of two-point mutations that distinguished *ERS* sequences of a gynoecious and hermaphrodite pair of lines, generated a molecular marker, tested a back-cross progeny and observed independent segregation of *M* and *ERS-1*. Other ethylene perception/transduction genes are presently being tested in our laboratory for linkage with sex loci.

6.3.3.7 Search for 'effector genes' that carry out the developmental program following sex determination

A number of approaches have been taken to isolate cucumber genes and study their role in the determination and development of male and female flowers. Some studies looked at genes whose roles are known and could be related to sex expression or flower development. Others set out to 'fish' for novel, unknown genes that are differentially expressed between sex genotypes, or during chemically induced sex reversal, trying to relate them to the sex expression program.

Among the 'known genes', the MADS-box gene family attracted much attention (see Section 6.3.1.1. earlier). Perl-Treves *et al.* (1998) set to study the expression of *AGAMOUS* homologues during cucumber flower development. Floral bud cDNA libraries were prepared from two cucumber lines, a gynoecious and an androecious one. Morphologically bisexual buds of 1 mm size were used, where important developmental 'decisions' regarding sex expression may take place. The libraries were screened with an *AGAMOUS* probe, and three homologues, *CAG-1*,2, 3 were isolated and shown to belong to the *AGAMOUS* sub-family of sequences. Northern analysis indicated that *CAG-1*,2, 3 were expressed during bud development. The *CAG-3* and *CAG-1* transcripts were present in both female and male buds, as well as in the stamens, carpel and nectaries of mature flower organs. Kater *et al.* (1998) also isolated two *AGAMOUS* homologues from cucumber, designated *CUM-1* and *CUM-10*; *CUM-1* is identical with *CAG-3* and *CUM-10* with *CAG-1*. *CUM-1* and *CUM-10* were ectopically expressed in transgenic petunia under the control of the 35S promoter. Their over-expression in all floral whorls indicated that the two are functional *AGAMOUS* orthologues; the homeotic transformation observed in petunia were consistent with the ABC model of organ identity.

The third cDNA, *CAG2*, had a carpel-specific expression pattern. The same sequence was also cloned from a cucumber embryogenic callus cDNA library (Filipecki *et al.*, 1997) and designated *CUS1*. *CUS1* was expressed in the carpel, ovules and developing fruits, as well as in somatic embryos, suggesting a role for this gene in embryo development. Female-specific genes of the *AGAMOUS* sub-family have been isolated also from other plants and implicated in ovule determination and development (Angenent *et al.*, 1995).

Are cucumber's unisexual flowers 'designed' by modulating the expression of floral organ-identity genes? Would ethrel and gibberellin treatments that affect sex expression induce a change in transcript levels of *AGAMOUS*-homologues? For example, would a feminizing treatment elevate the *CAG-2* female specific transcript, or decrease *CAG-1* or *CAG-3* transcript levels in male buds? We did not observe a change in the levels of *CAG-1, CAG-2* and *CAG-3* transcripts after such treatments, in either the male or female genotypes (Perl-Treves *et al.*, 1998). *CAG* genes are required for stamen and carpel development, but we could not implicate them in the 'sex expression cascade' that responds to sex-hormone stimuli. However, this conclusion is tentative, because a transient or strictly localized change in transcript level following the hormonal treatment could have passed unnoticed. Moreover, MADS-box gene products could have been affected by sex hormones at the post transcriptional level, e.g. by affecting proteins that interacts with CAG proteins.

Another MADS-box gene that was recently cloned correlates more significantly with cucumber sex determination. Ando *et al.* (2001) performed differential display of mRNAs from ethylene-induced versus control apices of monoecious cucumbers. They isolated a MADS-box gene, *ERAF-17*, with homology to tomato *TM8*, which is phylogenetically separate from the A, B and C MADS gene sub-families. Its expression was strong in gynoecious apices and developing female flowers throughout development to anthesis and it was induced by ethrel and inhibited by amino-ethoxyvinyl glycine (AVG), an ethylene synthesis inhibitor. This pattern makes *ERAF-17* an interesting candidate for studying differentiation events that follow the ethylene sex determining signal. Another gene isolated in the same study is *ERAF-16* (Ando and Sakai, 2002). It encodes a methyl-transferase of unknown physiological function, whose expression is ethylene-inducible and stronger in young female buds. Other studies that aimed at the isolation of differentially expressed transcripts from male and female buds were reported. Goldberg *et al.* (2000) performed differential display and differential library screens of young floral buds from a gynoecious and an androecious line. The screen resulted in several transcripts that were preferentially expressed in male buds, including serine-, threonine- and proline-rich unknown proteins that could have cell-wall related functions. Another cDNA represents a gene encoding a non-translatable RNA, and another clone represents a gene encoding a glutaredoxin homolog; both genes showed stronger expression in young and mature male buds. Two genes that were preferentially expressed in more mature male buds encoded a ubiquitine-like homologue and a ribosomal protein. Przybecki *et al.* (2003) performed differential screening of subtracted bud libraries from four sex genotypes and isolated several genes showing sex-related patterns of expression. These clones contained genes encoding unknown proteins, as well as known ones, such as chaperonins and ubiquitin-pathway proteins. None of them genetically segregated with sex genes. Some clones were studied by in situ-PCR analysis, revealing high expression in arrested stamens or pistils. These transcript-profiling studies identify many targets for future studies, although selecting those clones that play a relevant role in sex expression is not a trivial task.

The possible relationship between the sex determination machinery responsible for arresting inappropriate sex organs and MADS- box genes was further explored by Kater *et al.* (2001). The authors showed that *CUM-1*, the C-function gene, and *CUM-26*, a B-function gene, are expressed in wild-type flowers according to the ABC model-domains, not only in the developing sex organs but also in the arrested organs. This shows that the inappropriate organs are not arrested by suppressing their B or C functions. The authors then examined male and female flower development in three genotypes mutated in the A, B and C functions, respectively. A cucumber mutant *green petals* was defective in expression of *CUM-26*, a B-function gene. Its phenotype conformed to the ABC model predictions, with male flowers displaying a *sepal-sepal-carpel-no organs* structure, and females having a *sepal-sepal-no organ-carpel* architecture (Figure 6.7). The fact that the inhibited stamens could not form a functional carpel, despite the expression of C function in the third whorl, shows that the organ-arrest machinery blocks the formation of

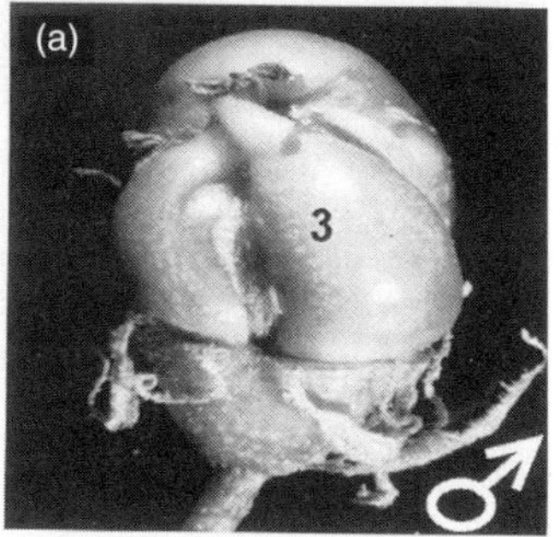

(b)

Male flower

	1	2	3	4
WT	sepals	petals	stamens	—
A mutant	carpels	stamens	stamens	—
B mutant	sepals	sepals	carpels	—
C mutant	sepals	petals	flower	flower

Female flower

	1	2	3	4
WT	sepals	petals	—	carpel
A mutant	carpels	stamens	—	carpel
B mutant	sepals	sepals	—	carpel
C mutant	sepals	petals	petals	flower

Figure 6.7 Interaction of the sex determination machinery in cucumber with the ABC functions that determine floral organ identity. (a) The green *petal mutant*, defective in the B-function gene *CUM-26*, forms three functional carpels instead of stamens in its third whorl. (b) Table summarizing homeotic conversions that were observed in cucumber mutants and transgenic plants, indicating that sex determination acts on specific floral whorls and not on the whole flower. Reproduced from Kater *et al.* (2001), with permission and copyrighted by the American Society of Plant Biologists.

any reproductive organ at that location. On the other hand, the male flower stamens were converted into carpels (Figure 6.7a), showing that sex determination acts specifically on the arrested whorl and not on the flower as a whole, as carpels could be produced in male flowers by conversion of non-blocked whorls. Elimination of the A function was obtained by genetically engineering cucumber to express *CUM-1* (class C) in all whorls, because the C function represses the A-class genes. A bisexual flower would have a *carpel-stamen-stamen-carpel* structure, whereas male cucumber flowers had a *carpel-stamen-stamen-no organ* structure, again showing that in non-arrested organs male flowers can produce a carpel, but the inhibited pistil cannot be de-repressed by supplying additional C function. A few of the transgenic plants underwent co-suppression (i.e. silencing) of the endogenous *CUM-1* gene, exhibiting a C-class mutant phenotype. Such flowers were indeterminate (Figure 6.7b). Here, the authors noted that the arrested organs were de-repressed and formed flowers or petals, suggesting that only sex organs are arrested, whereas non-reproductive organs can be induced. The organ-arresting machinery probably has to interact with homeotic gene products to recognize the identity of the whorls and inhibit only sex-organ programs.

A different approach was taken by Hao *et al.* (2003) to elucidate the downstream events that execute organ arrest in cucumber flowers. They focused on the rudimentary stamens of female flowers, dissected them and found signs of DNA degradation by the TUNEL assay (TdT mediated dUTP nick end labeling). DNA laddering typical of PCD was not seen, but a DNAase activity, specific to the staminoid region of female flowers, appeared. Ultrastructural changes in the staminoids included chromatin condensation and vacuolization, while nuclei and mitochondria remained intact. Cells did not disintegrate till anthesis, showing that the mechanism does not lead to PCD and is different from that described in maize flowers (Section 6.3.2.1). Isolation of the staminoid-specific DNAase and study of its regulation could provide an important link in the sex expression cascade.

6.4 Concluding remarks

Monoecious plants raise several interesting questions regarding their reproductive strategy and the development of unisexual flowers. The studies on the phenology, pollination and population dynamics of wild plants with varying sex habits offer a means of looking at sex evolution and experimentally testing specific models. In maize, mutants in sex determination of either the tassel or the ear are available, and initial progress in their characterization has been accomplished. Most of the mutants have not been cloned yet, and the biochemical function of the TS2 protein is not understood yet. Another important aspect that still needs to be examined relates to the endogenous hormone levels of the various mutants, and their responses to exogenous hormones: only the *anthereear/dwarf* mutations were physiologically characterized. In cucumber, molecular characterization of sex determining loci has only begun, but the genetic, molecular and physiological data fit into a tentative model, encouraging researchers to experimentally test its predictions. The model mostly covers the role of ethylene in sex expression, while other aspects such as the downstream events leading to organ arrest, and the interactions with other hormones, remain largely unexplored.

Abbreviations

ACC – 1-aminocyclopropane carboxylate
ACO – ACC oxidase
ACS – ACC synthase
AVG – aminoethoxyvinyl glycine
DAPI – 4,6-diamine-2-phenylindole dihydrochloride
GA – gibberellic acid
IPT – isopentenyl transferase
PCD – programmed cell death
PCR – polymerase chain reaction
RT-PCR – reverse transcription PCR
SK, sk – silkless
TS, ts – tasselseed
TUNEL – TdT mediated dUTP nick end labeling

References

Ando, S. and Sakai, S. (2002) Isolation of an ethylene-responsive gene (ERAF16) for a putative methyltransferase and correlation of ERAF16 gene expression with female flower formation in cucumber plants (*Cucumis sativus*), *Physiologica Plantarum*, **116**, 213–222.

Ando, S., Sato, Y., Kamachi, S. and Sakai, S. (2001) Isolation of a MADS-box gene (ERAF17) and correlation of its expression with the induction of formation of female flowers by ethylene in cucumber plants (*Cucumis sativus* L.), *Planta*, **213**, 943–952.

Angenent, G. C., Franken, J., Busscher, M. *et al.* (1995) A novel class of MADS box genes is involved in ovule development in *Petunia*, *The Plant Cell*, **7**, 1569–1582.

Atsmon, D. and Galun, E. (1960) A morphogenetic study of staminate, pistillate and hermaphrodite flowers in *Cucumis sativus* L, *Phytomorphology*, **10**, 110–115.

Atsmon, D. and Tabbak, C. (1979) Comparative effects of gibberellin, silver nitrate and aminoethoxyvinyl glycine on sexual tendency and ethylene evolution in the cucumber plant (*Cucumis sativus* L.), *Plant Cell Physiology*, **20**, 1547–1555.

Atsmon, D., Lang, A. and Light, E. N. (1968) Contents and recovery of gibberellins in monoecious and gynoecious cucumber plants, *Plant Physiology*, **43**, 806–810.

Barrett, S. C. H. (2002) The evolution of plant sexual diversity, *Nature Reviews Genetics*, **3**, 274–284.

Becker, A., Winter, K. U., Meyer, B., Saedler, H. and Theissen, G. (2000) MADS-box gene diversity in seed plants 300 million years ago, *Molecular Biology and Evolution*, **17**, 1425–1434.

Bensen, R. J., Johal, G. S., Crane, V. C. *et al.* (1995) Cloning and characterization of the maize *An1* gene, *The Plant Cell*, **7**, 75–84.

Bleecker, A. B. and Kende, H. (2000) Ethylene: a gaseous signal molecule in plants, *Annual Review of Cell and Developmental Biology*, **16**, 1–18.

Buckner, B., Johal, G. S. and Janick-Buckner, D. (2000) Cell death in maize, *Physiologica Plantarum*, **108**, 231–239.

Byers, R. E., Baker, L. R., Sell, H. M., Herner, R. C. and Dilley, D. R. (1972) Ethylene: a natural regulator of sex expression of *Cucumis melo* L, *Proceedings of the National Academy of Sciences of the United States of America*, **69**, 717–720.

Calderon-Urrea, A. and Dellaporta, S. L. (1999) Cell death and cell protection genes determine the fate of pistils in maize, *Development*, **126**, 435–441.

Charlesworth, D. and Guttman, D. S. (1999) The evolution of dioecy and plant sex chromosome systems, in *Sex Determination in Plants* (ed. C. Ainsworth), BIOS Scientific Publishers, Oxford, UK, pp. 183–188.

Cheng, P. C., Gryson, R. I. and Walden, D. B. (1983) Organ initiation and the development of unisexual flowers in the tassel and ear of *Zea mays*, *American Journal of Botany*, **70**, 450–462.

Costich, D. E. and Meagher, T. R. (2001) Impacts of floral gender and whole-plant gender on floral evolution in *Ecballium elaterium* (Cucurbitaceae), *Biological Journal of the Linnean Society*, **74**, 475–487.

Dellaporta, S. L. and Calderon-Urrea, A. (1993) Sex determination in flowering plants, *The Plant Cell*, **5**, 1241–1251.

De-long, A., Calderonurrea, A. and Dellaporta, S. L. (1993) Sex determination gene *TASSELSEED2* of maize encodes a short-chain alcohol-dehydrogenase required for stage-specific floral organ abortion, *Cell*, **74**, 757–768.

Dorken, M. E. and Barrett, S. C. H. (2003) Gender plasticity in *Sagittaria sagittifolia* (Alismataceae), a monoecious aquatic species, *Plant Systematics and Evolution*, **237**, 99–106.

Dorken, M. E. and Barrett, S. C. H. (2004) Phenotypic plasticity of vegetative and reproductive traits in monoecious and dioecious populations of *Sagittaria latifolia* (Alismataceae): a clonal aquatic plant, *Journal of Ecology*, **92**, 32–44.

Filipecki, M. K., Sommer, H. and Malepszy, S. (1997) The MADS-box gene *CUS1* is expressed during cucumber somatic embryogenesis, *Plant Science*, **125**, 63–74.

Fluhr, R. and Mattoo, A. K. (1996) Ethylene – biosynthesis and perception, *Critical Reviews in Plant Sciences*, **15**, 479–523.

Frankel, R. and Galun, E. (1977) *Pollination Mechanisms, Reproduction and Plant Breeding*, Springer-Verlag, Berlin, Heidelberg, New York.

Friedlander, M., Atsmon, D. and Galun, E. (1977a) Sexual differentiation in cucumber: abscisic acid and gibberellic acid contents of various sex genotypes, *Plant Cell Physiology*, **18**, 681–691.

Friedlander, M., Atsmon, D. and Galun, E. (1977b) The effect of grafting on sex expression in cucumber, *Plant Cell Physiology*, **18**, 1343–1350.

Galan, F. (1946) Sur la génétique de la monoecie et la dioecie zygotique chez *Ecballium elaterium* Rich, *Comptes rendus de l'Académie des Sciences Paris*, **222**, 1130–1131.

Galun, E. (1959a) Effect of gibberellic acid and naphtaleneacetic acid in sex expression and some morphological characters in the cucumber plant, *Phyton*, **13**, 1–8.

Galun, E. (1959b) The role of auxin in the sex expression of the cucumber, *Physiologica Plantarum*, **12**, 48–61.

Galun, E. (1961a) Gibberellic acid as a tool for the estimation of the time interval between physiological and morphological bisexuality of cucumber floral buds, *Phyton*, **16**, 57–62.

Galun, E. (1961b) Study of the inheritance of sex expression in the cucumber, the interactions of major genes with modifying genetic and non-genetic factors, *Genetica*, **32**, 134–163.

Galun, E., Jung, Y. and Lang, A. (1963) Morphogenesis of floral buds of cucumber cultured *in vitro*, *Developmental Biology*, **6**, 370–387.

Goffinet, M. C. (1990) Comparative ontogeny of male and female flowers of *Cucumis sativus*, in *Biology and Utilization of the Cucurbitaceae* (eds. D. M. Bates, R. W. Robinson and C. Jeffrey), Cornell University Press, New York. 288–304.

Goldberg, A., Kahana, A., Silberstein, L. and Perl-Treves, R. (2000) Markers for cucumber male flower development isolated by differential display and differential hybridization, *Plant Molecular Biology Conference*, Quebec, S25–S27.

Goldstein, P. (1997) Controlling cell death, *Science*, **275**, 1081–1082.

Grant, S., Houben, A., Vyskot, B. *et al.* (1994) Genetics of sex determination in flowering plants, *Developmental Genetics*, **15**, 214–230.

Hao, Y. J., Wang, D. H., Peng, Y. B. *et al.* (2003) DNA damage in the early primordial anther is closely correlated with stamen arrest in the female flower of cucumber (*Cucumis sativus* L.), *Planta*, **217**, 888–895.

Hirayama, T. and Alonso, J. M. (2000) Ethylene captures a metal! Metal ions are involved in ethylene perception and signal transduction, *Plant and Cell Physiology*, **41**, 548–555.

Hirosaka, S. and Sugiyama, N. (2004) Characterization of flower and fruit development of multipistillate type cucumbers, *Journal of Horticultural Science and Biotechnology*, **79**, 219–222.

Huang, S., Cerny, R. E., Qi, Y. L. *et al.* (2003) Transgenic studies on the involvement of cytokinin and gibberellin in male development, *Plant Physiology*, **131**, 1270–1282.

Huang, S. Q. (2003) Flower dimorphism and the maintenance of andromonoecy in *Sagittaria guyanensis ssp lappula* (Alismataceae), *New Phytologist*, **157**, 357–364.

Irish, E. E. (1999) Maize sex determination, in *Sex Determination in Plants* (ed. C. Ainsworth), BIOS Scientific Publishers, Oxford, UK, pp. 183–188.

Irish, E. E. and Nelson, T. M. (1993) Development of Tassel Seed-2 inflorescences in maize, *American Journal of Botany*, **80**, 292–299.

Irish, E. E., Langdale, J. A. and Nelson, T. M. (1994) Interactions between tassel seed genes and other sex-determining genes in maize, *Developmental Genetics*, **15**, 155–171.

Kahana, A., Silberstein, L., Kessler, N., Goldstein, R. S. and Perl-Treves, R. (1999) Expression of ACC oxidase genes differs among sex genotypes and sex phases in cucumber, *Plant Molecular Biology*, **41**, 517–528.

Kamachi, S., Mizusawa, H., Matsuura, S. and Sakai, S. (2000) Expression of two 1-aminocyclopropane-1-carboxylate synthase genes, *CS-ACS1* and *CS-ACS2*, correlated with sex phenotypes in cucumber plants (*Cucumis sativus* L.), *Plant Biotechnology*, **17**, 69–74.

Kamachi, S., Sekimoto, H., Kondo, N. and Sakai, S. (1997) Cloning of a cDNA for a 1-aminocyclopropane-1-carboxylate synthase that is expressed during development of female flowers at the apices of *Cucumis sativus* L, *Plant and Cell Physiology*, **38**, 1197–1206.

Kater, M. M., Colombo, L., Franken, J. *et al.* (1998) Multiple *AGAMOUS* homologs from cucumber and petunia differ in their ability to induce reproductive organ fate, *The Plant Cell*, **10**, 171–182.

Kater, M. M., Franken, J., Carney, K. J., Colombo, L. and Angenent, G. C. (2001) Sex determination in the monoecious species cucumber is confined to specific floral whorls, *The Plant Cell*, **13**, 481–493.

Krupnick, G. A., Avila, G., Brown, K. M. and Stephenson, A. G. (2000) Effects of herbivory on internal ethylene production and sex expression in *Cucurbita texana*, *Functional Ecology*, **14**, 215–225.

Krupnick, G. A., Brown, K. M. and Stephenson, A. G. (1999) The influence of fruit on the regulation of internal ethylene concentrations and sex expression in *Cucurbita texana*, *International Journal of Plant Sciences*, **160**, 321–330.

Kubicki, B. (1969a) Investigations on sex determination in cucumber (*Cucumis sativus* L.). V. Genes controlling intensity of femaleness, *Genetica Polonica*, **10**, 69–85.

Kubicki, B. (1969b) Investigations on sex determination in cucumber (*Cucumis sativus* L.). VII. Andromonoecism and hermaphroditism, *Genetica Polonica*, **10**, 101–120.

Le Corff, J., Agren, J. and Schemske, D. W. (1998) Floral display, pollinator discrimination, and female reproductive success in two monoecious Begonia species, *Ecology*, **79**, 1610–1619.

Le Roux, L. G. and Kellogg, E. A. (1999) Floral development and the formation of unisexual spikelets in the Andropogoneae (Poaceae), *American Journal of Botany*, **86**, 354–366.

Lebel-Hardenack, S., Ye, D., Koutnikova, H., Saedler, H. and Grant, S. R. (1997) Conserved expression of a TASSELSEED2 homolog in the tapetum of the dioecious *Silene latifolia* and *Arabidopsis thaliana*, *The Plant Journal*, **12**, 515–526.

Li, D., Blakey, C. A., Dewald, C. and Dellaporta, S. L. (1997) Evidence for a common sex determination mechanism for pistil abortion in maize and in its wild relative *Tripsacum*, *Proceedings of the National Academy of Sciences of the United States of America*, **94**, 4217–4222.

Li, J. M., Nagpal, P., Vitart, V., McMorris, T. C. and Chory, J. (1996) A role for brassinosteroids in light-dependent development of Arabidopsis, *Science*, **272**, 398–401.

Makus, D. J., Pharr, D. M. and Lower, R. L. (1975) Some morphogenic differences between monoecious and gynoecious cucumber seedlings as related to ethylene production, *Plant Physiology*, **55**, 352–355.

Mibus, H. and Tatlioglu, T. (2004) Molecular characterization and isolation of the F/f gene for femaleness in cucumber (*Cucumis sativus* L.), *Theoretical and Applied Genetics*, **109**, 1669–1676.

Mibus, H., Vural, I. and Tatlioglu, T. (2000) Molecular characterization and isolation of the F/f gene for femaleness in cucumber (*Cucumis sativus* L.), *Theortical and Applied Genetics*, **109**, 1669–1676.

Miller, J. S. and Venable, D. L. (2000) Polyploidy and the evolution of gender dimorphism in plants, *Science*, **289**, 2335–2338.

Mitchell, W. D. and Wittwer, S. H. (1962) Chemical regulation of sex expression and vegetative growth in *Cucumis sativus*, *Science*, **136**, 880–881.

Papadopoulou, K. and Grumet, R. (2005) Brassinosteroid- induced femaleness in cucumber and relationship to ethylene production. *HortScience*, **40**, 1763–1767.

Papadopoulou, E., Little, H. A., Hammar, S. A. and Grumet, R. (2005) Effect of modified endogenous ethylene production on hermaphrodite flower development and fruit production in melon (*Cucumis melo* L.), *Sex Plant Reprod*. **18**, 131–142

Papadopoulou, E., Little, H. A., Hammar, S. A. and Grumet, R. (2002) Effect of modified endogenous ethylene production and perception on sex expression in melon (*Cucumis melo* L.), in *Cucurbitaceae '02* (ed. D. Maynard), ASHS Press, Virginia, VS, pp. 157–164.

Perl-Treves, R. (1999) Male to female conversion along the cucumber shoot: approaches to studying sex genes and floral development in *Cucumis sativus*, in *Sex Determination in Plants* (ed. C.C. Ainsworth), Bios Scientific Publishers, Oxford, UK, pp. 189–215.

Perl-Treves, R., Kahana, A., Rosenman, N., Xiang, Y. and Silberstein, L. (1998) Expression of multiple AGAMOUS-like genes in male and female flowers of cucumber (*Cucumis sativus* L), *Plant and Cell Physiology*, **39**, 701–710.

Podolsky, R. D. (1992) Strange floral attractors – pollinator attraction and the evolution of plant sexual systems, *Science*, **258**, 791–793.

Przybecki, Z., Kowalczyk, M. E., Siedlecka, E., Urbanczyk-Wochniak, E. and Malepszy, S. (2003) The isolation of cDNA clones from cucumber (*Cucumis sativus* L.) floral buds coming from plants differing in sex, *Cell and Molecular Biology Letters*, **8**, 421–438.

Rajagopalan, P., Saraf-Levy, T., Lizhe, A. and Perl-Treves, R. (2004) Increased femaleness in transgenic cucumbers that overexpress an ethylene receptor, in (eds A. Lebeda and H. S. Paris) *Proceedings of Cucurbitaceae 2004*, Palacky University, Olomouc, Czech Republic, pp. 525–531.

Reusch, T. (2001) Fitness-consequences of geitonogamous selfing in a clonal marine angiosperm (*Zostera marina*), *Journal of Evolutionary Biology*, **14**, 129–138.

Rood, S. B., Pharis, R. P. and Major, D. J. (1980) Changes in endogenous gibberellin-like substances with sex reversal of the apical inflorescence in corn, *Plant Physiology*, **66**, 793–796.

Rudich, J., Halevy, A. H. and Kedar, N. (1969) Increase in femaleness of three cucurbits by treatment with Ethrel, an ethylene-releasing compound, *Planta*, **86**, 69–76.

Rudich, J., Halevy, A. H. and Kedar, N. (1972a) Ethylene evolution from cucumber plants as related to sex expression, *Plant Physiology*, **49**, 998–999.

Rudich, J., Halevy, A. H. and Kedar, N. (1972b) The level of phytohormones in monoecious and gynoecious cucumbers as affected by photoperiod and ethephon, *Plant Physiology*, **50**, 585–590.

Saraf-Levy, T., Kahana, A., Kessler, N. *et al.* (2000) Genes involved in ethylene synthesis and perception in cucumber, in *Proceedings of Cucurbitaceae 2000* (eds N. Katzir and H. S. Paris), Jerusalem, Israel, 7th Eucarpia meeting on cucurbit genetics and breeding.

Shannon, S. and de La Guardia, M. D. (1969) Sex expression and the production of ethylene induced by auxin in the cucumber (*Cucumis sativus* L.), *Nature*, **223**, 186.

Sun, T. P., Goodman, H. M. and Ausubel, F. M. (1992) Cloning the Arabidopsis Ga1 locus by genomic subtraction, *The Plant Cell*, **4**, 119–128.

Takahashi, H., Saito, T. and Suge, H. (1982) Intergeneric translocation of floral stimulus across a graft in monoecious Cucurbitaceae with special reference to the sex expression of flowers, *Plant Cell Physiology*, **23**, 1–9.

Talamali, A., Bajji, M., Le Thomas, A., Kinet, J. M. and Dutuit, P. (2003) Flower architecture and sex determination: how does Atriplex halimus play with floral morphogenesis and sex genes? *New Phytologist*, **157**, 105–113.

Tanurdzic, M. and Banks, J. A. (2004) Sex-determining mechanisms in land plants, *The Plant Cell*, **16**, S61-S71.

Trebitsh, T., Staub, J. E. and Oneill, S. D. (1997) Identification of a 1-aminocyclopropane-1-carboxylic acid synthase gene linked to the *female* (*F*) locus that enhances female sex expression in cucumber, *Plant Physiology*, **113**, 987–995.

Trebitsh, T., Rudich, J. and Riov, J. (1987) Auxin, biosynthesis of ethylene and sex expression in cucumber (*Cucumis sativus*), *Plant Growth Regulation*, **5**, 105–113.

Wang, K. L. C., Li, H. and Ecker, J. R. (2002) Ethylene biosynthesis and signaling networks, *The Plant Cell*, **14**, S131–S151.

Winkler, R. G. and Helentjaris, T. (1995) The maize *Dwarf3* gene encodes a cytochrome P450-mediated early step in gibberellin biosynthesis, *The Plant Cell*, **7**, 1307–1317.

Winter, K. U., Weiser, C., Kaufmann, K. *et al.* (2002) Evolution of class B floral homeotic proteins: obligate heterodimerization originated from homodimerization, *Molecular Biology and Evolution*, **19**, 587–596.

Yamasaki, S., Fujii, N., Matsuura, S., Mizusawa, H. and Takahashi, H. (2001) The *M* locus and ethylene-controlled sex determination in andromonoecious cucumber plants, *Plant and Cell Physiology*, **42**, 608–619.

Yamasaki, S., Fujii, N. and Takahashi, H. (2000) The ethylene-regulated expression of *CS-ETR2* and *CS-ERS* genes in cucumber plants and their possible involvement with sex expression in flowers, *Plant and Cell Physiology*, **41**, 608–616.

Yamasaki, S., Fujii, N. and Takahashi, H. (2003a) Characterization of ethylene effects on sex determination in cucumber plants, *Sexual Plant Reproduction*, **16**, 103–111.

Yamasaki, S., Fujii, N. and Takahashi, H. (2003b) Photoperiodic regulation of *CS-ACS2, CS-ACS4* and *CS-ERS* gene expression contributes to the femaleness of cucumber flowers through diurnal ethylene production under short-day conditions, *Plant Cell and Environment*, **26**, 537–546.

Yin, T. and Quinn, J. A. (1995) Tests of a mechanistic model of one hormone regulating both sexes in *Cucumis sativus* (Cucurbitaceae), *American Journal of Botany*, **82**, 1537–1546.

Young, T. E., Giesler-Lee, J. and Gallie, D. R. (2004) Senescence-induced expression of cytokinin reverses pistil abortion during maize flower development, *The Plant Journal*, **38**, 910–922.

7 Cytoplasmic male sterility

Françoise Budar, Pascal Touzet and Georges Pelletier

7.1 Introduction

Cytoplasmic male sterility (CMS) results from the dual genetic control of plant sex. Cytoplasmic genes provoking pollen abortion render the plant effectively female. Another set of nuclear genes — restorers of fertility (*Rf*) — overcomes the effects of the CMS genes, restoring the hermaphrodite condition. The effective sex of the individual is, therefore, determined by both its cytoplasmic and nuclear genotypes.

Unlike nuclear genes, which display Mendelian inheritance, cytoplasmic factors are generally inherited maternally. CMS in plant species therefore appears as a non-Mendelian variation of reproductive biology. The different modes of inheritance of the two key determinants underlie the considerable interest in CMS that has existed for many years, for both hybrid seed production in crops and for the study of genomic conflict in natural populations. CMS has also provided the first examples of mitochondrial involvement in plant sexual reproduction. For many years, the various scientific groups interested in different aspects of CMS worked on unrelated systems and communicated little. The progress made in the last few years, particularly in terms of the identification of sterility-inducing and restorer genes, provides an useful basis for the establishment of a more integrated view of this fascinating feature of plant reproductive biology.

This chapter provides an overview of current knowledge about CMS, focusing on genetic determinism, raising new open questions and proposing possible lines of research to extend our knowledge in the light of discoveries from the last decade.

7.2 Pioneering work in plant sexuality and hybridisation

Rudolph Jacob Camerarius (1665–1721) was the first to demonstrate plant sexuality; he described the different sexual forms and suggested that pollen was the male fertilising agent in plants in a 'Letter on the Sexuality of Plants' (*Epistolae de Sexu Plantarum*) in 1694. Many plant breeders and gardeners have since carried out crosses in plants. Joseph Gottlieb Koelreuter (1733–1806) carried out a large number of intra- and interspecific crosses and was the first to observe the phenomenon now known as heterosis, in which hybrid plants are more vigorous than their parents. Thomas Andrew Knight (1759–1838) used pea as an experimental model and probably understood the dominance of some characters. Gregor Mendel

(1822–1884) then had the brilliant idea of applying mathematical analysis to the results of his crosses and was perspicacious enough to interpret his results correctly (Mendel, 1865).

Darwin himself dedicated no less than two entire books to plant reproduction (Darwin, 1876, 1877), not counting his work on the pollination of orchids by insects. Darwin's reasoning was based not only on his own observations, but also on the outcome of many previous crosses performed by individuals such as Koelreuter. Unfortunately, he did not consider the work of Mendel. In 'The Different Forms of Flowers on Plants of the Same Species' (Darwin, 1877), Darwin described the occurrence of different sexual forms in related species and defined gynodioecious species as those that 'consist of hermaphrodites and females without males'. He suggested that dioecy evolved via the selection of separate sexes from hermaphroditism, which he considered to be the 'normal' form. About two- thirds of angiosperm species are hermaphrodites.

The value of hybridisation in plants rapidly became obvious to many plant breeders, as demonstrated by the extensive review published on 'Hybrids and their Utilization in Plant Breeding' (Swingle and Webber, 1897). Hybrid vigour, or heterosis, was commonly observed and was considered highly valuable in plant improvement. In 1917, Jones pointed out in the introduction to a paper on heterosis, 'The increased growth as the result of crossing is so common an occurrence that it is probably familiar to everyone who has made any hybridization experiments' (Jones, 1917).

7.3 Early studies of CMS

Less than ten years after the rediscovery of Mendel's laws by deVries (de Vries, 1900a, b), Baur and Correns independently noted that colour variegation was maternally inherited in *Pelargonium* and *Mirabilis*, respectively (Baur, 1909; Correns, 1909). The genetic autonomy of chloroplasts seems to have been accepted rapidly, but was regarded as a peculiarity of the organelle. In 1921, Bateson and Gairdner reported an unusual pattern of segregation of male sterility in flax crosses, and suggested that alleles were unevenly distributed in male and female gametes via somatic segregation (Bateson and Gairdner, 1921). In 1927, Chittenden and Pellew reviewed these results and suggested the involvement of a combination of Mendelian segregation of nuclear genes and maternal inheritance of cytoplasmic factors (Chittenden and Pellew, 1927). Subsequently, Chittenden elegantly demonstrated the validity of their hypothesis in a series of experiments published in the same year (Chittenden, 1927). He suggested that a component of the cytoplasm other than the chloroplasts affected reproductive phenotype via interaction with the nucleus, and that this constituent could be variable and different between species or races and was inherited through the female line (Chittenden, 1927). In the early 1930s, Rhoades provided the first description of CMS in maize (Rhoades, 1931, 1933). In papers published in 1931 and 1933, he stressed the limits of the 'chromosome theory of heredity'. It should be borne in mind that there was much controversy concerning the respective roles of the cytoplasm and of nuclear genes or chromosomes in the formation of an organism from an egg at that time, especially among embryologists. The paper published by Rhoades in 1933 provides an impressive demonstration of the maternal inheritance of complete or partial male sterility, and its independence from chromosomal genetic markers. The stance taken by Rhoades in his papers shows clearly that he did not believe that chromosomal genes could account for male sterility. As he was convinced that nuclear factors were not involved, he insisted that this phenomenon should be called 'cytoplasmic male sterility'. This may explain why he cited a number of previously described cases of cytoplasmic inheritance of male sterility, but explicitly disregarded previous work on flax and provided only short comments on his own results for crosses in which clear Mendelian segregation of restorer genes occurred. Undoubtedly, Chittenden and Pellew would have preferred the name 'nuclear-cytoplasmic male sterility'. In subsequent work on CMS induced by *iojap*, Rhoades suggested that mitochondria were the cytoplasmic factors responsible for male sterility (Rhoades, 1950). The nature of the heritable factors present in the cytoplasm was a subject of debate over several decades. In the 1960s, nucleic acids were clearly shown to be present in both chloroplasts (in unicellular algae) and mitochondria (in yeast) (Gibor and Granick, 1964).

Jones and his colleagues were the first to suggest that CMS could be used to produce hybrid seed; they developed and formalised a method for onion (Jones and Emsweller, 1937; Jones and Clarke, 1943). This led to the production of hybrid onions on a commercial scale by means of CMS. Hybridisation was already known to be of great value for plant breeding and improvement. Breeders knew that the vigour and homogeneity of hybrids was lost in subsequent generations and the advantages of large-scale hybrid production were obvious. Hybrid maize had been produced since the 1920s, but at high cost, by removing the tassels of the plants. In 1944, Mangelsdorf initiated efforts to develop male sterile inbred lines for use in hybrid production, based on a male sterility-inducing cytoplasm found in the Texan varieties Mexican June, Golden June and Honey June. This work led to extensive use of the so-called Texas CMS in the commercial production of hybrid maize seeds (Rogers and Edwardson, 1952). In 1971, Beckett proposed a classification for male sterility-inducing cytoplasms in maize, based on the genetics of fertility restoration (Beckett, 1971). Two male sterility-inducing cytoplasms are identical, or at least belong to the same group, if fertility is restored by the same nuclear genotype. This classification criterion is still used and is entirely valid.

7.4 Definition of CMS as a system with two genetic determinants

CMS should really be called nucleo-cytoplasmic male sterility as it is controlled by two genetic determinants with different modes of inheritance. It involves variations (i) at the cytoplasmic level, with at least a sterility-inducing cytoplasm (S) and a cytoplasm with no sterility effect (N) and (ii) at the nuclear level, with a dominant

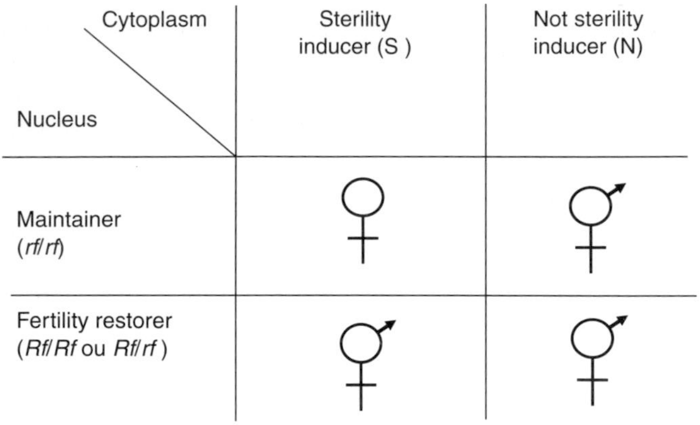

Nucleus \ Cytoplasm	Sterility inducer (S)	Not sterility inducer (N)
Maintainer (*rf*/*rf*)	♀	⚥
Fertility restorer (*Rf*/*Rf* ou *Rf*/*rf*)	⚥	⚥

Figure 7.1 CMS is determined by two genetic components.

restorer allele (*Rf*) that enables plants with the S cytoplasm to produce pollen and a recessive allele (*rf*), also called a maintainer of sterility, maintaining the male sterile phenotype induced by the S cytoplasm. The only combination resulting in male sterility is, therefore, a combination of the S cytoplasm with a nuclear genome not carrying the appropriate *Rf* allele, as shown in Figure 7.1.

If one of the variants in one of the genetic compartments is not identified, the system is not complete and may be misinterpreted. For instance, if no *rf* allele is present in the genotypes considered then no male sterility will be detected. Conversely, if no *Rf* allele is present, then the observed sterility appears to be entirely cytoplasmic (as suggested by Rhoades). If no S cytoplasm is present in the considered genotypes, then *Rf* genes cannot be detected and if no N cytoplasm is present then the male sterility phenotype appears to be entirely determined by the nucleus and *rf* alleles are likely to be considered male sterility mutations.

We will therefore consider ‘CMS’ to be a genetic system with two genetic components: a male sterility-inducing cytoplasm (S), and a nuclear locus carrying either restorer (*Rf*) or maintainer (*rf*) alleles. In some cases, several loci for the restoration of fertility have been detected, only some or all of which may be necessary for fertility restoration. This definition is purely genetic and implies no functional hypothesis for male sterility or fertility restoration. Indeed, total ignorance of the two mechanisms involved has not precluded the widespread use of CMS for hybrid seed production over many years.

7.5 Use of CMS in hybrid seed production

In most crops with hermaphrodite flowers, hybrid seed production is difficult to envisage in the absence of a genetic castration system, given that the use of chemicals for such purposes remains very limited. It is easy to see why CMS is popular with hybrid seed producers and is largely due to the genetic specificities of these systems.

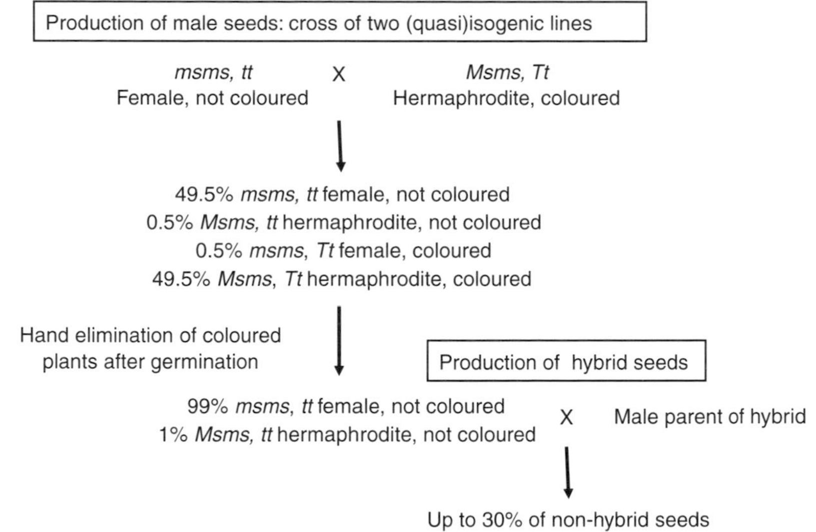

Figure 7.2 The use of a genic male sterility (ms) linked to an early phenotypic marker (T: synthesis of anthocyane leading to coloured seedlings), for the production of hybrid seeds in sunflower.

The main problem for hybrid seed producers is how to produce, as cheaply as possible, male sterile plants to serve as the female parent of the hybrid seed (final product). In all processes involving male sterility with Mendelian inheritance, the production of females involves screening to eliminate hermaphrodites directly in the hybrid seed production field, entailing additional costs. In sunflower, for instance, a Mendelian recessive gene for male sterility has been used for hybrid seed production, with screening for a linked locus for anthocyanin synthesis used to eliminate hermaphrodites (Leclercq, 1966). As shown in Figure 7.2, the main drawbacks of this system are (i) the need for very dense sowing and manual elimination of half the plantlets after germination and (ii) the residual 1% hermaphrodites, resulting from recombination between the Ms and T loci, giving up to 30% non-hybrid seeds in the harvest (Vear, 1992). After the discovery of a CMS, *pet1*, in the back-cross progeny of an interspecific hybrid between *Heliantus petiolaris* and *Heliantus annuus* (Leclercq, 1969), the nuclear system for generating male sterility was abandoned, and hybrid sunflowers are now produced with the *pet1* cytoplasm.

The production of females remains a major problem for genetically engineered nuclear male sterilities (Perez-Prat and van Lookeren Campagne, 2002). Nevertheless, some elegant solutions have been proposed, based on reversible transgene-induced male sterilities. Such systems are based on a transgene that induces gametophytic male sterility in normal conditions (for hybrid production), the expression or effects of which are abolished by the application of a chemical (for the production of females by selfing). Restoration is obtained in F1 hybrids due to their heterozygous state for the transgene, allowing production of 50% of non-transgenic pollen, which is sufficient for correct pollination (Drouaud *et al.*, 1999).

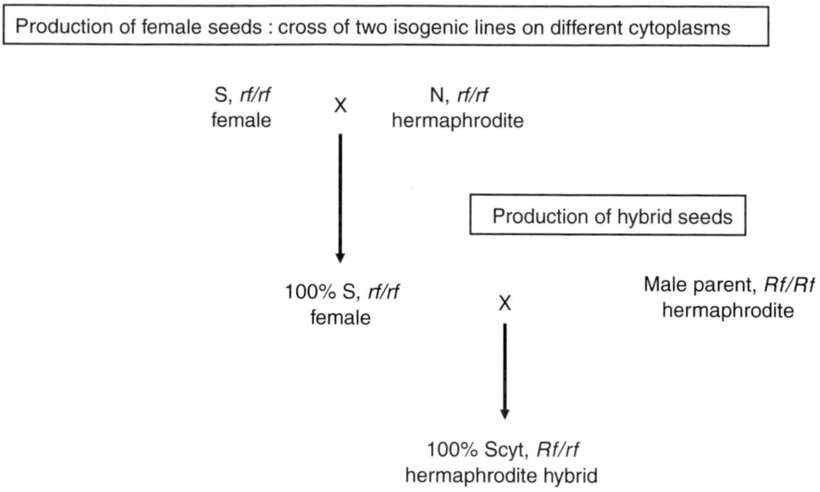

Production of female seeds : cross of two isogenic lines on different cytoplasms

S, *rf/rf* female X N, *rf/rf* hermaphrodite

Production of hybrid seeds

100% S, *rf/rf* female X Male parent, *Rf/Rf* hermaphrodite

100% Scyt, *Rf/rf* hermaphrodite hybrid

Figure 7.3 Use of a CMS system for the production of hybrid seeds.

Maternally inherited male sterility is an ideal solution as crosses between a male sterile plant and a hermaphrodite plant homozygous for maintainer alleles give rise to 100% female plants (Figure 7.3). The use of CMS to produce the female parent leads to the production of hybrid seeds, all of which carry the sterility-inducing cytoplasm. Consequently, unless a dominant restorer gene is incorporated into the nuclear genotype from the male parent, the hybrids are male sterile. The restoration of male fertility in hybrids is not an absolute necessity if crop production does not involve pollination, as is the case for many vegetables (onion, carrot, radish, cabbage, turnip etc.) or certain other crops such as beets. However, the restoration of male fertility is much more important in crops in which the product results from fertilisation (fruit or grain crops). In such cases, a complete system is required for hybrid seed production. In some cases, the lack of restorers, or the association of such restorers with undesirable agronomic traits, has been overcome by the production of 'hybrid and line composite varieties'. In such varieties, a small proportion of the male fertile parent line is mixed with the male sterile hybrid to facilitate pollination in the crop field. Varieties of this type have been registered for winter and spring oilseed rape carrying the 'Ogu-INRA' CMS in Europe (Budar *et al.*, 2004).

The convenience associated with the use of CMS for the production of hybrid seeds has led plant breeders to search for such systems in almost all crops. They have been successful in many cases and significant quantities of hybrid seed are produced using CMS systems in many cultivated species, including beet, oilseed rape, sunflower, onion, carrot, radish, cabbage and rice. In rice, hybrid seed production is based on the use of CMS or recessive and 'reversible' Mendelian sterilities. These Mendelian sterilities are influenced by environmental factors, such as temperature and photoperiod. Depending on the environmental conditions, plants may be male sterile and used for hybrid production, or male fertile and self-pollinated for the propagation of females (Xue *et al.*, 2003). However, such environmentally modified male sterilities are very difficult to use unless the females and hybrids can be produced in very different environments.

7.6 The search for CMS: intraspecific crosses, alloplasmic situations and cybrids

Kaul distinguished spontaneous CMS from CMS arising following intraspecific, interspecific or intergeneric crosses (Kaul, 1988). The apparent difference between a spontaneous CMS and a CMS resulting from an intraspecific cross may depend entirely on the generation evaluated for male sterility, especially in cases where the genetic composition of the first-described male sterile plant is unknown.

A number of cases of maternally inherited male sterilities arising in particular nuclear backgrounds, such as *iojap* in maize (Rhoades, 1950), or after *in vitro* culture, such as 'CMSI' and 'CMSII' in *Nicotiana sylvestris* (De Paepe *et al.*, 1990), have been reported. However, in such cases, the male sterility is exclusively determined by the cytoplasm. In the case of *N. sylvestris* cytoplasmic male sterile mutants, deletions in mitochondrial genes encoding complex I subunits, leading to the impairment of complex I assembly, have been implicated in the male sterile phenotype (Sabar *et al.*, 2000). These *N. sylvestris* mutants produce pollen in some environmental conditions (strong light) and generally develop more slowly than the wild type (Li *et al.*, 1988; Budar *et al.*, 2003). Indeed, due to the lack of the nuclear genetic component, such sterilities cannot be considered to be genuine CMS systems as defined earlier.

In his extensive review on male sterility, Kaul estimated that whereas most Mendelian male sterilities arise from spontaneous mutations, CMSs (gc-msts in Kaul's nomenclature) arise predominantly from hybridisations between varieties or species (Kaul, 1988). He counted 71 species displaying intraspecific CMS and 271 inter-species crosses in which CMS emerged (Kaul, 1988). When male sterility occurs in the progeny of an intraspecific cross, the parents are often, if not always, unrelated, or of 'distant genotypes'. This situation is essentially similar to that of spontaneous CMS. The male fertility of the female parent of the cross results from the presence of fertility restorers in the nuclear genotype. The male parent, genetically distant from the female, carries maintainer alleles. Therefore, male sterility induced by the female parent cytoplasm appears, generally, in the second generation. This seems to be the case for a number of CMSs reported to arise after intraspecific crosses. For instance, the flax CMS described by Bateson and Gairdner (Bateson and Gairdner, 1921) resulted from a cross between two genotypes, the cytoplasm donor being described as a 'peculiar plant whose origin is unknown' (Chittenden, 1927). The simple Mendelian segregation of male fertility in the female descent of this plant after crossing with flax is entirely consistent with the presence of an *Rf* gene in the genotype of the female parent and an *rf* allele in the genotype of the recurrent male parent. The male sterility was not generated by the cross, but was instead already potentially present in the cytoplasm donor, masked by the *Rf* gene. The cross simply brought the inherent potential sterility to light by combining the male sterility-inducing cytoplasm with maintainer alleles. This is also the case for the *nap* CMS in oilseed rape (*Brassica napus*), which was detected following crosses in which particular cultivars ('Bronowski' or 'Hokuriku 23') were used as the recurrent male parent. All oilseed rape varieties except these two male parents carry restorers for a male sterility induced by the *B. napus* cytoplasm (Budar *et al.*, 2004). Currently, the main problem facing breeders trying to use an intraspecific CMS is the limited number of good maintainer genotypes. This is the case for the *pol* and *nap* CMSs in oilseed (Delourme and Budar, 1999), the Ogura CMS in radish (Bonnet, 1975), the C CMS in maize (Laughnan and Gabay-Laughnan, 1983).

In 1950, Clayton showed that CMS emerged in *Nicotiana*, after an interspecific cross between *N. debneyi* and *N. tabacum* followed by repeated back-crosses with the male parent. Morphological abnormal flowers appeared after the second back-cross and male sterility was complete when seven-eighths of the *debneyi* chromosomes had been eliminated (Clayton, 1950). Clayton suggested that male sterility was due to incompatibility between the *debneyi* cytoplasm and *tabacum* chromosomes, and proposed to use this situation to produce hybrids or to avoid topping. He also suggested that this incompatibility could provoke unexpected problems in breeding programs when such programs involve interspecific crosses followed by recurrent back-crosses with the cultivated species if the wild and subsequent hybrids are used as female parents. This subsequently proved to be the case in breeding programs (Bannerot *et al.*, 1977). Cytoplasmic male sterilities arising in interspecific crosses are often linked to the confrontation of the cytoplasm of one species with the nuclear genome of the other parent. This so-called alloplasmic situation is obtained by using the nuclear donor as the recurrent male parent in successive back-crosses on the first interspecific hybrid. It should be noted that although the hybrid plants resulting from the interspecific cross are often sterile or partially sterile, this is mostly due to abnormal meioses. Fertility generally increases after the first back-cross, and CMS appears with the progressive replacement of the chromosomes by those from the recurrent male parent. In some taxa, considerable effort has been made to detect CMS after interspecific crosses, mostly with the aim of identifying a CMS that can be used in hybrid seed production. A large number of alloplasmic situations have been generated in two dicotyledonous genera, *Brassica* and *Nicotiana*, and in the wheat family, using *Aegilops* cytoplasms (Kaul, 1988). CMS has been actively sought, with some success, in crops of the Brassicaceae family by interspecific crosses (reviewed in Delourme and Budar, 1999). In the *Nicotiana* genus, no less than 27 different interspecific crosses have been shown to result in CMS (Kaul, 1988; Nikova and Vladova, 2002). However, the alloplasmic situation frequently results in the modification of characters others than male sterility, and is therefore highly undesirable for hybrid seed production. There may be a deficiency in photosynthesis, of various degrees (linked to the cytoplasm donor's plastids), and modifications to flower morphology, sometimes associated with female sterility. These modifications may be considered homeotic; they may be limited to modification of the anthers (petaloid or carpelloid anthers, sometimes with naked ovules) or may affect all organs of the flower. They have been linked to the mitochondria of the cytoplasm donor. The alloplasmic situation is thought to reflect impaired co-operation between the mitochondrial and nuclear genomes. This impairment occurs because, unlike the nuclear and mitochondrial genomes of a given species, these genomes did not co-evolve in the alloplasmic situation. The lack of co-evolution of the nuclear and plastid genomes in alloplasmic situations may also account for the observed defects in photosynthesis. In all cases, changes to flower morphology, including male sterility, result directly from the alloplasmic situation. Some alloplasmic situations have been extensively studied in *Nicotiana*. In the alloplasmic line of *N. tabacum* carrying the *N. repanda* (*rep*) cytoplasm, a low ATP/ADP ratio and defective cell proliferation in the flower meristem have been reported (Bergman *et al.*, 2000, Farbos *et al.*, 2001). Moreover, the ectopic expression of a cadastral gene controlling organ identity led to partial correction of the homeotic phenotype (Bereterbide *et al.*, 2002). The recent implication of the cytoplasm in floral differentiation has been tentatively integrated into our global understanding of nuclear–cytoplasmic cross-talk (Zubko, 2004).

When crosses are not possible between species, somatic hybridisation can be used.

Rearrangements in the mitochondrial genome after protoplast fusion were first demonstrated in *Nicotiana* (Belliard *et al.*, 1979). Following protoplast fusion, the mitochondrial genomes of the two parents undergo extensive rearrangements. Chloroplast genomes are not affected in the regenerated plants, with each plant carrying one of the two parental plastid types. The very different behaviour of the plastid and mitochondrial genomes in such experiments makes it possible (i) to replace the plastids, facilitating the recovery of normal photosynthetic performance (ii) to produce new, recombined mitochondrial genomes, in some cases breaking the linkage between mitochondrially encoded features (Belliard *et al.*, 1979; Bonnett *et al.*, 1991; Kofer *et al.*, 1991, 1992). Plants with rearranged cytoplasms are called cybrids. Protoplast fusion provided a new means of searching for CMS in two ways. First, this made it possible to correct undesirable phenotypes in some alloplasmic situations. This was the case for the introduction of the Ogura radish cytoplasm (Ogura, 1968) into *Brassica* crops by sexual hybridisation. The undesirable traits, such as cold sensitivity of photosynthesis and reduced nectar and seed production, present in the alloplasmic lines were corrected by protoplast fusion (Bannerot *et al.*, 1974, 1977; Pelletier *et al.*, 1983). In some of the resulting cybrids, *Brassica* chloroplasts were recovered, and rearranged mitochondrial genomes were associated with male sterility, good female fertility and nectar production (Pelletier *et al.*, 1987). The cybrids subsequently chosen for hybrid seed production displayed even greater vegetative vigour than normal isogenic fertile lines (M. Renard, personal communication). Second, protoplast fusion can be used to generate somatic hybrids between species that cannot be hybridised sexually, making it possible to set up new alloplasmic situations, in some cases modified by mitochondrial recombination. This method was used to generate CMS in chicory by protoplast fusion between chicory and a male sterile sunflower carrying the *pet1* cytoplasm (Rambaud *et al.*, 1993). It was also used to produce new alloplasmic combinations in *Nicotiana* (Zubko *et al.*, 2003) and in *Brassica* (Kirti *et al.*, 1992, 1995; Stiewe *et al.*, 1995; Leino *et al.*, 2003). In the case of Ogura CMS, protoplast fusions made it possible to transfer the mitochondrial sterility determinant originally identified in *Raphanus* to *Brassica*, as shown by identification of the sterility gene in both the *Brassica* cybrids and the donor species (Bonhomme *et al.*, 1992; Krishnasamy and Makaroff, 1993). However, in chicory, the mitochondrial gene associated with *pet1* sterility in sunflower does not seem to be involved in the male sterility of some chicory cybrids (Dubreucq *et al.*, 1999).

Thus, somatic hybridisation makes it possible to create new alloplasmic situations and to transfer mitochondrial sterility determinants from one species to another. These two situations are not mutually exclusive and are difficult to distinguish on the basis of the phenotype of the plant. However, although somatic hybridisation is a promising way to generate new cytoplasmic–nuclear combinations leading to CMS, this method is restricted to plant families in which regeneration from protoplasts is technically feasible.

In alloplasmic situations, whether resulting from sexual or somatic hybridisation, and in cases of transfer of sterility determinants between unrelated species, the main problem after identification and characterisation of the male sterile plants is the identification of restorer genotypes. Generally, the cytoplasm donor species from which the sterility originates also provides the nuclear restorers. The transfer of these genes to the recipient species may represent a major challenge (Heyn, 1976; Pellan-Delourme and Renard, 1988; Delourme *et al.*, 1991, 1995; Suzuki *et al.*, 1994; Prakash *et al.*, 1998).

7.7 Spontaneous CMSs result from genomic conflict between nuclear and cytoplasmic genes

Gynodioecy can be seen as an intermediate state towards dioecy or a stable plant mating system. An important question remains unresolved if this is the case: how do male sterile plants, which produce no progeny of male descent, avoid being selected against in populations in which natural selection occurs? (Charlesworth, 2002).

Lewis showed mathematically that the conditions for the presence of females 'at equilibrium' was that females produced at least twice as many seeds as hermaphrodites in the case of a dominant or recessive nuclear male sterility gene, but only slightly more seeds than hermaphrodites in the case of maternally inherited male sterlity (Lewis, 1941). However, he considered only maternally inherited male sterility and disregarded situations in which genetic control did 'not satisfy any simple genic scheme', as in *Silene inflata*, *Cirsium pratense* or *Plantago lanceolata*, which have clearly nucleo-cytoplasmic male sterlity. Lewis believed that the main role of CMS in gynodioecious populations was avoiding self-fertilisation to maintain optimum hybridity. He did raise the question of cytoplasmic variability (which he called 'cytoplasmic differentiation') and hypothesised that cytoplasmic changes arise more frequently than would have been predicted from the available data. He also suggested that the exchange of cytoplasm between related species by interspecific hybridisation could lead to females and hermaphrodites within a population having different cytoplasms. However, CMS must be more than simply a mechanism for maintaining cross-fertilisation. More recent models evoke a genomic conflict between cytoplasmic and nuclear genes (Gouyon and Couvet, 1987; Frank, 1989). Nuclear and cytoplasmic genes are subject to natural selection and their different modes of inheritance lead to different situations of selective value with respect to male descent. In the vegetative growth phase and in female descent, improvements to the genotype improve the transmission of both nuclear and cytoplasmic genes. This results in the co-evolution of nuclear and cytoplasmic genes, with vegetative growth and seed production becoming more efficient. In contrast, pollen production may be increased or made more efficient through selection acting on nuclear genes, but this has no effect on cytoplasmic genes, which are inherited exclusively from the maternal plant. A cytoplasmic gene inducing male sterility increases the selective value of cytoplasmic genomes when female plants produce slightly more seeds because all the progeny of the female plant carry its cytoplasm. This has been described as the 'female advantage'. However, the nuclear genes associated with the male sterility-inducing cytoplasm lose much of their selective value in this situation as they are no longer transmitted through the male line. Therefore, any nuclear gene restoring male fertility appearing in plants carrying male sterility-inducing cytoplasm is likely to be efficiently selected, even if less pollen is produced in the restored plant than in the original hermaphrodite. However, in this case, *Rf* genes should become fixed in the population, leading to a complete disappearance of functionally female plants. The observed persistence of females in natural populations has been explained in theoretical models in two non-mutually exclusive ways. First, there is a 'cost of restoration', resulting in the restorer nuclear genotype having lower selective value in the absence of the male sterility-inducing cytoplasm. In this case, the maintainer alleles may be maintained and male sterile plants observed in each generation (Charlesworth, 1981; Delannay *et al.*, 1981; Frank, 1989; Gouyon *et al.*, 1991). Second, a population structure out of equilibrium due to founder effects, allows the maintenance of females (Figure 7.4) (Frank, 1989; McCauley and Taylor, 1997; Pannell, 1997; Couvet *et al.*, 1998; McCauley *et al.*, 2000). In all cases, the female advantage is required to favour selection of the male sterility-inducing cytoplasm in the first place.

This theoretical model of genomic conflict is attractive and a number of its features can be tested experimentally. If this model holds true then:

(1) The female reproductive success of female plants should be greater than that of hermaphrodites.
(2) CMS should not be associated with deleterious effects on vegetative growth and seed production.

(3) Gynodioecious populations should display cytoplasmic variability.
(4) Fertility-restoring genes should arise from maintainer genes in the presence of male sterility-inducing cytoplasms.
(5) In the absence of the male sterility-inducing cytoplasm, there should be selection against nuclear restorers.

Studies on gynodioecious species have focused on *Thymus vulgaris*, *Beta maritima*, *Plantago lanceolata* and *Silene* sp. In *Silene*, the occurrence of closely related gynodioecious and dioecious species has raised particular interest in studies of the evolution of sex determinism (Desfeux *et al.*, 1996; Matsunaga and Kawano, 2001). In *T. vulgaris*, both significantly higher levels of seed production by females than by hermaphrodites and structured populations have been observed, and a cost of restoration has been suggested (Manicacci *et al.*, 1996; Thompson and Tarayre, 2000). Female advantage has also been described in *P. lanceolata* (Poot, 1997), but no clear female advantage could be detected for *B. maritima* (Boutin *et al.*, 1988). In all studied gynodioecious species, cytoplasmic variability is demonstrated by the occurrence of several types of CMS (Cuguen *et al.*, 1994). These CMSs are differentiated by nuclear restorer genotypes, but also, in some cases, by morphological differences. The use of molecular markers has recently facilitated the efficient characterisation

Stage 0: Arrival of founders

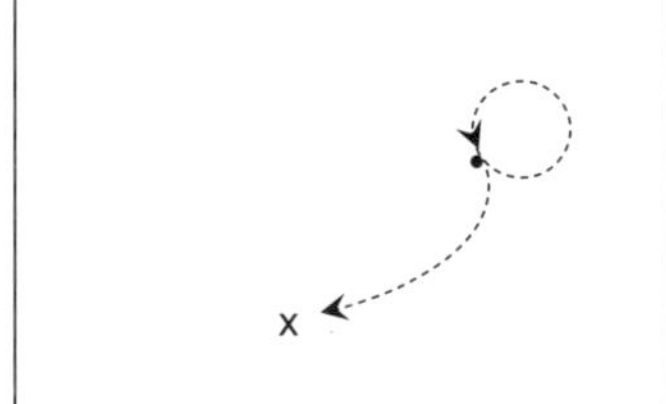

Stage 1: Patch of females
Females are the major contributors for seeds, the S cytoplasm spreads in the population

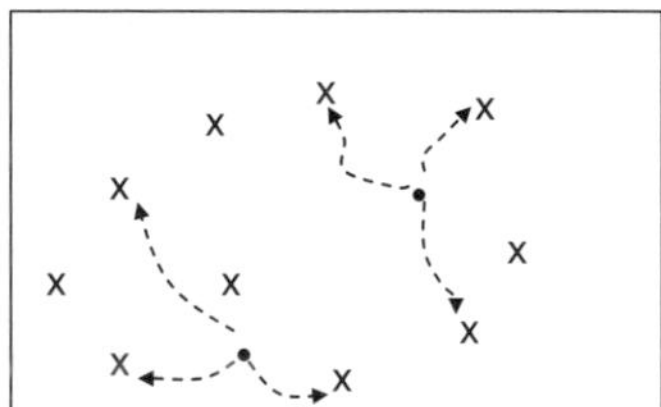

Stage 2: Arrival of nuclear restorer(s)

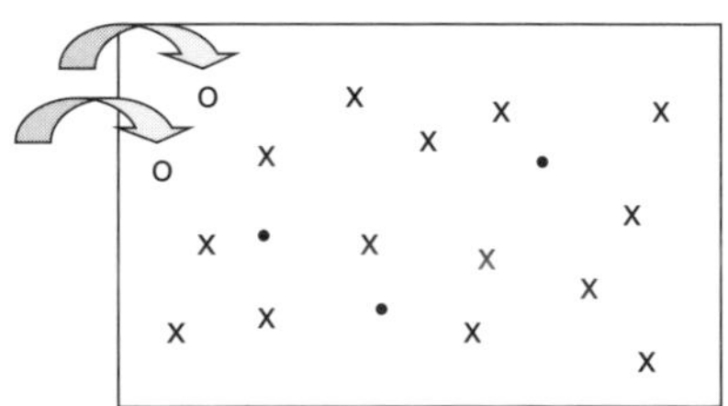

Stage 3: The restorer invades the population
Frequency of females decreases

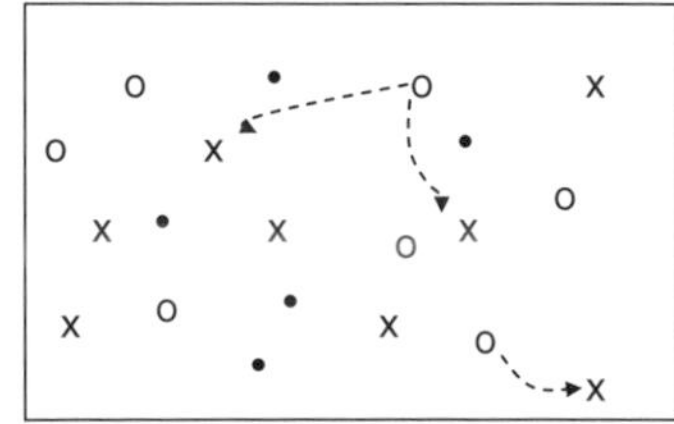

Figure 7.4 CMS and founder effect in thyme (Atlan, 1991). •, rf/rf hermaphrodites; o, restoring hermaphrodites; x, females; --→, pollination.

of cytoplasms and has made possible systematics studies based on mitochondrial variability (Saumitou-Laprade *et al.*, 1993; De Haan *et al.*, 1997; Desplanque *et al.*, 2000; Ingvarsson and Taylor, 2002; Stadler and Delph, 2002; Olson and McCauley, 2002; McCauley and Olson, 2003). The mitochondrial DNA of *T. vulgaris* has been shown to be highly polymorphic (Belhassen *et al.*, 1993). Interestingly, multiple CMSs have also been described in several cultivated species such as sunflower (Crouzillat *et al.*, 1991, 1994; Horn and Friedt, 1999; Horn, 2002), maize (Laughnan and Gabay-Laughnan, 1983; Beckett, 1971), *Brassica* crops (Delourme and Budar, 1999), sorghum (Pring *et al.*, 1995; Xu *et al.*, 1995), onion. (Havey, 2000) and rice (Virmani and Shinjyo, 1988). Some of the CMSs used in hybrid seed production and/or studied at a molecular level probably evolved over a long period of time in the species considered, as they have been detected in natural populations of ancestral species (Hervieu *et al.*, 1993, 1994; Yamagishi and Terachi, 1996, 1997; Horn *et al.*, 1996; Arrieta-Montiel *et al.*, 2001). Moreover, molecular evolution has been observed for Ogura CMS in *Raphanus*, as several molecular variants of the sterility gene have been described (Yamagishi and Terachi, 2001), including the gene associated with so-called kosena CMS (Iwabuchi *et al.*, 1999), which is restored by the same nuclear restorer (Koizuka *et al.*, 2000, 2003; Desloire *et al.*, 2003; Brown *et al.*, 2003), and can therefore be considered to belong to the same system.

Studies on gynodioecious species and theoretical models have not focused on the CMSs used for hybrid seed production, but their results and predictions are nonetheless useful to plant breeders. Evolutionary constraints should result in the selection of CMSs with characteristics highly suitable for practical use. In particular, natural CMSs should not be associated with deleterious drawbacks for vegetative growth or seed production. However, a male sterility cytoplasm that has evolved over a significant period of time in a species would be expected to have selected a number of restorer genes. Maintainer genotypes should therefore be sought in unrelated populations, if possible. This is entirely consistent with the observation that maternally inherited male sterility arises in crosses between genetically distant genotypes.

Conclusions drawn from theoretical models also suggest that male sterility due to poor co-operation between the nuclear and mitochondrial genomes, as suggested for alloplasmic CMSs, is very unlikely to be selected in natural conditions. This implies that occasional alloplasmic situations, arising from spontaneous interspecific hybridisation, as suggested by Lewis (1941), are unlikely to be at the origin of natural CMSs.

7.8 Molecular studies on CMSs: a step towards understanding or confusion?

7.8.1 The Texas story

Molecular studies on CMSs were initially driven by the problems encountered by maize breeders with the Texas cytoplasm in maize. Wise and colleagues have described the fascinating story of T-CMS in maize in their excellent and complete review (Wise *et al.*, 1999). The Texas cytoplasm was extensively used for hybrid seed production in maize until the 1970s. In 1970, a severe epidemic of southern corn leaf blight, caused by race T of *Cochliobolus heterostrophus*, was responsible for very heavy losses in maize production, especially in the USA. At that time, 85% of maize hybrids were produced with the Texas cytoplasm. Race T of the leaf blight fungus was first detected in the Philippines, and a link between the Texas cytoplasm and high susceptibility to this disease had already been established. The Texas cytoplasm also confers high susceptibility to *Mycosphaerella zeae-maydis*, which causes yellow leaf blight, and to the insecticide S-methyl-N-[(methyl-carbamoyl)oxy]thioacetimididate (methomyl). The 1970 epidemic had two consequences: (i) in less than two years, the Texas cytoplasm was almost completely replaced by tassel removal for hybrid production and (ii) a number of molecular and physiological studies were undertaken with the aim of identifying the susceptibility determinant and dissociating it from male sterility. It soon became obvious that a single gene was responsible for male sterility and fungal toxin sensitivity. All plants obtained from mutagenic treatment or *in vitro* culture of Texas CMS plants appeared to recover male fertility as soon as they became toxin resistant. The gene concerned, T-*urf13*, is a mitochondrial gene found only in the Texas cytoplasm, and seems to have been generated by mitochondrial genome recombination. It is the first chimeric mitochondrial gene to be described (Dewey *et al.*, 1986) and encodes a 13-kDa protein, T-URF13, constitutively produced in plants carrying the Texas cytoplasm and first detected by means of *in organello* synthesis experiments (Forde *et al.*, 1978). T-URF13 is associated with the inner mitochondrial membrane. Paradoxically, the toxin sensitivity conferred by T-URF13 has been extremely useful for biochemical studies of the protein and its possible function. Sensitivity to toxin and methomyl has been transferred, by ectopic expression of the T-*urf13* gene, into heterologous systems: the bacterium *Escherichia coli* (Dewey *et al.*, 1988), the yeast *Saccharomyces cerevisiae* (Huang *et al.*, 1990; Glab *et al.*, 1990), insect cells (Korth and Levings, 1993) and tobacco plants (Chaumont *et al.*, 1995). The mechanism of T-URF13-associated toxin sensitivity was elucidated by studying the structure of the protein and its interaction with toxin molecules in *E. coli* membranes (Korth *et al.*, 1991; Levings and Siedow, 1992; Rhoads *et al.*, 1995; Siedow *et al.*, 1995). T-URF13 tetramers are assembled in the membrane, with each monomer anchored by three *trans*-membrane domains. The resulting structure constitutes a pore that opens when specific fungal toxins or methomyl bind to exposed residues. The opening of the pore leads to electrolyte leakage and the uncoupling of oxidative phosphorylation followed by mitochondrial swelling and rupture, and resulting in the death in *E. coli*.

The T-URF13 protein of Texas maize is not only the first CMS-inducing protein to be identified, it also remains the best-studied protein of this type in terms of biochemistry. This is undoubtedly because this protein confers susceptibility to a toxin, a property unique among all CMS systems studied to date. Is the Texas CMS an archetype for other CMSs — natural CMSs in particular — and what features of this fascinating model are shared by other CMSs?

Maize breeders also observed that hybrids carrying the Texas cytoplasm were slightly less vigorous than their isogenic equivalents with a normal cytoplasm (Duvick, 1965). As the T-URF13 protein is constitutively produced in Texas maize mitochondria, it seems likely that the presence of T-URF13 in vegetative tissues has a slight, but significant, deleterious effect on vigour, even in the absence of specific pathogens. However, male sterile plants are reported to yield more grain than male fertile plants, indicating a real female advantage (Wise *et al.*, 1999). Restorers have been detected in South American varieties (Edwardson, 1955), and this is consistent with a natural origin for Texas CMS, with restorers selected by the presence of the sterility-inducing cytoplasm in an area in which the species is diversifying.

7.8.2 Identification of CMS genes

After Texas CMS, a number of other systems were investigated to identify the cytoplasmic factor causing sterility. Research efforts focused on the mitochondrial genome, which appeared from very early studies to be the best candidate. Indeed, all but one of the CMS determinants studied to date are linked to the mitochondrial genome.

The unique exception is ‘447’ CMS in *Vicia faba*, in which male sterility is caused by a cytoplasmic double-stranded RNA molecule (Lefebvre *et al.*, 1990). Little work has been carried out on this system since the identification of the CMS-inducing viroid. Nevertheless, this case of ‘pathogenic’ CMS is interesting because it is similar to a situation described in many animals, especially arthropods, in which intracytoplasmic bacteria of the genus *Wolbachia* (closely related to the presumed bacterial ancestor of mitochondria) manipulate the sex of their host (Knight, 2001; Zimmer, 2001). Selection models relating to the manipulation of sex or sex ratios by parasites are very similar to those for CMSs.

The principle of genetic conflict with the nuclear genome applies equally to all maternally inherited genomes. However, no plastid sequence has yet been found to be associated with male sterility. A deletion variant of the *rpoC2* plastid gene, encoding RNA polymerase subunit β'', has been associated with a small anther phenotype in some CMSs in sorghum (Chen *et al.*, 1993, 1995). However, no link has been established between sterility phenotype and plastid variants.

A number of mitochondrial genes have been suggested or demonstrated to be responsible for CMS in some species. With two exceptions – the *orf224* of *pol* CMS related to the *orf222* of *nap* CMS in *B. napus*, and the *orf107* of A3 CMS in sorghum homologous to the *orf79* of ‘Boro’ CMS in rice – they appear to be completely unrelated but share some remarkable features. They all result from recombination events in the mitochondrial genome, and are co-transcribed with normal mitochondrial genes (Schnable and Wise, 1998; Budar and Pelletier, 2001; Hanson and Bentolila, 2004). Plant mitochondrial genomes have a high level of recombination activity, as illustrated by their behaviour following protoplast fusion. Recombination also seems to be one of the mechanisms by which plant mitochondrial genomes evolve (Palmer and Herbon, 1988; Small *et al.*, 1989). Plant mitochondrial genomes are also the largest and most variable among eukaryotes, although they carry only slightly more genes than those of other taxa (Knoop, 2004; Gray *et al.*, 1999). Indeed, the large size of plant mitochondrial genomes may be a direct consequence of their high levels of recombination activity (Small *et al.*, 1989; Fauron *et al.*, 1995). The high variability of mitochondrial genomes in gynodioecious species is also a direct result of this recombination activity (Palmer and Herbon, 1988; Albert *et al.*, 1998). These observations provide insight into the origin of sterility genes: they are products of the recombination activity of plant mitochondrial genomes. New open reading frames, resulting from recombination events, have been detected in the sequenced plant mitochondrial genomes (Marienfeld *et al.*, 1997; Kubo *et al.*, 2000; Notsu *et al.*, 2002; Handa, 2003; Clifton *et al.*, 2004). These data also provide an explanation for the lack of homology between male sterility genes: they result from independent events in different species. However, not all open reading frames generated by recombination in a plant mitochondrial genome are likely to cause male sterility. The conditions under which a new mitochondrial gene is likely to cause male sterility are unclear and are unlikely to be elucidated until the mechanisms underlying male sterility are resolved.

Most identified male sterility genes display co-transcription with essential mitochondrial genes. Indeed, the corresponding proteins are generated by translation from the resulting bicistronic messenger RNAs (mRNA). Even if monocistronic mRNAs are observed for the partner gene, this is not the case for the gene associated with sterility. Given their origins, it is not surprising that male sterility genes use promoters and 3′ stabilising sequences from essential mitochondrial genes for their expression. However, co-expression does not at first glance appear necessary, as recombination can duplicate sequences, and therefore expression signals. Studies on *Brassica* cybrids carrying Ogura male sterility shed light on this feature. Most cybrids carried the original radish male sterility locus, in which the male sterility gene *orf138* is co-expressed with the *orfB* (*atp8*) gene. However, some cybrids carried a recombined male sterility locus in which *orf138* was expressed as a monocistronic transcript, or carried a second copy of the *orfB* gene, unlinked to *orf138*. In both these types of cybrid, male sterility was often frequently reversed by recombination. This process deleted the male sterility gene or modified its genetic environment, leading to the deletion of 3′ stabilising sequences (Bonhomme *et al.*, 1991, 1992; Grelon *et al.*, 1994; Bellaoui *et al.*, 1997, 1998). The ‘permanent restorer’ of Sprite CMS in bean (*Phaseolus vulgaris*) provides another example. Elimination of the *pvs-orf239* gene by recombination occurs in some genetic backgrounds, which have been named ‘permanent restorers’(Mackenzie *et al.*, 1988; Mackenzie and Chase, 1990; Janska and Mackenzie, 1993; He *et al.*, 1995). Interestingly, male sterile plants of *Phaseolus* carry two copies of the *atpA* gene, one of which is linked to and co-transcribed with pvs-orf239 (Janska and Mackenzie, 1993; Chase, 1994). In fertile revertants, the large co-transcript carrying both *atpA* and *pvs-orf239* is lost, but the normal monocistronic *atpA* mRNA remains unchanged

(Chase, 1994). Co-expression with an essential mitochondrial gene seems to be required for male sterility genes to avoid elimination or silencing by the same phenomenon that creates them. They are thus protected by selection for the expression of their essential partner.

These two features – a chimaeric gene resulting from recombination and co-transcription with an essential mitochondrial gene – are so widely observed that the search for male sterility genes often involves systematically searching for new open reading frames expressed on extended mRNAs for essential mitochondrial genes (Singh and Brown, 1991; Landgren *et al.*, 1996; de la Canal *et al.*, 2001; Engelke and Tatlioglu, 2002; Dieterich *et al.*, 2003).

The most convincing data demonstrating the involvement of a mitochondrial gene in male sterility come from the identification of fertile revertants, generated by modifications to the mitochondrial genome. These modifications may be spontaneous, induced by *in vitro* culture or even provoked by nuclear genes. This is the case for T-*urf13* in Texas CMS in maize (Gengenbach *et al.*, 1981; Kemble *et al.*, 1982; Umbeck and Gengenbach, 1983; Fauron *et al.*, 1987), for *pvs-orf239* in 'Sprite' CMS in bean (Mackenzie *et al.*, 1988; Mackenzie and Chase, 1990; Janska and Mackenzie, 1993), and for *orf138* in Ogura CMS in *Brassica* (Bonhomme *et al.*, 1991, 1992). In the case of CMS-S in maize, spontaneous reversion has been associated with the integration of linear mitochondrial episomes into the main genome via the recombination of terminal repeated sequences (Pring *et al.*, 1977; Levings *et al.*, 1980; Laughnan and Gabay-Laughnan, 1983). However, in the vast majority of cases, no fertile revertant has been identified, and the causal link between a mitochondrial gene and male sterility is difficult to establish. In such cases, the evidence implicating a mitochondrial gene in male sterility is derived primarily from differential expression in maintainer and restorer genetic backgrounds. This is the case for *Pcf* in the CMS of *Petunia* (Nivison and Hanson, 1989; Rasmussen and Hanson, 1989; Pruitt and Hanson, 1991), *orfH522* in the *pet1* CMS of sunflower (Laver *et al.*, 1991; Horn *et al.*, 1991; Kohler *et al.*, 1991; Smart *et al.*, 1994; Monéger *et al.*, 1994), *orf193* in the '*tournefortii*-Stiewe' CMS of oilseed rape (Dieterich *et al.*, 2003), *orf263* in the *tour* CMS of oilseed rape (Landgren *et al.*, 1996), *orf107* in the A3 CMS of sorghum (Tang *et al.*, 1996), *urf-rmc* in the 'Boro' CMS of rice (Kadowaki *et al.*, 1990; Akagi *et al.*, 1994), *orf224* in the *pol* CMS of oilseed rape (Menassa *et al.*, 1999; Yuan *et al.*, 2003), *orf355-orf77* in the CMS-S of maize (Zabala *et al.*, 1997), and the new open reading frame detected in CMS1 of chives (Engelke and Tatlioglu, 2004).

Comparison of the expression patterns of the original open reading frames between genetic backgrounds differing only in alleles at the restorer locus provides convincing evidence implicating the mitochondrial gene concerned in male sterility. However, in alloplasmic situations in which chimeric genes are thought to cause male sterility, comparisons have often been made between the nuclear genotypes of the different species. As mitochondrial gene expression is tightly controlled by nuclear genetic background, the inference of a causal link between male sterility and expression of a chimaeric open reading frame not expressed in the cytoplasm donor parent remains speculative in the absence of any other evidence. Nevertheless, theoretical models suggest that cryptic male sterility genes may be present in the mitochondrial genome of a species, if the corresponding restorer alleles have been completely fixed. In such cases, the introduction of a new genetic background, devoid of restorer alleles, may bring the cryptic male sterility to light by allowing expression of the chimeric mitochondrial gene. This may be the case for *orf263* in *tour* CMS of *Brassica* (Landgren *et al.*, 1996) and for *orf274* in alloplasmic tobacco carrying *repanda* cytoplasm (Bergman *et al.*, 2000).

Alloplasmic situations therefore combine two possible sources of male sterility: revelation of male sterility that is cryptic in the cytoplasm donor species and a lack of co-operation between the nuclear and mitochondrial genomes for flower differentiation and reproductive functions. The revelation of cryptic CMS may provide a source of CMS that can be used in hybrid seed production whereas nuclear–mitochondrial incompatibility is likely to be associated with undesirable traits. Although attractive, and consistent with data obtained relating to mitochondrial recombination after protoplast fusion (Primard *et al.*, 1988), this formal difference between 'natural CMSs' and 'alloplasmic CMSs' remains speculative. It could be formally tested if the physiological mechanisms underlying these two types of CMS were known and could be compared.

7.8.3 The mechanisms underlying male sterility

Investigation of the molecular and physiological mechanisms underlying CMS involves the analysis of apparently contradictory data. 'Nuclear–mitochondrial incompatibility' is thought to lead to male sterility in most cases, as is often observed in alloplasmic situations. In such situations, male sterility is often accompanied by homeotic floral modifications, which are also attributed to the poor interaction between the nucleus and the cytoplasm. However, in natural gynodioecious populations, CMS is likely to occur only if it increases female reproductive success, even only slightly. Indeed in some widely used CMSs, male-sterile female plants appear to be more vigorous than their male-fertile counterparts. However, this is not the case for all CMSs (e.g. Texas CMS in maize). In addition, the male sterility-inducing genes identified to date have been found to be expressed throughout the plant, whereas the observed phenotype is restricted to the microgametogenesis developmental programme. This apparent paradox is shared by all studied CMS systems for which data are available, except for the 'Sprite' CMS in bean, in which a specific protease degrades the male sterility protein in all tissues of the plants except anthers (Abad *et al.*, 1995; Sarria *et al.*, 1998). Finally, it remains unclear whether a common mechanism is shared by most, if not all, CMSs or whether each CMS involves a specific mechanism. The absence of homology between unrelated male sterility genes suggests that specific mechanisms are involved. However, it is difficult to imagine how a multitude of different mechanisms could lead to microspore abortion without impairing other functions given that the mitochondrial protein concerned is produced constitutively. Indeed, it is hard to come up with even one such mechanism!

Very few male sterility gene products have been studied to date, and none has been studied in as much detail as T-URF13. However, a few common characteristics can be tentatively identified based on the available data. These proteins contain at least one hydrophobic domain, and some have been shown to be associated with mitochondrial membranes to varying degrees. This is the case for the ORF138 protein of Ogura CMS in *Brassica* (Grelon *et al.*, 1994) and radish (Krishnasamy and Makaroff, 1994), and the PCF protein of CMS in petunia (Nivison and Hanson, 1989). A remarkable exception is again provided by the 'Sprite' CMS gene product in bean, which has been detected outside mitochondria, and even outside cells, in the callose layer and primary cell wall of developing microspores (Abad *et al.*, 1995).

Several possible mechanisms of male sterility have been proposed. The first was formulated by Flavell based on the initial results obtained for T-URF13 (Flavell, 1974). An unknown 'X factor', present only in the anthers, may interact with T-URF13, as do the specific toxins. This interaction provokes the same cascade of events in the cells concerned as does interaction with toxins, resulting in microspore abortion. This hypothesis elegantly solves the apparent paradox of the specificity of the phenotype: the deleterious effect of the sterility protein is exerted only in the presence of a tissue-specific partner. Although this hypothesis was formulated specifically for the Texas CMS of maize and by analogy with the mechanism of disease susceptibility, a similar hypothesis could be put forward for any system in which the male sterility gene is constitutively expressed. However, although this hypothesis was formulated 30 years ago, no 'X factor' has yet been isolated from the anthers of maize or any other species.

Alternatively, the male sterility gene product may interfere with mitochondrial physiology, decreasing the efficiency of respiration or ATP production. This deficiency would prevent the anther from reaching an energy supply threshold, resulting in the inhibition of microspore development. This second hypothesis was also proposed for Texas CMS (Levings, 1993). It is applicable both to alloplasmic situations and to 'natural CMSs'. In alloplasmic situations, a lack of co evolution of the nuclear-encoded and mitochondrially encoded subunits of respiratory complexes would account for the defects in mitochondrial function. In cases of natural CMS, male sterility genes would be expected to interfere with the normal subunits of the respiratory complexes, either by directly interacting with complexes or affecting the normal expression of essential genes they are linked to (Akagi *et al.*, 1994; Sabar *et al.*, 2003; Hanson and Bentolila, 2004). If this was the case, male sterility genes could be considered as 'dominant deleterious mutants' of the normal mitochondrial genes, as suggested by Sabar (Sabar *et al.*, 2003) for *pet1* CMS in sunflower and by Dieterich (Dieterich *et al.*, 2003) for '*tour*' CMS in *Brassica*. Such a dominant mutant was produced in the construction of male sterile tobacco plants producing a modified form of the ATP9 subunit targeted to mitochondria (Hernould *et al.*, 1993). To our knowledge, this is the only example of a dominant mutation in a respiratory complex subunit that has been shown to induce male sterility. However, this is an artificial situation and some transgenic plants displayed only partial male sterility (Hernould *et al.*, 1993). The male sterility displayed by complex I deletion mutants of *N. sylvestris* is conditional and some viable pollen may be produced under strong illumination (Budar *et al.*, 2003). Indeed, in some CMSs, male sterility has been described as dependent on environmental factors, temperature and photoperiod in particular (Kaul, 1988). Some alloplasmic situations also lead to a sort of incomplete or condition-dependent male sterility (McCollum, 1981). In these cases, inefficient respiration in male reproductive tissues may result in impaired pollen production. Some recent reports have provided physiological data consistent with this hypothesis. In alloplasmic tobacco carrying the *N. repanda* cytoplasm, the ATP/ADP ratio is low, suggesting a deficiency in mitochondrial function (Bergman *et al.*, 2000). In *pet1* male sterile sunflower, a decrease in ATP synthase activity is observed in in-gel assays on mitochondrial protein extracts (Sabar *et al.*, 2003). Similarly, mitochondrial protein extracts of beet plants with *G*CMS have low levels of cytochrome c oxidase activity (Ducos *et al.*, 2001). In these examples, the impaired activity measured *in vitro* suggests a deficiency *in vivo* that is responsible for the CMS phenotype. It is currently unclear why male sterile plants display normal vegetative growth and female reproductive function, and, in the case of *G* CMS in beet, why such plants are successfully selected in natural populations.

High levels of alternative oxidase (AOX) activity have been reported in male sterile plants for *G* CMS in beet (Ducos *et al.*, 2001), maize CMSs (Karpova *et al.*, 2002), *Petunia* CMS (Hanson and Bentolila, 2004), Ogura CMS in *Brassica* (Y. Duroc, personal communication) and CMS mutants of *N. sylvestris* (Sabar *et al.*, 2000). In *N. sylvestris* mutants, increased AOX activity has been linked to redox homeostasis in the leaf (Dutilleul *et al.*, 2003). However, the significance of AOX enhanced activity in natural CMSs remains unclear.

It has been suggested that programmed cell death plays a role in CMS (Balk and Leaver, 2001). As tapetal cells enter a cell death programme at a precise stage of normal anther development, the premature triggering of this mechanism might account for male sterility, at least in CMSs in which the tapetum is the first tissue affected in male sterile plants. This is the case in a number of CMSs, including Texas CMS (Wise *et al.*, 1999) and Ogura CMS (Gourret *et al.*, 1992). Moreover, mitochondria are known to play a central role in programmed cell death (Hoeberichts and Woltering, 2003; Yao *et al.*, 2004). Even if future experimental data provide further evidence for the involvement of a cell death programme in the final steps of the sterility mechanism, how and why such a programme is triggered in plants with CMS, remains unclear.

Finding the answers to these questions will necessarily involve thorough biochemical analyses of sterility-inducing proteins, physiological studies of the mitochondria producing them and the use of newly obtained cytological markers to improve characterisation of the cascade of events preceding microspore abortion.

7.8.4 Nuclear restorers and mechanisms of restoration to fertility

In early molecular studies on CMS, it was thought that the identification of restorers would help us to understand the mechanism of male sterility. However, as more mitochondrial genes for male sterility were identified, it became clear that nuclear restorers acted by impairing expression of the male sterility gene. Efforts to identify male sterility genes have even made use of this fact (see above § 7.8.2). In most cases, nuclear restorers impair the expression of mitochondrial male sterility genes in a post-transcriptional manner (Budar *et al.*, 2003). The effects of restorers can provide valuable information, even if the genes involved are not formally identified. This is the case when nuclear restorers affect expression of the sterility gene only in certain cells or tissues. In particular, in all CMSs in which restoration is gametophytic, the physiological events leading to male sterility are likely to occur after meiosis in developing microspores. This is the case for A3 CMS in sorghum (Tang *et al.*, 1999), CMS-HL and CMS BoroII in rice (Shinjyo, 1969; Liu *et al.*, 2004) and CMS-S in maize (Buchert, 1961; Wen and Chase, 1999b). In plants heterozygous for the restorer allele, only pollen carrying this allele is produced. The maintainer allele at this locus is therefore never transmitted via the pollen, even in restored plants. This will probably modify both evolution scheme and equilibrium as has been proposed for sporophytic restoration, as the increase of the frequency of the gametophytic restorer allele, while being expected to be slower, will tend to fixation more easily in populations.

The case of CMS-S in maize is particularly fascinating. Natural restorer alleles have been identified in Mexican maize and teosinte, indicating that this system developed early in the evolution of these species (Gabay-Laughnan *et al.*, 2004). Some of these alleles are probably allelic to the *Rf3* restorer, which has been shown to affect the accumulation of CMS-associated mitochondrial transcripts (Wen and Chase, 1999a). Some spontaneous 'restorer mutants' have also been isolated in the field. These mutants have been named *rfl* because they are lethal if homozygous (Laughnan and Gabay, 1973; Wen *et al.*, 2003). A recent study of one of these restorers revealed that the mutation decreased levels of the CMS-associated *orf355-orf77* mitochondrial transcript and of the *atpA* transcript. Pollen grains carrying the *rfl* allele are rescued from male sterility by the decrease in levels of the *orf355-orf77* transcript, and appear to tolerate the low levels of *atpA* transcript. However, the impairment of *atpA* gene expression and accumulation of the ATPA protein is lethal in plants homozygous for *rfl* (Wen *et al.*, 2003). These results strongly suggest that the haploid pollen may have more flexible requirements for mitochondrial essential proteins than sporophytic tissues.

To our knowledge, only two restorer genes have no effect on expression of the mitochondrial sterility gene. One is the 'permanent' restorer gene of bean, the action of which leads to the loss of the sterility gene (Mackenzie *et al.*, 1988; Mackenzie and Chase, 1990; Janska and Mackenzie, 1993; He *et al.*, 1995). The second is *Rf2*, one of the two genes necessary for restoration of male fertility in Texas cytoplasm carrying plants. Unlike *Rf1*, it has no effect on *T-urf13* transcript processing or accumulation. *Rf2* was the first restoration gene to be cloned and encodes a mitochondrial aldehyde dehydrogenase (Cui *et al.*, 1996). Normal mitochondrial aldehyde dehydrogenase activity is necessary for male fertility in all cytoplasm types (Liu *et al.*, 2001). This observation led to some discussion about its specificity as a restorer for

T-CMS (Touzet, 2002; Schnable, 2002). Plants carrying the Texas cytoplasm seem to be much more susceptible to the impairment of mitochondrial dehydrogenase activity than plants with other types of cytoplasm. This finding may provide some insight into the biochemical mechanism underlying sterility (Moller, 2001).

All other restorers for which experimental data are available seem to have an effect on accumulation of the mitochondrial sterility gene product, after transcription. It should be kept in mind that most recent studies have used this observation as indicating the involvement of a given mitochondrial gene in male sterility. This may have biased our vision, exaggerating the number of examples following this rule. Nevertheless, the recent identification of certain *Rf* genes as encoding pentatricopeptide repeat (PPR) proteins in Petunia (Bentolila *et al.*, 2002), in Ogura and kosena radish (Brown *et al.*, 2003; Desloire *et al.*, 2003; Koizuka *et al.*, 2003) and in Boro rice (Kazama and Toriyama, 2003; Akagi *et al.*, 2004; Komori *et al.*, 2004) has provided insight into the genomic conflict underlying CMSs (Touzet and Budar, 2004). As the mitochondrial genomes of plants use recombination to create new genes, some of which may lead to male sterility, the source of variability of the nuclear component seems to lie in the impressive expansion of the PPR protein family in plants (Small and Peeters, 2000; Lurin *et al.*, 2004). PPR proteins are thought to be specific for particular organelle transcripts, directly or indirectly modulating their processing, accumulation and/or translation (Lurin *et al.*, 2004). Restorer alleles encoding PPR proteins may be variants that have changed target. In this regards, the *pol* and *nap* CMSs in rapeseed are an interesting case. Their specific restorers, *Rfp* and *Rfn* are allelic and also co-localise with a modifier of mitochondrial transcript locus (*Mmt*) (Li *et al.*, 1998; Brown, 1999). This could reflect the allelic variation of a single derivated PPR coding gene, or the existence of a complex restorer locus composed of diverged PPR genes, each of them being specifically ‘resistant’ to a given CMS (Touzet and Budar, 2004). Nevertheless, the co-localisation of these nuclear restorers and the observed homology between the mitochondrial sterility genes strongly suggest a common origin for these two CMSs of oilseed rape. To our knowledge, this is the only example of such a situation, and its study should provide interesting information concerning the evolution of CMSs. In CMS-T in maize, several restorers have been described that affect *T-urf13* transcript processing and can at least partially substitute for *Rf1*, but are not allelic with *Rf1* (Dill *et al.*, 1997). This suggests that different PPR loci may be able to produce new alleles affecting the processing of a given mitochondrial transcript. It also suggests that these restorers appeared independently during evolution. If we could understand the molecular features underlying the specificity of a PPR protein for a given mitochondrial gene product, and the target cells of the sterility phenotype, we would be able to design restorer genes.

7.9 Conclusion and prospects

CMS is not simply a peculiarity of plant reproduction. It reflects the genomic conflict between nucleus and cytoplasm. This conflict seems to be possible because the mitochondrial genome, although highly dependent on the nucleus for its maintenance and expression, has retained the possibility of creating new genes by recombination. This is a remarkable feature specific to plant mitochondrial genomes and it is currently completely unknown why these genomes evolved so differently from those in other groups from the same common ancestor that gave rise to all mitochondria in eukaryotes (Gray *et al.*, 1999; Andersson *et al.*, 2003). Moreover, the nuclear genomes of plants seem to use a specific type of protein, PPR proteins, to 'correct' male sterility induced by the cytoplasm and these proteins seem to have evolved by a process similar to that for the products of resistance genes (Touzet and Budar, 2004). Evolutionary biologists might even consider that the specific features of the co-evolution of plant mitochondrial and nuclear genomes merit closer study.

It remains unclear how some mitochondrial genes, created by recombination, impair pollen production without affecting vegetative and female function, and how populations of plants with male sterility manage to spread. There is increasing evidence to suggest that mitochondrial biogenesis and function display a certain level of specificity in male reproductive organs. Our knowledge of the possibly specific role of mitochondria in pollen development remains extremely limited and further studies are required (Berthomé *et al.*, 2003; Hanson and Bentolila, 2004).

Conflicting data are available concerning possible mechanisms of male sterility in CMSs. Even in the best-known case CMS, Texas CMS, the mechanism responsible is not really understood. Many experimental data suggest that there are deficiencies in oxidative phosphorylation and energy supply in male sterile plants but no real causal link between biochemical observations and male sterility has been established, except in transgenic plants in which the impairment of mitochondrial function led to male sterility (Hernould *et al.*, 1993; Heiser *et al.*, 1997; Kitashiba *et al.*, 1999; Yui *et al.*, 2003). However, total pollen abortion and normal female reproductive success are rarely observed if essential mitochondrial function is impaired. The 'paradox of CMS', namely the specific effect on pollen production of the constitutive expression of a mitochondrial protein, remains unsolved. However, tapetal cells seem to be the most frequent targets of male sterility mechanisms, probably due to the dramatic increase in mitochondrial activity in those cells during microgametogenesis. The fate of these cells, which undergo programmed cell death before the release of mature pollen, has also received some attention as the timing of this event is of crucial importance for pollen maturation and survival. We nevertheless require more precise and reliable information concerning the specific function of mitochondria in the tapetal cells, and the molecular and cytological events triggering the cascade ending with pollen abortion.

Gametophytic CMSs seem to be less frequent than sporophytic CMSs. It is possible that the molecular mechanisms provoking pollen abortion are harder to trigger in the gametophyte than in the tapetal cells. This remains to be shown, and requires an understanding of both gametophytic and sporophytic CMS mechanisms. In addition, gametophytic CMSs may be difficult to observe because in this case restorer alleles are easily fixed.

Until we manage to construct a comprehensive model of CMS molecular mechanisms, our search for systems suitable for hybrid seed production will remain somewhat speculative. Nevertheless, we have already acquired valuable knowledge that could help to direct our efforts. Studies of the assembly of new nuclear and mitochondrial genomes will undoubtedly remain the best way of discovering new CMSs. The most promising CMSs will be 'cryptic', silent in their original background but brought to light by the absence of restorers in the new genetic background. Some of these CMSs may have all the qualities required for a good hybrid seed production system, the induction of pollen abortion with no effect on vegetative growth and seed set in particular. The tremendous variability of plant mitochondrial genomes is encouraging and suggests that many more types of CMSs are probably awaiting discovery.

Acknowledgements

The precious help from the librarians of the Centre de Versailles (INRA) is gratefully acknowledged.

References

Abad, A. R., Mehrtens, B. J. and Mackenzie, S. A. (1995) Specific expression in reproductive tissues and fate of a mitochondrial sterility-associated protein in cytoplasmic male-sterile bean, *The Plant Cell*, **7**, 271–285.

Akagi, H., Nakamura, A., Yokozeki-Misono, Y., Inagaki, H., Mori, K. and Fujimura, T. (2004) Positional cloning of the rice Rf-1 gene, a restorer of BT-type cytoplasmic male sterility that encodes a mitochondria-targeting PPR protein, *Theoretical and Applied Genetics*, **108**, 1449–1457.

Akagi, H., Sakamoto, M., Shinjyo, C., Shimada, H. and Fujimura, T. (1994) A unique sequence located downstream from the rice mitochondrial atp6 may cause male sterility, *Current Genetics*, **25**, 52–58.

Albert, B., Godelle, B. and Gouyon, P. (1998) Evolution of the plant mitochondrial genome: dynamics of duplication and deletion of sequences, *Journal of Molecular Evolution*, **46**, 155–158.

Andersson, S. G., Karlberg, O., Canback, B. and Kurland, C. G. (2003) On the origin of mitochondria: a genomics perspective, *Philosophical Transactions of the Royal Society of London, series B, Biological Sciences*, **358**, 165–177; discussion 177–179.

Arrieta-Montiel, M., Lyznik, A., Woloszynska, M., Janska, H., Tohme, J. and Mackenzie, S. (2001) Tracing evolutionary and developmental implications of mitochondrial stoichiometric shifting in the common bean, *Genetics*, **158**, 851–864.

Balk, J. and Leaver, C. J. (2001) The PET1-CMS mitochondrial mutation in sunflower is associated with premature programmed cell death and cytochrome c release, *The Plant Cell*, **13**, 1803–1818.

Bannerot, H., Boulidard, L., Cauderon, Y. and Tempé, J. (1974) Transfer of cytoplasmic male sterility from *Raphanus sativus* to *Brassica oleracea*, *Proceedings Eucarpia Crop Section, Cruciferae*, **25**, 52–54.

Bannerot, H., Boulidard, L. and Chupeau, Y. (1977) Unexpected difficulties met with the radish cytoplasm in *Brassica oleracea*, *Eucarpia Cruciferae Newsletter*, **2**, 16.

Bateson, W. and Gairdner, A. E. (1921) Male-sterility in flax, subject of two types of segregation, *Journal of Genetics*, **11**, 269–275.

Baur, E. (1909) Das Wesen und die Erblichkeitsverhältnisse der Varietates albomarginatae hort von Pelargonium zonale, *Z. Vererbungslehre*, **1**, 330–351.

Beckett, J. B. (1971) Classification of male-sterile cytoplasms in maize (*Zea mays* L.), *Crop Science*, **11**, 724–727.

Belhassen, E., Atlan, A., Couvet, D., Gouyon, P.-H. and Quétier, F. (1993) Mitochondrial genome of *Thymus vulgaris* L. (Labiate) is highly polymorphic between and among natural populations, *Heredity*, **71**, 462–472.

Bellaoui, M., Martin-Canadell, A., Pelletier, G. and Budar, F. (1998) Low-copy-number molecules are produced by recombination, actively maintained and can be amplified in the mitochondrial genome of Brassicaceae: relationship to reversion of the male sterile phenotype in some cybrids, *Molecular & General Genetics*, **257**, 177–185.

Bellaoui, M., Pelletier, G. and Budar, F. (1997) The steady-state level of mRNA from the Ogura cytoplasmic male sterility locus in Brassica cybrids is determined post-transcriptionally by its 3′ region, *EMBO Journal*, **16**, 5057–5068.

Belliard, G., Vedel, F. and Pelletier, G. (1979) Mitochondrial recombination in cytoplasmic hybrids of *Nicotiana tabacum* by protoplast fusion, *Nature*, **281**, 401–403.

Bentolila, S., Alfonso, A. A. and Hanson, M. R. (2002) A pentatricopeptide repeat-containing gene restores fertility to cytoplasmic male-sterile plants, *Proceedings of the National Academy of Sciences of the United States of America*, **99**, 10887–10892.

Bereterbide, A., Hernould, M., Farbos, I., Glimelius, K. and Mouras, A. (2002) Restoration of stamen development and production of functional pollen in an alloplasmic CMS tobacco line by ectopic expression of the Arabidopsis thaliana SUPERMAN gene, *The Plant Journal*, **29**, 607–615.

Bergman, P., Edqvist, J., Farbos, I. and Glimelius, K. (2000) Male-sterile tobacco displays abnormal mitochondrial atp1 transcript accumulation and reduced floral ATP/ADP ratio, *Plant Molecular Biology*, **42**, 531–544.

Berthomé, R., Froger, N., Hiard, S., Balasse, H., Martin-Canadell, A. and Budar, F. (2003) The involvement of organelles in plant sexual reproduction: a post-genomic approach, *Acta Biologica Cracoviensia Series Botanica*, **45**, 119–124.

Bonhomme, S., Budar, F., Férault, M. and Pelletier, G. (1991) A 2.5 kb *Nco*I fragment of Ogura radish mitochondrial DNA is correlated with cytoplasmic male-sterility in *Brassica* cybrids, *Current Genetics*, **19**, 121–127.

Bonhomme, S., Budar, F., Lancelin, D., Small, I., Defrance, M.-C. and Pelletier, G. (1992) Sequence and transcript analysis of the *Nco2.5* Ogura-specific fragment correlated with cytoplasmic male sterility in *Brassica* cybrids, *Molecular & General Genetics*, **235**, 340–348.

Bonnet, A. (1975) Introduction et utilisation d'une stérilité mâle cytoplasmique dans des variétés précoces européennes de radis *Raphanus sativus* L., *Annales de l'Amélioration des Plantes*, **25**, 381–397.

Bonnett, H. T., Kofer, W., Hakansson, G. and Glimelius, K. (1991) Mitochondrial involvement in petal and stamen development studied by sexual and somatic hybridization of *Nicotiana* species, *Plant Science*, **80**, 119–130.

Boutin, V., Jean, R., Valero, M. and Vernet, P. (1988) Gynodioecy in *Beta maritima*, *Oecologia Plantarum*, **9**, 61–66.

Brown, G. G. (1999) Unique aspects of cytoplasmic male sterility and fertility restoration in *Brassica napus*, *Journal of Heredity*, **90**, 351–356.

Brown, G. G., Formanova, N., Jin, H. *et al.* (2003) The radish Rfo restorer gene of Ogura cytoplasmic male sterility encodes a protein with multiple pentatricopeptide repeats, *The Plant Journal*, **35**, 262–272.

Buchert, J. G. (1961) The stage of the genome–plasmon interaction in the restoration of fertility to cytoplasmically pollen-sterile maize, *Proceedings of the National Academy of Sciences of United States of America*, **47**, 1436–1440.

Budar, F. and Pelletier, G. (2001) Male sterility in plants: occurrence, determinism, significance and use, *Comptes rendus de l'Académie des sciences Série III*, **324**, 543–550.

Budar, F., Delourme, R. and Pelletier, G. (2004) Male sterility, in *Brassica*, Vol. 54 (eds E. C. Pua, and C. J. Douglas) Springer-Verlag, Berlin, pp. 43–65.

Budar, F., Touzet, P. and De Paepe, R. (2003) The nucleo-mitochondrial conflict in cytoplasmic male sterilities revisited, *Genetica*, **117**, 3–16.

Charlesworth, D. (1981) A further study of the problem of the maintenance of females in gynodioecious species, *Heredity*, **46**, 27–39.

Charlesworth, D. (2002) What maintains male-sterility factors in plant populations?, *Heredity*, **89**, 408–409.

Chase, C. D. (1994) Expression of CMS-unique and flanking mitochondrial DNA sequences in *Phaseolus vulgaris* L, *Current Genetics*, **25**, 245–251.

Chaumont, F., Bernier, B., Buxant, R., Williams, M. E., Levings, C. S., 3rd and Boutry, M. (1995) Targeting the maize T-urf13 product into tobacco mitochondria confers methomyl sensitivity to mitochondrial respiration, *Proceedings of the National Academy of Sciences of the United States of America*, **92**, 1167–1171.

Chen, Z., Muthukrishnan, S., Liang, G. H., Schertz, K. F. and Hart, G. E. (1993) A chloroplast deletion located in RNA polymerase gene *rpoC2* in CMS lines of sorghum, *Molecular & General Genetics*, **236**, 251–259.

Chen, Z., Schertz, K. F., Mullet, J. E., DuBell, A. and Hart, G. E. (1995) Characterization and expression of *rpoC2* in CMS and fertile lines of sorghum, *Plant Molecular Biology*, **28**, 799–809.

Chittenden, R. J. (1927) Cytoplasmic inheritance in flax, *Journal of Heredity*, **18**, 337–343.

Chittenden, R. J. and Pellew, C. (1927) A suggested interpretation of certain cases of anisogeny, *Nature*, **119**, 10–12.

Clayton, E. E. (1950) Male sterile tobacco, *Journal of Heredity*, **41**, 171–175.

Clifton, S. W., Minx, P., Fauron, C. M. *et al.* (2004) Sequence and comparative analysis of the maize NB mitochondrial genome, *Plant Physiology*, **136**, 3486–3503.

Correns, C. (1909) Vererbungsversuche mit blass (gelb) grünen und buntblättrigen Sippen bei Mirabilis, Urtica, und Lunaria, *Zeitschrift fiir Induktive Abstammungsund Vererbungslehre*, **1**, 291–329.

Couvet, D., Ronce, O. and Gliddon, C. (1998) Maintenance of nucleo-cytoplasmic polymorphism in a metapopulation: the case of gynodioecy, *American Naturalist*, **152**, 59–70.

Crouzillat, D., de la Canal, L., Perrault, A., Ledoigt, G., Vear, F. and Serieys, H. (1991) Cytoplasmic male sterility in sunflower: comparison of molecular biology and genetic studises, *Plant Molecular Biology*, **16**, 415–426.

Crouzillat, D., de la Canal, L., Vear, F., Serieys, H. and Ledoigt, G. (1994) Mitochondrial DNA RFLP and genetical studies of cytoplasmic male sterility in the sunflower (*Helianthus annuus*), *Current Genetics*, **26**, 146–152.

Cuguen, J., Wattier, R., Saumitou-laprade, P., Forcioli, D., Mörchen, M., Van-Dijk, H. and Vernet, P. (1994) Gynodioecy and mitochondrial DNA polymorphism in natural populations of *Beta vulgaris* ssp *maritima*, *Genetics Selection Evolution*, **26**, 87–101.

Cui, X., Wise, R. P. and Schnable, P. S. (1996) The *rf2* nuclear restorer gene of male-sterile T-cytoplasm maize, *Science*, **272**, 1334–1336.

Darwin, C. (1876) *The Effects of Cross and Self-Fertilisation in the Vegetable Kingdom*, John Murray, London.

Darwin, C. (1877) *The Different Forms of Flowers on Plants of the Same Species*, John Murray, London.

De Haan, A. A., Mateman, A. C., Van Dijk, P. J. and Van Damme, J. M. M. (1997) New CMS types in *Plantago lanceolata* and their relatedness, *Theoretical and Applied Genetics*, **94**, 539–548.

de la Canal, L., Crouzillat, D., Quetier, F. and Ledoigt, G. (2001) A transcriptional alteration in the atp9 gene is associated with a sunflower male-sterile cytoplasm., *Theoretical and Applied Genetics*, **102**, 1185–1189.

De Paepe, R., Chetrit, P., Vitart, V., Ambard-Bretteville, F., Prat, D. and Vedel, F. (1990) Several nuclear genes control both male sterility and mitochondrial protein synthesis in Nicotiana sylvestris protoclones, *Molecular & General Genetics*, **222**, 206–10.

de Vries, H. (1900a) Sur la loi de disjonction des hybrides, *Comptes rendus de l'Académie des Sciences de Paris*, **130**, 845–847.

de Vries, H. (1900b) Sur les unités des caractères spécifiques et leur application à l'étude des hybrides, *Revue générale de botanique*, **12**, 257–271.

Delannay, X., Gouyon, P. H. and Valdeyron, G. (1981) Mathematical study of the evolution of gynodioecy with cytoplasmic inheritance under the effect of a nuclear gene, *Genetics*, **99**, 169–181.

Delourme, R. and Budar, F. (1999) Male sterility, in *Biology of Brassica Coenospecies* (Ed., C. Gomez-Campo) Elsevier, Amsterdam, pp. 185–216.

Delourme, R., Eber, F. and Renard, M. (1991) Radish cytoplasmic male sterility in rapeseed: breeding restorer lines with a good female fertility, in *Proccedings of 8th International Rapeseed Conference*, Saskatoon, Saskatechewan, Canada, 1056.

Delourme, R., Eber, F. and Renard, M. (1995) Breeding double low restorer lines in radish cytoplasmic male sterility of rapeseed (*Brassica napus* L.), in *Proccedings of 9th International Rapeseed Conference*, Cambridge, UK, 6–8.

Desfeux, C., Maurice, S., Henry, J.-P., Lejeune, B. and Gouyon, P.-H. (1996) Evolution of reproductive systems in the genus *Silene*, *Proceedings of the Royal Society of London, series B-Biological Sciences*, **263**, 409–414.

Desloire, S., Gherbi, H., Laloui, W., Marhadour, S., Clouet, V., Cattolico, L., Falentin, C., Giancola, S., Renard, M., Budar, F., Small, I., Coboche, M., Delourme, R. and Bendahmane, A. (2003) Identification of the fertility restoration locus, Rfo, in radish, as a member of the pentatricopeptide-repeat protein family, *EMBO Reports*, **4**, 588–594.

Desplanque, B., Viard, F., Bernard, J., Forcioli, D., Saumitou-Laprade, Cuguen, J. and Van Dijk, H. (2000) The linkage disequilibrium between chloroplast DNA and mitochondrial DNA haplotypes in Beta vulgaris ssp. maritima (L.): the usefulness of both genomes for population genetic studies, *Molecular Ecology*, **9**, 141–154.

Dewey, R. E., Levings, C. S., 3rd and Timothy, D. H. (1986) Novel recombinations in the maize mitochondrial genome produce a unique transcriptional unit in the Texas male-sterile cytoplasm, *Cell*, **44**, 439–449.

Dewey, R. E., Siedow, J. N., Timothy, D. H. and Lewings, C. S. I. I. I. (1988) A 13-kilodalton maize mitochondrial protein in *E. coli* confers sensitivity to *Bipolaris maydis* toxin, *Science*, **239**, 293–294.

Dieterich, J. H., Braun, H. P. and Schmitz, U. K. (2003) Alloplasmic male sterility in *Brassica napus* (CMS 'Tournefortii-Sticwe') is associated with a special genc arrangement around a novel atp9 gene, *Molecular & General Genetics*, **269**, 723–731.

Dill, C. L., Wise, R. P. and Schnable, P. S. (1997) Rf8 and Rf* mediate unique T-urf13-transcript accumulation, revealing a conserved motif associated with RNA processing and restoration of pollen fertility in T-cytoplasm maize, *Genetics*, **147**, 1367–1379.

Drouaud, J., Fourgoux, A., Pelletier, G. and Guerche, P. (1999) Microspore specific promoter and method for producing hybrid plants, Patent No 2768745, 26/03/99 France.

Dubreucq, A., Berthe, B., Asset, J.-F., Boulidard, L., Budar, F., Vasseur, J. and Ramband, C. (1999) Analyses of mitochondrial DNA structure and expression in three cytoplasmic male-sterile chicories originating from somatic hybridisation between fertile chicory and CMS sunflower protoplasts, *Theoretical and Applied Genetics*, **99**, 1094–1105.

Ducos, E., Touzet, P. and Boutry, M. (2001) The male sterile G cytoplasm of wild beet displays modified mitochondrial respiratory complexes, *The Plant Journal*, **26**, 171–180.

Dutilleul, C., Garmier, M., Noctor, G., Mathieu, C., Chetrit, P., Foyer, C.H. and De Paepe, R. (2003) Leaf mitochondria modulate whole cell redox homeostasis, set antioxidant capacity, and determine stress resistance through altered signaling and diurnal regulation, *The Plant Cell*, **15**, 1212–1226.

Duvick, D. N. (1965) Cytoplasmic pollen sterility in corn, in *Advances in Genetics*, Vol. 13 (eds, E. W. Caspari and J. M. Thoday), Academic Press, New York, pp. 1–56.

Edwardson, J. R. (1955) The restoration of fertility to cytoplasmic male-sterile corn, *Agronomy Journal*, **47**, 457–461.

Engelke, T. and Tatlioglu, T. (2002) A PCR-marker for the CMS1 inducing cytoplasm in chives derived from recombination events affecting the mitochondrial gene *atp9*, *Theoretical and Applied Genetics*, **104**, 698–702.

Engelke, T. and Tatlioglu, T. (2004) The fertility restorer genes X and T alter the transcripts of a novel mitochondrial gene implicated in CMS1 in chives (*Allium schoenoprasum* L.), *Molecular Genetics and Genomics*, **271**, 150–160.

Farbos, I., Mouras, A., Bereterbide, A. and Glimelius, K. (2001) Defective cell proliferation in the floral meristem of alloplasmic plants of *Nicotiana tabacum* leads to abnormal floral organ development and male sterility, *The Plant Journal*, **26**, 131–142.

Fauron, C. M.-R., Abbott, A. G., Brettell, R. I. S. and Gesteland, R. F. (1987) Maize mitochondrial DNA rearrangements between the normal type, the Texas male sterile cytoplasm, and a fertile revertant cms-T regenerated plant, *Current Genetics*, **11**, 339–342.

Fauron, C. M.-R., Moore, B. and Casper, M. (1995) Maize as a model of higher plant mitochondrial genome plasticity, *Plant Science*, **112**, 11–32.

Flavell, R. (1974) A model for the mechanism of cytoplasmic male sterility in plants, with special reference to maize, *Plant Science Letters*, **3**, 259–263.

Forde, B. G., Oliver, R. J. C. and Leaver, C. J. (1978) Variation in mitochondrial translation products associated with male-sterile cytoplasms in maize, *Proceedings of the National Academy of Sciences of the United States of America*, **77**, 418–422.

Frank, S. A. (1989) The evolutionary dynamics of cytoplasmic male sterility, *American Naturalist*, **133**, 345–376.

Gabay-Laughnan, S., Chase, C. D., Ortega, V. M. and Zhao, L. (2004) Molecular–genetic characterization of CMS-S restorer-of-fertility alleles identified in Mexican maize and teosinte, *Genetics*, **166**, 959–970.

Gengenbach, B. G., Connelly, J. A., Pring, D. R. and Conde, M. F. (1981) Mitochondrial DNA variation in maize plants regenerated during tissue culture selection, *Theoretical and Applied Genetics*, **59**, 161–167.

Gibor, A. and Granick, S. (1964) Plastids and mitochondria: inheritable systems, *Science*, **145**, 890–897.

Glab, N., Wise, R. P., Pring, D. R., Jacq, C. and Slonimski, P. P. (1990) Expression in *Saccharomyce cerevisiae* of a gene associated with cytoplasmic male sterility from maize: respiratory dysfunction and uncoupling of yeast mitochondria, *Molecular & General Genetics*, **223**, 24–32.

Gourret, J., Delourme, R. and Renard, M. (1992) Expression of *ogu* cytoplasmic male sterility in cybrids of *Brassica napus*., *Theoretical and Applied Genetics*, **83**, 549–556.

Gouyon, P. H. and Couvet, D. (1987) A conflict between two sexes, females and hermaphrodites, *Experientia Supplementum*, **55**, 245–261.

Gouyon, P. H., Vichot, F. and Van Damme, J. M. M. (1991) Nuclear–cytoplasmic male sterility: single point equilibria versus limit cycles, *American Naturalist*, **137**, 498–514.

Gray, M. W., Burger, G. and Lang, B. F. (1999) Mitochondrial evolution, *Science*, **283**, 1476–1481.

Grelon, M., Budar, F., Bonhomme, S. and Pelletier, G. (1994) Ogura cytoplasmic male-sterility (CMS)-associated *orf138* is translated into a mitochondrial membrane polypeptide in male-sterile *Brassica* cybrids, *Molecular & General Genetics*, **243**, 540–547.

Handa, H. (2003) The complete nucleotide sequence and RNA editing content of the mitochondrial genome of rapeseed (*Brassica napus* L.): comparative analysis of the mitochondrial genomes of rapeseed and Arabidopsis thaliana, *Nucleic Acids Research*, **31**, 5907–5916.

Hanson, M. R. and Bentolila, S. (2004) Interactions of mitochondrial and nuclear genes that affect male gametophyte development, *The Plant Cell*, **16** Suppl., S154–S169.

Havey, M. J. (2000) Diversity among male-sterility-inducing and male-fertile cytoplasms of onion, *Theoretical and Applied Genetics*, **101**, 778–782.

He, S., Lyznik, A. and Mackenzie, S. (1995) Pollen fertility restoration by nuclear gene *Fr* in CMS bean: nuclear-directed alteration of a mitochondrial population, *Genetics*, **139**, 955–962.

Heiser, V., Rasmusson, A. G., Thieck, O. and Brennicke, A. (1997) Antisense repression of the mitochondrial NADH-binding subunit of complex I in transgenic potato plants affects male fertility, *Plant Science*, **127**, 61–69.

Hernould, M., Suharsono, S., Litvak, S., Araya, A. and Mouras, A. (1993) Male-sterility induction in transgenic tobacco plants with an unedited *atp9* mitochondrial gene from wheat, *Proceedings of the National Academy of Sciences of the United States of America*, **90**, 2370–2374.

Hervieu, F., Bannerot, H. and Pelletier, G. (1994) A unique cytoplasmic male sterility (CMS) determinant is present in three *Phaseolus* species characterized by different mitochondrial genomes, *Theoretical and Applied Genetics*, **88**, 314–320.

Hervieu, F., Charbonnier, L., Bannerot, H. and Pelletier, G. (1993) The cytoplasmic male-sterility (CMS) determinant of common bean is widespread in *Phaseolus coccineus* L. and *Phaseolus vulgaris* L, *Current Genetics*, **24**, 149–155.

Heyn, F. W. (1976) Transfer of restorer genes from *Raphanus* to cytoplasmic male-sterile *Brassica napus*, *Cruciferae Newsletters*, **1**, 15–16.

Hoeberichts, F. A. and Woltering, E. J. (2003) Multiple mediators of plant programmed cell death: interplay of conserved cell death mechanisms and plant-specific regulators, *Bioessays*, **25**, 47–57.

Horn, R. (2002) Molecular diversity of male sterility inducing and male-fertile cytoplams in the genus *Heliantus*, *Theoretical and Applied Genetics*, **104**, 562–570.

Horn, R. and Friedt, W. (1999) CMS sources in sunflower: different origin but same mechanism? *Theoretical and Applied Genetics*, **98**, 195–201.

Horn, R., Hustedt, J., Horstmeyer, A., Hahnen, J., Zetsche, K. and Friedt, W. (1996) The CMS-associated 16 kDa protein encoded by *orfH522* in the PET1 cytoplasm is also present in other male-sterile cytoplasms of sunflower., *Plant Molecular Biology*, **30**, 523–538.

Horn, R., Kohler, R. H. and Zetsche, K. (1991) A mitochondrial 16 kDa protein is associated with cytoplasmic male sterility in sunflower, *Plant Molecular Biology*, **17**, 29–36.

Huang, J., Lee, S.-H., Lin, C., Medici, R., Hack, E. and Myers, A. M. (1990) Expression in yeast of the T-URF13 protein from Texas male-sterile maize mitochondria confers sensitivity to methomyl and to Texas-cytoplasm-specific fungal toxins, *EMBO Journal*, **9**, 339–347.

Ingvarsson, P. K. and Taylor, D. R. (2002) Genealogical evidence for epidemics of selfish genes, *Proceedings of the National Academy of Sciences of the United States of America*, **99**, 11265–11269.

Iwabuchi, M., Koizuka, N., Fujimoto, H., Sakai, T. and Imamura, J. (1999) Identification and expression of the kosena radish (*Raphanus sativus* cv. Kosena) homologue of the Ogura radish CMS-associated gene, *orf138*, *Plant Molecular Biology*, **39**, 183–188.

Janska, H. and Mackenzie, S. A. (1993) Unusual mitochondrial genome organization in cytoplasmic male sterile common bean and the nature of cytoplasmic reversion to fertility, *Genetics*, **135**, 869–879.

Jones, D. F. (1917) Dominance of linked factors as a means of accounting for heterosis, *Genetics*, **2**, 466–479.

Jones, H. A. and Clarke, A. E. (1943) Inheritance of male sterility in the onion and the production of hybrid seed, *Proceedings of the American Society of Horticultural Science*, **43**, 189–194.

Jones, H. A. and Emsweller, S. L. (1937) A male-sterile onion, *Proceedings of the American Society of Horticultural Science*, **34**, 582–585.

Kadowaki, K., Suzuki, T. and Kazama, S. (1990) A chimeric gene containing the 5′ portion of atp6 is associated with cytoplasmic male-sterility of rice, *Molecular & General Genetics*, **224**, 10–16.

Karpova, O. V., Kuzmin, E. V., Elthon, T. E. and Newton, K. J. (2002) Differential expression of alternative oxidase genes in maize mitochondrial mutants, *The Plant Cell*, **14**, 3271–3284.

Kaul, M. L. H. (1988) *Male Sterility in Higher Plants*, Springer-Verlag, Berlin Heidelberg.

Kazama, T. and Toriyama, K. (2003) A pentatricopeptide repeat-containing gene that promotes the processing of aberrant atp6 RNA of cytoplasmic male-sterile rice, *FEBS Letters*, **544**, 99–102.

Kemble, R. J., Flavell, R. B. and Brettell, R. I. S. (1982) Mitochondrial DNA analysis of fertile and sterile maize plants derived from tissue culture with the Texas male sterile cytoplasm, *Theoretical and Applied Genetics*, **62**, 213–217.

Kirti, P. B., Mohapatra, T., Baldev, A., Prakash, S. and Chopra, V. L. (1995) A stable male sterile line of *Brassica juncea* carrying restructured organelle genomes from the somatic hybrid + *B. juncea*, *Plant Breeding*, **114**, 434–438.

Kirti, P. B., Narasimhulu, S. B., Prakash, S. and Chopra, V. L. (1992) Somatic hybridization between *Brassica juncea* and *Moricandia arvensis* by protoplast fusion, *Plant Cell Reports*, **11**, 318–321.

Kitashiba, H., Kitazawa, E., Kishitani, S. and Toriyama, K. (1999) Partial male sterility in transgenic tobacco carrying an antisense gene for alternative oxidase under the control of a tapetum-specific promoter, *Molecular Breeding*, **5**, 209–218.

Knight, J. (2001) Meet the Herod bug, *Nature*, **412**, 12–14.

Knoop, V. (2004) The mitochondrial DNA of land plants: peculiarities in phylogenetic perspective, *Current Genetics*, **46**, 123–139.

Kofer, W., Glimelius, K. and Bonnett, H. T. (1991) Restoration of normal stamen development and pollen formation by fusion of different cytoplasmic male-sterile cultivars of *Nicotiana tabacum*, *Theoretical and Applied Genetics*, **81**, 390–396.

Kofer, W., Glimelius, K. and Bonnett, H. T. (1992) Fusion of male sterile tobacco causes modifications of mtDNA leading to changes in floral morphology and restoration of fertility in cybrid plants, *Physiologia Plantarum*, **85**, 334–338.

Kohler, R. H., Horn, R., Lossl, A. and Zetsche, K. (1991) Cytoplasmic male sterility in sunflower is correlated with the co-transcription of a new open reading frame with the atpA gene, *Molecular & General Genetics*, **227**, 369–376.

Koizuka, N., Imai, R., Fujimoto, H., Hayakawa, T., Kimura, Y., Kohno-Murase, J., Sakai, T., Kawasaki, S. and Imamura, J. (2003) Genetic characterization of a pentatricopeptide repeat protein gene, orf687, that restores fertility in the cytoplasmic male-sterile Kosena radish, *The Plant Journal*, **34**, 407–415.

Koizuka, N., Imai, R., Iwabuchi, M., Sakai, T. and Imamura, J. (2000) Genetic analysis of fertility restoration and accumulation of ORF125 mitochondrial protein in the kosena radish (*Raphanus sativus* cv. Kosena) and a *Brassica napus* restorer line, *Theoretical and Applied Genetics*, **100**, 949–955.

Komori, T., Ohta, S., Murai, N. *et al.* (2004) Map-based cloning of a fertility restorer gene, Rf-1, in rice (*Oryza sativa* L.), *The Plant Journal*, **37**, 315–325.

Korth, K. L. and Levings, C. S. I. (1993) Baculovirus expression of the maize mitochondrial protein URF13 confers insectidal activity in cell cultures and larvae, *Proceedings of the National Academy of Sciences of the United States of America*, **90**, 3388–3392.

Korth, K. L., Kaspi, C. I., Siedow, J. N. and Levings, C. S. I. (1991) URF13, a maize mitochondrial pore-forming protein, is oligomeric and has a mixed orientation in *Escherichia coli* plasma membranes, *Proceedings of the National Academy of Sciences of the United States of America*, **88**, 10865–10869.

Krishnasamy, S. and Makaroff, C. A. (1993) Characterization of the radish mitochondrial *orfB* locus: possible relationship with male sterility in Ogura radish, *Current Genetics*, **24**, 156–163.

Krishnasamy, S. and Makaroff, C. A. (1994) Organ-specific reduction in the abundance of a mitochondrial protein accompanies fertility restoration in cytoplasmic male-sterile radish, *Plant Molecular Biology*, **26**, 935–946.

Kubo, T., Nishizawa, S., Sugawara, A., Itchoda, N., Estiati, A. and Mikami, T. (2000) The complete nucleotide sequence of the mitochondrial genome of sugar beet (*Beta vulgaris* L.) reveals a novel gene for tRNA(Cys)(GCA), *Nucleic Acids Research*, **28**, 2571–2576.

Landgren, M., Zetterstrand, M., Sundberg, E. and Glimelius, K. (1996) Alloplasmic male-sterile *Brassica* lines containing *B. tournefortii* mitochondria express an ORF 3′ of the *atp6* gene and a 32kDa protein., *Plant Molecular Biology*, **32**, 879–890.

Laughnan, J. R. and Gabay, S. J. (1973) Mutations leading to nuclear restoration of fertility in S male-sterile cytoplasm in maize, *Theoretical and Applied Genetics*, **43**, 109–116.

Laughnan, J. R. and Gabay-Laughnan, S. (1983) Cytoplasmic male sterility in maize, *Annual Review of Genetics*, **17**, 27–48.

Laver, H. K., Reynolds, S. J., Moneger, F. and Leaver, C. J. (1991) Mitochondrial genome organization and expression associated with cytoplasmic male sterility in sunflower (*Helianthus annuus*), *The Plant Journal*, **1**, 185–193.

Leclercq, P. (1966) Une stérilité mâle utilisable pour la production d'hybrides simples de tournesol, *Annales de l'Amélioration des Plantes*, **16**, 135–144.

Leclercq, P. (1969) Une stérilité mâle cytoplasmique chez le tournesol, *Annales de l'Amélioration des Plantes*, **19**, 99–106.

Lefebvre, A., Scalla, R. and Pfeiffer, P. (1990) The double-stranded RNA associated with the '447' cytoplasmic male sterility in Vicia faba is packaged together with its replicase in cytoplasmic membranous vesicles, *Plant Molecular Biology*, **14**, 477–490.

Leino, M., Teixeira, R., Landgren, M. and Glimelius, K. (2003) *Brassica napus* lines with rearranged Arabidopsis mitochondria display CMS and a range of developmental aberrations, *Theoretical and Applied Genetics*, **106**, 1156–1163.

Levings, C. S. I. and Siedow, J. N. (1992) Molecular basis of disease susceptibility in the Texas cytoplasm of maize, *Plant Molecular Biology*, **19**, 135–147.

Levings, C. S. 3rd (1993) Thoughts on cytoplasmic male sterility in maize, *The Plant Cell*, **5**, 1285–1290.

Levings, C. S. r., Kim, B. D., Pring, D. R. *et al.* (1980) Cytoplasmic reversion of *cms-S* in maize: association with a transpositional event, *Science*, **209**, 1021–1023.

Lewis, D. (1941) Male sterility in natural populations of hermaphrodite plants. The equilibrium between females and hermaphrodites to be expected with different types of inheritance, *New Phytologist*, **40**, 56–63.

Li, X. Q., Chetrit, P., Mathieu, C., Vedel, F., De Paepe, De Paepe, R., Remy, R. and Ambard-Bretteville, F. (1988) Regeneration of cytoplasmic male sterile protoclones of *Nicotiana sylvestris* with mitochondrial variations, *Current Genetics*, **13**, 261–266.

Li, X. Q., Jean, M., Landry, B. S. and Brown, G. G. (1998) Restorer genes for different forms of *Brassica* cytoplasmic male sterility map to a single nuclear locus that modifies transcripts of several mitochondrial genes, *Proceedings of the National Academy of Sciences of the United States of America*, **95**, 10032–10037.

Liu, F., Cui, X., Horner, H. T., Weiner, H. and Schnable, P. S. (2001) Mitochondrial aldehyde dehydrogenase activity is required for male fertility in maize, *The Plant Cell*, **13**, 1063–1078.

Liu, X. Q., Xu, X., Tan, Y. P., Li, S. Q., Hu, J., Huang, J. Y., Yang, D. C., Li, Y.S. and Zhu, Y.G. (2004) Inheritance and molecular mapping of two fertility-restoring loci for Honglian gametophytic cytoplasmic male sterility in rice (*Oryza sativa* L.), *Molecular Genetics & Genomics*, **271**, 586–594.

Lurin, C., Andres, C., Aubourg, S., Bellaoui, M., Bitton, F., Bruyére, C., Caboche, M., Debast, C., Gualberto, J., Hoffmann, B., Lechamy, A., Le Ret, M., Martin-Magniette, L., Mireau, H., Peeters, N., Renou, J.-P., Szurek, B., Teconnat, L. and Small, I. (2004) Genome-wide analysis of Arabidopsis pentatricopeptide repeat (PPR) proteins reveals their essential role in organelle biogenesis, *The Plant Cell*, **16**, 2089–2103.

Mackenzie, S. A. and Chase, S. (1990) Fertility restoration is associated with a loss of a portion of the mitochondrial genome in cytoplasmic male sterile common bean, *The Plant Cell*, **2**, 905–912.

Mackenzie, S. A., Pring, D. R., Bassett, M. and Chase, C. D. (1988) Mitochondrial DNA rearrangement associated with fertility restoration and reversion to fertility in cytoplasmic male sterile *Phaseolus vulgaris*, *Proceedings of the National Academy of Sciences of the United States of America*, **85**, 2714–2717.

Manicacci, D., Couvet, D., Belhassen, E., Gouyon, P.-H. and Atlan, A. (1996) Founder effects and sex ratio in the gynodioecious *Thymus vulgaris* L, *Molecular Ecology*, **5**, 63–72.

Marienfeld, J. R., Unseld, M., Brandt, P. and Brennicke, A. (1997) Mosaic open reading frames in the Arabidopsis thaliana mitochondrial genome, *Biological Chemistry*, **378**, 859–862.

Matsunaga, S. and Kawano, S. (2001) Sex determination by sex chromosomes in dioecious plants, *Plant Biology*, **3**, 481–488.

McCauley, D. E. and Olson, M. S. (2003) Associations among cytoplasmic molecular markers, gender, and components of fitness in Silene vulgaris, a gynodioecious plant, *Molecular Ecology*, **12**, 777–787.

McCauley, D. E. and Taylor, D. R. (1997) Local population structure and sex ratio: evolution in gynodioecious plants, *American Naturalist*, **150**, 406–419.

McCauley, D. E., Olson, M. S., Emery, S. N. and Taylor, D. R. (2000) Population and structure influences sex ratio evolution in a gynodioecious plant, *American Naturalist*, **155**, 814–819.

McCollum, G. D. (1981) Induction of an alloplasmic male-sterile *Brassica oleracea* by substituting cytoplasm from 'Early Scarlet Globe' radish (*Raphanus sativus*), *Euphytica*, **30**, 855–859.

Menassa, R., L'Homme, Y. and Brown, G. G. (1999) Post-transcriptional and developmental regulation of a CMS-associated mitochondrial gene region by a nuclear restorer gene, *The Plant Journal*, **17**, 491–499.

Mendel, G. (1865) *Experiments in Plant Hybridization* (ed. Blumberg Roger B (http://www.mendelweb.org/)) Edition 97.1 1997.

Moller, I. M. (2001) A general mechanism of cytoplasmic male fertility? *Trends in Plant Science*, **6**, 560.

Monéger, F., Smart, C. J. and Leaver, C. J. (1994) Nuclear restoration of cytoplasmic male sterility in sunflower is associated with the tissue-specific regulation of a novel mitochondrial gene, *EMBO Journal*, **13**, 8–17.

Nikova, V. and Vladova, R. (2002) Wild *Nicotiana* species as a source of cytoplasmic male sterility in *Nicotiana tabacum*, *Contributions to Tobacco Research*, **20**, 301–311.

Nivison, H. T. and Hanson, M. R. (1989) Identification of a mitochondrial protein associated with cytoplasmic male sterility in petunia, *The Plant Cell*, **1**, 1121–1130.

Notsu, Y., Masood, S., Nishikawa, T. *et al.* (2002) The complete sequence of the rice (*Oryza sativa L.*) mitochondrial genome: frequent DNA sequence acquisition and loss during the evolution of flowering plants, *Molecular Genetics and Genomics*, **268**, 434–445.

Ogura, H. (1968) Studies on the new male sterility in japanese radish, with special references to utilization of this sterility towards the practical raising of hybrid seeds, *Memories of the Faculty of Agriculture, Kagoshima University*, **6**, 39–78.

Olson, M. S. and McCauley, D. E. (2002) Mitochondrial DNA diversity, population structure, and gender association in the gynodioecious plant *Silene vulgaris*, *Evolution; International Journal of Organic Evolution*, **56**, 253–262.

Palmer, J. D. and Herbon, L. A. (1988) Plant mitochondrial DNA evolve rapidly in structure but slowly in sequence, *Journal of Molecular Evolution*, **28**, 87–97.

Pannell, J. (1997) The maintenance of gynodioecy and androdioecy in a metapopulation, *Evolution*, **51**, 10–20.

Pellan-Delourme, R. and Renard, M. (1988) Cytoplasmic male sterility in rapeseed (*Brassica napus* L.): female fertility of restored rapeseed with 'Ogura' and cybrid cytoplasms, *Genome*, **30**, 234–238.

Pelletier, G., Primard, C., Vedel, F., Chétrit, P., Rény, R., Rousselle, P. and Renard, M. (1983) Intergeneric cytoplasmic hybridization in Cruciferae by protoplast fusion, *Molecular Genetics and Genomics*, **191**, 244–250.

Pelletier, G., Primard, C., Vedel, F., Chétrit, P., Renard, M., Pellan-Delourme, R. and Mesquida, J. (1987) Molecular, phenotypic and genetic characterization of mitochondrial recombinants in rapeseed, in *Proccedings of 7th International Rapeseed Conference*, Poznan, Poland, 113–118.

Perez-Prat, E. and van Lookeren Campagne, M. M. (2002) Hybrid seed production and the challenge of propagating male-sterile plants, *Trends in Plant Science*, **7**, 199–203.

Poot, P. (1997) Reproductive allocation and resource compensation in male-sterile and hermaphroditic plants of *Plantago lanceolata* (Plantaginaceae), *American Journal of Botany*, **84**, 1256–1265.

Prakash, S., Kirti, P. B., Bhat, S. R., Gaikwad, K., Kumar, V. D. and Chopra, V. L. (1998) A *Moricandia arvensis*-based cytoplasmic male sterility and fertility restoration system in *Brassica juncea.*, *Theoretical and Applied Genetics*, **97**, 488–492.

Primard, C., Bonhomme, S., Budar, F. and Pelletier, G. (1988) Segregation of two types of cytoplasmic male sterility after regeneration of rapeseed plants derived from fusion between normal and Ogura male-sterile rapeseed, in *Proccedings of 2nd International Congress of Plant Molecular Biology*, Jerusalem, Israel, 351.

Pring, D. R., Levings, C. S. r., Hu, W. W. L. and Timothy, D. H. (1977) Unique DNA associated with mitochondria in the 'S'-type cytoplasm, *Proceedings of the National Academy of Sciences of the United States of America*, **74**, 2904–2908.

Pring, D. R., Tang, H. V. and Schertz, K. F. (1995) Cytoplasmic male sterility and organelle DNAs of sorghum, in *The Molecular Biology of Plant Mitochondria* (eds C. S. I. Levings and I. K. Vasil), Kluwer Academic, Dordrecht, Netherlands, pp. 461–495.

Pruitt, K. D. and Hanson, M. R. (1991) Transcription of the petunia mitochondrial CMS-associated *Pcf* locus in male sterile and fertility-restored lines, *Molecular Genetics and Genomics*, **227**, 348–355.

Rambaud, C., Dubois, J. and Vasseur, J. (1993) Male-sterile chicory cybrids obtained by intergeneric protoplast fusion, *Theoretical and Applied Genetics*, **87**, 347–352.

Rasmussen, J. and Hanson, M. R. (1989) A NADH dehydrogenase subunit gene is co-transcribed with the abnormal *Petunia* mitochondrial gene assciated with cytoplasmic male sterility, *Molecular and General Genetics*, **215**, 332–336.

Rhoades, M. M. (1931) Cytoplasmic inheritance of male sterility in *Zea mays*, *Science*, **73**, 340–341.

Rhoades, M. M. (1933) The cytoplasmic inheritance of male sterility in *Zea mays*, *Journal of Genetics*, **27**, 71–93.

Rhoades, M. M. (1950) Gene induced mutation of a heritable cytoplasmic factor producing male sterility in maize, *Proceedings of the National Academy of Sciences of the United States of America*, **36**, 634–635.

Rhoads, D., Levings, C. 3rd and Siedow, J. (1995) URF13, a ligand-gated, pore-forming receptor for T-toxin in the inner membrane of cms-T mitochondria, *Journal of Bioenergetics and Biomembranes*, **27**, 437–445.

Rogers, J. S. and Edwardson, J. R. (1952) The utilization of cytoplasmic male-sterile inbreds in the production of corn hybrids, *Agronomy Journal*, **44**, 8–13.

Sabar, M., De Paepe, R. and de Kouchkovsky, Y. (2000) Complex I impairment, respiratory compensations, and photosynthetic decrease in nuclear and mitochondrial male sterile mutants of *Nicotiana sylvestris*, *Plant Physiology*, **124**, 1239–1250.

Sabar, M., Gagliardi, D., Balk, J. and Leaver, C. J. (2003) ORFB is a subunit of F(1)F(O)-ATP synthase: insight into the basis of cytoplasmic male sterility in sunflower, *EMBO Reports*, **4**, 1–6.

Sarria, R., Lyznik, A., Vallejos, C. E. and Mackenzie, S. A. (1998) A cytoplasmic male sterility-associated mitochondrial peptide in common bean is post-translationally regulated, *The Plant Cell*, **10**, 1217–1228.

Saumitou-Laprade, P., Rouwendal, G. J. A., Cuguen, J., Krens, F. A. and Michaelis, G. (1993) Different CMS sources found in *Beta vulgaris* ssp. *maritima*: mitochondrial variability in wild populations revealed by a rapid rapid screening procedure, *Theoretical and Applied Genetics*, **85**, 529–535.

Schnable, P. S. (2002) Is rf2 a restorer gene of CMS-T in maize? Answer, *Trends in Plant Science*, **7**, 434.

Schnable, P. S. and Wise, R. P. (1998) The molecular basis of cytoplasmic male sterility and fertility restoration, *Trends in plant science*, **3**, 175–180.

Shinjyo, C. (1969) Cytoplasmic genetic male sterility in cultivated rice, *Oryza sativa* L. II The inheritance of male sterility, *Japanese Journal of Genetics*, **44**, 149–156.

Siedow, J. N., Rhoads, D. M., Ward, G. C. and Lewings, C. S. 3rd (1995) The relationship between the mitochondrial gene T-*urf13* and fungal pathotoxin sensitivity in maize, *Biochimica et Biophysica Acta*, **1271**, 235–240.

Singh, M. and Brown, G. G. (1991) Suppression of cytoplasmic male sterility by nuclear genes alters expression of a novel mitochondrial gene region, *The Plant Cell*, **3**, 1349–1362.

Small, I. and Peeters, N. (2000) The PPR motif: A TPR-related motif prevalent in plant organellar proteins, *Trends in Biochemical Sciences*, **25**, 46–47.

Small, I., Suffolk, R. and Leaver, C. J. (1989) Evolution of plant mitochondrial genomes via substoichiometric intermediates, *Cell*, **58**, 69–76.

Smart, C. J., Moneger, F. and Leaver, C. J. (1994) Cell-specific regulation of gene expression in mitochondria during anther development in sunflower, *The Plant Cell*, **6**, 811–825.

Stadler, T. and Delph, L. F. (2002) Ancient mitochondrial haplotypes and evidence for intragenic recombination in a gynodioecious plant, *Proceedings of the National Academy of Sciences of the United States of America*, **99**, 11730–11735.

Stiewe, G., Sodhi, Y. S. and Robbelen, G. (1995) Establishment of a new CMS system in Brassica napus by mitochondrial recombination with B. tournefortii, *Plant Breeding*, **113**, 294–304.

Suzuki, T., Nakamura, C., Mori, N., Isawa, Y. and Kaneda, C. (1994) Homeologous group-1 chromosomes of *Agropyron* restore nucleus-cytoplasmic compatibility in alloplasmic common wheat with *Agropyron* cytoplasms, *Japanese Journal of Genetics*, **69**, 41–51.

Swingle, W. T. and Webber, H. J. (1897) Hybrids and their utilization in plant breeding, in *Yearbook of Department of Agriculture* http://www.bulbnrose.com/Heredity/Swingle/swingle.html, pp. 383–420.

Tang, H. V., Chen, W. and Pring, D. R. (1999) Mitochondrial *orf107* transcription, editing, and nucleolytic cleavage conferred by the*Rf3* are expressed in sorghum pollen, *Sexual Plant Reproduction*, **12**, 53–59.

Tang, H. V., Pring, D. R., Shaw, L. C., Salazar, R. A., Muza, F. R., Yan, B. and Schertz, K. F. (1996) Transcript processing internal to a mitochondrial open reading frame is correlated with fertility restoration in male-sterile sorghum, *The Plant Journal*, **10**, 123–133.

Thompson, J. D. and Tarayre, M. (2000) Exploring the genetic basis and proximate causes of female fertility advantage in gynodioecious *Thymus vulgaris*, *Evolution International Journal of Organic Evolution*, **54**, 1510–1520.

Touzet, P. (2002) Is the *rf2* a restorer gene of CMS-T in maize? *Trends in Plant Science*, **7**, 434.

Touzet, P. and Budar, F. (2004) Unveiling the molecular arms race between two conflicting genomes in cytoplasmic male sterility? *Trends in Plant Sciences*, **9**, 568–570.

Umbeck, P. F. and Gengenbach, B. G. (1983) Reversion of male-sterile T-cytoplasm maize to male fertility in tissue culture, *Crop Science*, **23**, 584–588.

Vear, F. (1992) Le Tournesol, In *Amélioration des espèces végétales cultivées. Objectifs et critères de sélection* (eds A. Gallais and H. Bannerot), INRA Editions, Paris, pp. 146–160.

Virmani, S. S. and Shinjyo, C. (1988) Current status of analysis and symbols for male-sterile cytoplasms and fertility-restoring genes, *Rice Genetics Newsletter*, **5**, 9–15.

Wen, L. and Chase, C. D. (1999a) Pleiotropic effects of a nuclear restorer-of-fertility locus on mitochondrial transcripts in male-fertile and S male-sterile maize, *Current Genetics*, **35**, 521–526.

Wen, L., Ruesch, K. L., Ortega, V. M., Kamps, T. L., Gabay-Laughnan, S. and Chase, C. D. (2003) A nuclear restorer-of-fertility mutation disrupts accumulation of mitochondrial ATP synthase subunit alpha in developing pollen of S male-sterile maize, *Genetics*, **165**, 771–779.

Wen, L. Y. and Chase, C. D. (1999b) Mitochondrial gene expression in developing male gametophytes of male-fertile and S male-sterile maize, *Sexual Plant Reproduction*, **11**, 323–330.

Wise, R. P., Bronson, C. R., Schnable, P. S. and Horner, H. T. (1999) The genetics, pathology and molecular biology of T-cytoplasm male sterility in maize, *Advances in Agronomy*, **65**, 79–131.

Xu, G. W., Cui, Y.-X., Schertz, K. F. and Hart, G. E. (1995) Isolation of mitochondrial DNA sequences that distinguish male-sterility-inducing cytoplasms in *Sorghum bicolor* (L.) Moench, *Theoretical and Applied Genetics*, **90**, 1180–1187.

Xue, Y. B., Li, J. and Xu, Z. (2003) Recent highlights of the China rice functional genomics program, *Trends in Genetics*, **19**, 390–394.

Yamagishi, H. and Terachi, T. (1996) Molecular and biological studies on male-sterile cytoplasm in the Cruciferae. III. Distribution of Ogura-type cytoplasm among Japanese wild radishes and Asian radish cultivars, *Theoretical and Applied Genetics*, **93**, 325–332.

Yamagishi, H. and Terachi, T. (1997) Molecular and biological studies on male-sterile cytoplasm in the Cruciferae. IV. Ogura-type cytoplasm found in the wild radish, *Raphanus raphanistrum*, *Plant Breeding*, **116**, 323–329.

Yamagishi, H. and Terachi, T. (2001) Intra- and Inter-specific variations in the mitochondrial gene orf138 of Ogura-type male sterile cytoplasm from *Raphanus sativus* and *Raphanus raphanistrum*, *Theoretical and Applied Genetics*, **103**, 725–732.

Yao, N., Eisfelder, B. J., Marvin, J. and Greenberg, J. T. (2004) The mitochondrion – an organelle commonly involved in programmed cell death in *Arabidopsis thaliana*, *The Plant Journal*, **40**, 596–610.

Yuan, M., Yang, G. S., Fu, T. D. and Li, Y. (2003) Transcriptional control of orf224/atp6 by the pol CMS restorer Rfp gene in Brassica napus L, *Yi Chuan Xue Bao*, **30**, 469–473.

Yui, R., Iketani, S., Mikami, T. and Kubo, T. (2003) Antisense inhibition of mitochondrial pyruvate dehydrogenase E1alpha subunit in anther tapetum causes male sterility, *The Plant Journal*, **34**, 57–66.

Zabala, G., Gabay-Laughnan, S. and Laughnan, J. R. (1997) The nuclear gene Rf3 affects the expression of the mitochondrial chimeric sequence R implicated in S-type male sterility in maize, *Genetics*, **147**, 847–860.

Zimmer, C. (2001) Wolbachia. A tale of sex and survival, *Science*, **292**, 1093–1095.

Zubko, M. K. (2004) Mitochondrial tuning fork in nuclear homeotic functions, *Trends in Plant Science*, **9**, 61–64.

Zubko, M. K., Zubko, E. I., Adler, K., Grimm, B. and Gleba, Y. Y. (2003) New CMS-associated phenotypes in cybrids *Nicotiana tabacum* L. (+ *Hyoscyamus niger* L.), *Annals of Botany*, (London), **92**, 281–288.

8 The diversity and significance of flowering in perennials

Theresa Townsend, Maria Albani, Mike Wilkinson,
George Coupland and Nick Battey

8.1 Introduction

The subject of flowering in perennials is large and diverse. Previous reviews have tended to be concerned with the physiology of plants that are cultivated as perennials, their patterns of flowering and the practical manipulation of flowering for crop production (e.g. Jackson and Sweet, 1972; Sedgley, 1990). Furthermore, in recent years the focus on model (annual) plants has tended to diminish the worldwide effort into the study, at both fundamental and applied levels, of perennial plants and their flowering. Advances in the understanding of plant development at genctic and molecular levels have, in contrast, progressed very rapidly, and the time is now overdue for application and extension of that understanding into the challenging arena of perennial flowering (Battey and Tooke, 2002). Here, our intention is to provide a sketch map of the types of perennial flowering, their relationship to annual flowering and their integration within the other aspects of the perennial life cycle. But first we emphasise the importance of the perennial trait, both in natural and cultivated plant populations. This provides a backdrop against which the fundamentals of perennial flowering take on added significance.

8.2 The importance of perenniality

Perennial plants dominate natural plant communities in temperate zones, where deciduous forests and grassland are the prominent vegetation types. In contrast, the principal food crops in these regions are annuals, which are typically subject to high-input, intensive farming methods. The four most important food crops are rice, wheat, maize and soybean, whose wild progenitors have been bred to create high-yielding and disease-resistant cultivars adapted to a range of climates (Ladizinsky, 1998). The perennial forage grasses are intermediate between natural, perennial-dominated communities and the intensively farmed annual cereals: they are an important crop with (semi-)natural status. Flowering is a crucial process in this situation because it must be suppressed to allow vegetative biomass production, yet there is an agri-commercial requirement to make money from seed sales of forage perennials. Turf grass is an extreme case, where seed sales are very valuable but flowering can be a major problem.

In the tropics, in contrast to temperate regions, perennial food crops predominate. It has been argued for some time that the creation of a more sustainable future for temperate agriculture requires a general move away from the instability and unnaturalness of annual monoculture cultivation; and that agriculture should be modelled after natural (perennial) systems (Soule and Piper, 1992). Concerns over wind- and water-mediated soil erosion have meant that there has been increased interest in the conversion from conventional tillage to minimum or reduced tillage systems. Yet, on the other hand, tillage is an effective part of weed control procedures and so its loss increases reliance on other methods such as herbicide application. Indeed, the commercial release of GM crops that are resistant to broad range herbicides has profoundly altered weed management practices in North America and increased the adoption of reduced-tillage agriculture (Duke *et al.*, 2002). The appearance of weeds showing resistance to these herbicides threatens to compromise the economic validity of such strategies (Main *et al.*, 2004; Ng *et al.*, 2004). In the longer term, a change towards the cultivation of perennial crops could circumvent this problem, because acknowledged advantages associated with the use of perennials include reduced soil erosion, increased soil porosity and nutrient content, and a decrease in tilling and seeding (Lefroy *et al.*, 1999; Scheinost *et al.*, 2001).

Farming systems based on the use of perennial crops may, therefore, provide an environmentally sound and economically viable alternative to annual cropping systems, but there are also significant disadvantages. These vary between species but can include a prolonged establishment phase, variable crop height, more difficult mechanical harvesting, delayed maximum yields (because of the extended juvenile phase), greater diversity in breeding systems, extended breeding programmes for crop improvement and continual exposure of the crop to pests and diseases. Therefore, further work needs to be carried out to create economically viable perennial crop species. Some breeding efforts have examined the possibility of producing perennial crop cultivars of existing annual crops through interspecific crosses with perennial relatives, although effort was ultimately diverted into the production of improved annual cultivars (Cox *et al.*, 2002). The Soviet Union led in this area, trying to breed perennial wheat through the hybridisation of annual species with perennial relatives in the 1920s, and plant breeders at the Land Institute in the USA aimed to develop a 'domestic prairie' by domesticating promising wild perennials. As a result of this work valuable perennial germplasm has been created, even though perennial cultivars have yet to be developed and adopted (Wagoner, 1990; Cox *et al.*, 2002).

8.3 Flowering in the context of perenniality

Flowering behaviour in relation to life expectancy is central to the way we categorise plants. An 'annual' dies after flowering during the course of one growing season; in temperate regions this is usually 1 year. Winter annuals and biennials flower and die after 1–2 years and require a winter in their life cycle because this promotes the transition to flowering (and death). It is in the 'perennial' group of plants that the connection between flowering and death becomes more complex. Monocarpic perennials, like some bamboos, and cacti like *Agave* grow for years in the vegetative state before flowering and death (Battey and Tooke, 2002; Thomas, 2003). These perennials employ the same strategy as annuals and biennials: when they flower, the individual dies. Polycarpic perennials, however, repeatedly survive the flowering event and continue to live for many years. Flowering in this way (more than once) requires that the individual has a mechanism for renewed vegetative growth after flower initiation.

Brief consideration indicates that there is a variety of methods by which plants achieve this. An early classification was devised by Raunkiaer (1934), and has been usefully condensed into the form presented in Table 8.1 (Thomas, 2003). In this system, with the exception of the therophytes ('annuals'), the key variable is the location of the perennating buds that allow survival of post-flowering senescence and provide for future vegetative growth. These buds may be above ground – as in trees which produce buds that are dormant through unfavourable periods; or they may be at the soil level – aerial parts of the plant involved in reproduction die back and leave low-growing plant structures, such as rosettes, to survive unfavourable conditions; or the perennating buds may be below ground – the plant dies back and survives unfavourable conditions underground. These buds may be located on roots, rhizomes, bulbs, corms or stem or root tubers; examples are dock, nettles and potato.

8.4 Raunkiaer's five life form groups: deeper and deeper connections between flowering and perenniality

We have seen earlier that the pattern of perennial flowering is intimately related to the yearly growth pattern. To understand flowering in perennials therefore requires knowledge of the temporal controls on flowering (length of the juvenile phase and meristem dormancy, time of the floral transition and flower emergence), the factors that regulate plant architecture – shoot branching patterns, spatial location of new meristems (e.g. as root buds or tuber meristems), and the mechanisms that co-ordinate these two aspects of plant development. Some work is already conceived in this integrative manner, although not with the focus on perennial flowering (e.g. Beveridge *et al.*, 2002). Here, as a starting point, we describe examples of the range of possibilities for perennial flowering. They emphasize, at progressively more subtle levels, the intimate connections between flowering, senescence and dormancy.

8.4.1 The Ravenelle wallflower – timing is everything

The Ravenelle wallflower is a perennial variety of *Cheiranthus cheiri* and adopts a strikingly simple approach to the periodic need to renew flowering (Diomaiuto, 1988). The terminal meristem cycles back and forth between the vegetative and the inflorescence states, utilising inflorescence reversion to achieve

Table 8.1 The classification of plant life forms, based on Raunkiaer, C. (1934) *The Life Forms of Plants and Statistical Plant Geography*, Oxford University Press (after Thomas, 2003)

Life form	Definition	Types included
Phanerophytes	Generally tall plants visible throughout the year, carrying surviving buds or apices at least 25 cm up from the ground. Examples are trees, large shrubs and lianas	(a) Evergreens without bud covering (b) Evergreens with bud covering (c) Deciduous with bud covering (d) Less than 2 m high
Chamaephytes	Low growing plants visible all year round, bearing perennial buds between ground level and 25 cm up. Examples include shrubby tundra species	(a) Suffruticose chamaephytes, i.e. those bearing erect shoots which die back to the portion that bears the surviving buds (b) Passive chamaephytes with persistent weak shoots that trail on or near the ground (c) Active chamaephytes that trail on or near the ground because they are persistent and have horizontally directed growth (d) Cushion plants
Hemicryptophytes	The surviving buds or shoot apices are situated at or just below the soil surface. Includes perennial grasses, many forbs and ferns	(a) Protohemicryptophytes with aerial shoots that bear normal foliage leaves, but of which the lower ones are less perfectly developed (b) Partial rosette plants bearing most of their leaves (and the largest) on short internodes near ground level (c) Rosette plants bearing all their foliage leaves in a basal rosette
Cryptophytes	At the end of the growing season, die back to bulbs, corms, rhizomes or similar underground (in some species, underwater) structures. For example, lilies, onions, garlic, potatoes and similar forbs	(a) Geocryptophytes or geophytes which include forms with (i) rhizomes, (ii) bulbs, (iii) stem tubers and (iv) root tubers (b) Marsh plants (helophytes) (c) Aquatic plants (hydrophytes)
Therophytes	Plants that complete their life cycle from seed to seed and die within a season, or that germinate in fall and reproduce and die in the spring of the following year	

Figure 8.1 Perennial flowering in the Ravenelle wallflower. (a) Fruits (fr) followed by shoots after the point of reversion (R) of the inflorescence meristem. (b) Flowered, fruited regions (Fl_1) and vegetative regions (veg. br) of a reverted plant. Reproduced from Diomaiuto (1988).

polycarpic perenniality. The successive transitions are regulated by cold and warm temperatures, and Diomaiuto has linked these to the temperate climate to which the Ravenelle wallflower is adapted. The consequence of this growth pattern is a disorganised-looking bush, probably best described as a chamaephyte, composed of axes that show the evidence of successive vegetative and floral episodes (Figure 8.1). Interestingly, the wallflower is not unique in this growth pattern; *Callistemon*, the bottlebrush tree, adopts a similar strategy and so, in a more complex manner, does another member of the Myrtaceae, *Metrosideros* (Sreekantan *et al.*, 2001). Further discussion of reversion as a basis for perenniality can be found in Tooke *et al.* (2005).

8.4.2 Arabidopsis and its perennial relatives – timing with a spatial dimension

Arabidopsis thaliana is a typical therophyte, rapidly passing from vegetative growth to flowering, fruiting and finally whole plant senescence within a single growing season. Accessions grown in warm environments overwinter as seeds and germinate and flower in spring and summer (summer annuals). In these accessions floral induction is mainly promoted by increasing daylengths in the spring. The light receptors (phytochromes and cryptochromes) act together with the circardian clock components (e.g. CCA1/LHY and TOC1) to detect changes in photoperiod and activate downstream genes *CO* and *FT* to promote flowering (Searle and Coupland, 2004). Accessions grown in temperate regions overwinter as seedlings and flower the following spring after a period of cold winter temperatures (winter annuals), in which case the decision to flower is not only controlled by changes in photoperiod. *FRIGIDA* (*FRI*) and *FLOWERING LOCUS C* (*FLC*) act together to repress flowering of a non-vernalised seedling and chilling can rapidly promote flowering in these accessions by abolishing the effect of these two genes. Long days still promote flowering but compete with *FRI* and *FLC* action.Thus, in laboratory conditions, these accessions show late flowering phenotypes under continuous long days. The effect of chilling can also be reversible when seedlings experience inadequate chilling and heat treatment for several days immediately after the cold period (reviewed in Michaels and Amasino, 2000; Mouradov *et al.*, 2002).

The Skye ecotype is an example from *Arabidopsis* that shows a clear link between shoot architecture and flowering pattern (Grbić and Bleecker, 1996). In this ecotype the rosette structure is recapitulated in the axillary shoots (Figure 8.2), a phenotype that has been hypothesised to result from delayed competence of the axillary meristems to respond to floral signal (Grbić and Bleecker, 1996). Alternatively, it may be due to local deficiency of floral signal at these meristems, an interpretation consistent with Laibach's finding that aerial rosettes can form in response to daylength transfer treatments (see Koorneef *et al.*, 1994). The combined activity of dominant *FRI* and *FLC* alleles with a dominant allele at the *AERIAL ROSETTE 1* locus is required for the phenotype, indicating that modulation of flowering time genes can

Figure 8.2 The 'F' mutant of *Arabidopsis thaliana*, also known as the Skye ecotype (Grbić and Bleecker, 1996). Reproduced with permission from Koorneef *et al.* (1994).

Figure 8.3 *Arabidopsis lyrata* ssp. *petraea.*

lead to qualitative changes in life form (Poduska *et al.*, 2003). In this case, it is unclear why most flowering time mutants show variation in leaf number in the main rosette, but the Skye ecotype shows the specific character of axillary rosetting.

These mechanisms are all concerned with the timing of flowering, and they act within a life form that makes no provision for renewed growth after flowering. To understand how flowering in the *Arabidopsis* pattern might be integrated within the context of (polycarpic) perennial development, we can turn to the perennial relatives of *Arabidopsis*. *Arabidopsis lyrata* ssp. *petraea* is an herbaceous perennial and shows a protohemicryptophyte life form (Table 8.1). The part of the plant that persists for several years is situated close to the soil surface, where dead leaves protect the perennating meristems during the winter period (Figure 8.3). In the spring, reproductive shoots develop from the surviving meristems and die back to the soil level after flowering. Pollard and co-workers (2001) show that *A. lyrata* ssp. *petraea* under a 22-h daylength produces more basal (reproductive) shoots than under a 14-h daylength. They interpret this responsiveness as a measure of phenotypic plasticity, and it is interesting that both of the perennial *Arabidopsis* relatives they studied (*A. lyrata* ssp. *petraea* and *A. arenosa*) were much more plastic in this respect than *A. thaliana*. It could be that this trait is of significance, because these same basal shoots sustain perennial development. However, root buds also appear to be an important method of propagation (our unpublished observations).

Arabis alpina is a herbaceous perennial with a passive chamaephyte life form (Figure 8.4). Seedlings produce axillary shoots early in their life and under inductive conditions the apical meristems from the main stem and branches bolt, giving more leaves, axillary shoots and an inflorescence. These shoots are long and weak in proportion to their length and by the end of the seed set period they lie on the ground

Figure 8.4 *Arabis alpina.*

under their own weight. The reproductive parts in each shoot senesce after seed dispersal, whereas the newly grown vegetative parts rest on the soil with their shoot apices erect. These new shoots bear the perennating buds and produce adventitious roots which enable them to act as an independent module the following year. The flowering response of *A. alpina* accessions depends on their origin. Populations are found dispersed in various habitats from circumpolar regions in Scandinavia to the Pyrenees and Italian Alps and even in Mediterranean islands (Kirchner, 2002). Accessions from alpine regions of the Pyrenees initiate flowers in response to winter chilling, suggesting regulation of flower initiation by cold responsive genes such as *FLC* (Albani and Wang, unpublished data).

8.4.3 Apple – the mystery of the bourse

Apple, *Malus domestica*, adopts the phanerophyte perennial strategy – overwintering buds grow seasonally to generate the structure of the tree. The flowers of apple emerge in spring, just before the new leaves unfurl. The spurs, which bear both flowers and leaves, epitomise perennial flowering: they typically produce a terminal inflorescence whilst maintaining vegetative growth from a lateral meristem, known as the bourse shoot (Figure 8.5). The bourse shoot grows for a limited time in the spring and early summer initiating leaves, bud scales and (usually) a terminal inflorescence. The terminal inflorescence itself terminates in a flower (Foster *et al.*, 2003), but crucially for perennial development, at least one lateral bud below the terminal inflorescence remains vegetative, forming the bourse shoot for the next season's growth. The 'saving' of this meristem from the floral transition could be explained in terms of spatial factors (localized distribution of floral signal), or by timing – the apple shoot apical meristem may be juvenile for its first period of its development. Although many studies have been carried out relating to flowering

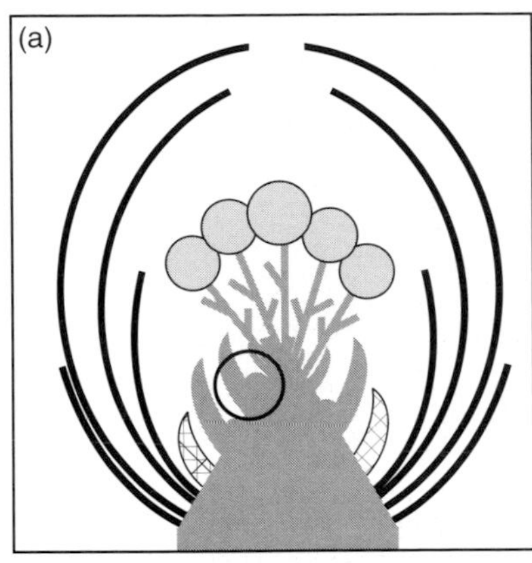

Figure 8.5 Apple floral bud before and after bud break. (a) The organisation of the winter dormant floral bud. In the previous season, the terminal meristem initiated leaves (not shown), then bud scales (curved lines), transition leaves (cross-hatched) and leaf primordia (filled), before terminating in an inflorescence. Subsequent vegetative growth is from the bourse shoot (circled) that develops from the axillary meristem of one of the leaf primordia. (b) Short shoot derived from the bud after bud break and bourse growth. Flowers and expanded leaves (VL) were initiated in the previous season; new leaves (L1–L3) were initiated by the bourse shoot meristem (arrowed). Reproduced with permission from Foster *et al.* (2003).

and fruiting patterns (see Buban and Faust, 1982; Gur, 1985), the answer to this question remains unclear.

The isolation of a *TERMINAL FLOWER 1* homologue from apple (Kotoda and Wada, 2004), and its ability to delay the floral transition in transgenic *Arabidopsis* are therefore of considerable interest. The apple *TFL1* gene is expressed in apical buds in a way that is generally consistent with a function in maintaining the vegetative (juvenile?) phase; it may also dictate the local fate of the bourse shoot. However, the modifying effects of spatial signals, possibly in the form of inhibition by gibberellic acid, are implied by physiological experiments (see Buban and Faust, 1982). That spatial signals modulate an underlying temporal control of meristem fates within the tree is also suggested by the natural tendency of many apple varieties towards irregular bearing, in which very heavy flowering years are interspersed by years with very few flowers; and of other varieties towards repeat flowering within one year, particularly where frost has damaged the first cohort of flowers. These appear to be responses to crop load and to be brought about by hormonal signals derived from the fruit (Dennis, 2003).

8.4.4 Strawberry – an overwintering crown and a summer of environmental sensing

Strawberry, *Fragaria* spp., is probably best considered a hemicryptophyte, with short internodes creating a partial rosette plant with an overwintering crown.

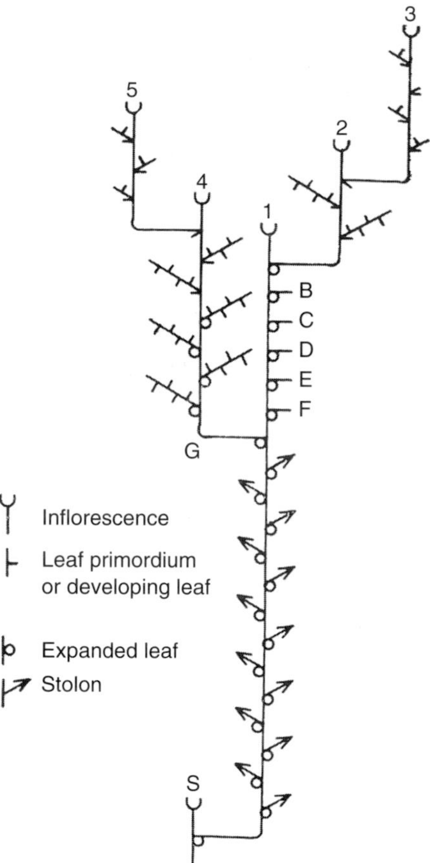

Figure 8.6 A diagram of the strawberry crown when dissected in winter. Inflorescence S flowered the previous spring, and the vegetative growth of the crown was continued by the axillary meristem below it. This meristem initiated 13 leaves subtending axillary stolons, followed by leaves with axillary branch crown buds (B–G) and finally inflorescence 1. Inflorescences 2–3 and 4–5 derive from axillary branch buds. Whereas stolons typically grow out in the same season as they are initiated, inflorescences grow out the following spring, and continuation of vegetative growth occurs from an axillary meristem arising below inflorescence 3. Branch crown buds that did not initiate inflorescences in the autumn may also break and form new branch crowns during the next season. Reproduced from Guttridge (1985).

Figure 8.6, taken from Guttridge (1985), contains all the information needed for an understanding of the perennial cycle of development in the octoploid, Junebearing strawberry. It requires careful study, however, because it includes developmental structures that have emerged (stolons) as well as those yet to appear from the buds (branch crowns and inflorescences). It shows the crown (shoot) at dissection in the winter. The stolons indicated as arrows grew out during the previous summer; the inflorescences (numbered 1–5) will not make their appearance until the coming spring, although they were initiated during the autumn. Branch crowns were initiated just before the inflorescences (with some differentiation continuing during inflorescence formation). These branch crowns may emerge after the main crown inflorescence in late spring, bearing a terminal inflorescence of their own. Alternatively, they may grow vegetatively alongside the main crown extension axis that arises from below one of the main crown inflorescences (e.g. inflorescence 3). The vegetative meristem thus goes through a series of developmental phases, initiating leaves subtending stolons, then branch crowns before finally converting to inflorescence development. Stolons typically grow out during the same season as initiation; branch crowns and inflorescences grow out the season after initiation.

One key to perennial growth in this plant is, therefore, the maintenance of axillary meristems in a vegetative condition while the terminal meristem differentiates an inflorescence. However, this sympodial mode of growth is shared by many plants whose development is not necessarily perennial – tomato being a familiar example. The maintenance of some vegetative axillary meristems therefore only offers the *potential* for perennial development. This is realised by means of the dormant phase, which is supported by the adaptations of the overwintering crown (Hancock, 1999) and which is broken by cold temperature that promotes vegetative growth (Battey, 2000). Additionally, there is a mechanism which confines senescence to the flowers and fruits, in contrast to the monocarpic annual in which senescence is all-consuming (see Thomas *et al.*, 2000).

Against this background, the duration of the flowering episode is tightly regulated in seasonal flowering ('Junebearing') varieties, because flower initiation only takes place in the autumn, in response to short days and cool temperatures. In the spring, flower initiation is repressed as a consequence of winter chilling (Battey *et al.*, 1998). This characteristic is controlled in diploid strawberry by a single gene, the *SEASONAL FLOWERING LOCUS* (*SFL*; Brown and Wareing, 1965; Albani *et al.*, 2004). In the perpetual flowering genotype *F. v. semperflorens*, inflorescences are formed regularly every two nodes (Brown and Wareing, 1965; Guttridge, 1985) in a sympodial growth pattern reminiscent of day-neutral tomato (see Battey, 2005). Physiological data are consistent with *SFL* encoding a floral repressor that prevents this regular transition to flowering. It acts in conjunction with the (independent) *RUNNERING* locus (Brown and Wareing, 1965; Albani *et al.*, 2001; Cekic *et al.*, 2001) to take the shoot meristem through the yearly succession of developmental phases (stolon, branch crown and inflorescence initiation), although the genetic control of branch crown formation (B–G in Figure 8.6) is currently unclear.

8.4.5 Perennial grasses, leafy spurge – cryptophytes where the rhizome is crucial

In members of the grass family, which can be considered hemicryptophytes or cryptophytes (Table 8.1), perenniality is typically a consequence of persistence of

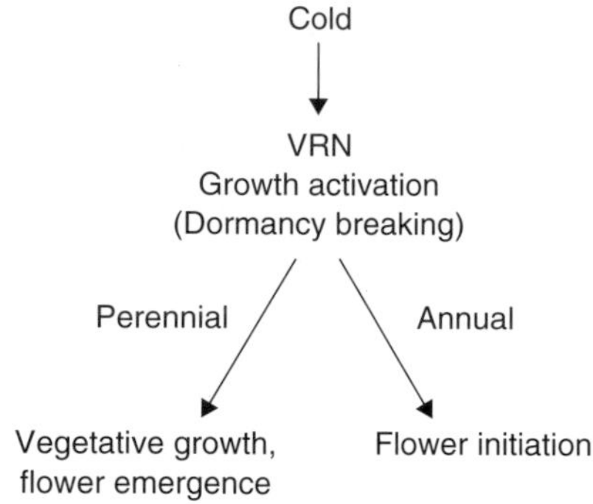

Figure 8.7 Scheme showing potential pathways of cold perception and action in perennials (adapted from Battey, 2000). VRN here suggests a cold-sensing pathway analogous to that in winter annual *Arabidopsis*, in which VIN3 represses *FLC* (Sung and Amasino, 2004) and the effects of cold are 'remembered' because *FLC* repression is maintained by the action of VRN1 (Levy *et al.*, 2002) and VRN2 (Gendall *et al.*, 2001).

tillers through the winter. This persistence can be achieved through the production of stolons in species such as *Agrostis stolonifera*; and through the presence of a persistent rhizome. Although we shall not discuss it further here, pseudovivipary can be another means of perennial growth in the grasses. This is a form of reversion and has clear connections to the control of flowering. Further details can be found in Heide (1994) and Tooke *et al.* (2005).

The overwintering of tillers means a vernalisation requirement in temperate grasses like *Lolium perenne* (Aitken, 1985). Although *FLC*/*FRI* homologues have not been described for this species, a *TFL1* homologue has been studied in some detail (Jensen *et al.*, 2001). The expression pattern was consistent with a role in maintaining tillers in the vegetative state while the shoot apical meristem flowered, and the branching pattern of the inflorescence was interpreted in terms of *TFL1* expression. However, a specific role in the perennial cycle has not been explored.

Rhizome formation is considered to be the key trait relating to perenniality in the Poaceae, because it is the principal means of persistence and vegetative propagation (Paterson *et al.*, 1995). In perennial rice (*Oryza longistaminata*) two major quantitative trait loci (QTLs) for the rhizome trait have been mapped (*Rhz2* and *Rhz3*) in an F2 population derived from a cross with *Oryza sativa* (Hu *et al.*, 2003). Both QTLs were found to be dominant, consistent with perenniality being ancestral to annuality, and were considered likely to encode regulatory genes. It will be of great interest to discover how the action of *Rhz2* and *Rhz3* leads to the development of the rhizome; in weedy *Sorghum halepense* vegetative buds become tillers or rhizomes according to the action of one of three QTLs that influence the number of rhizomes (Paterson *et al.*, 1995).

The research on perenniality in rice is motivated by the potential benefits of persistence in rice growing environments that have been damaged by soil erosion

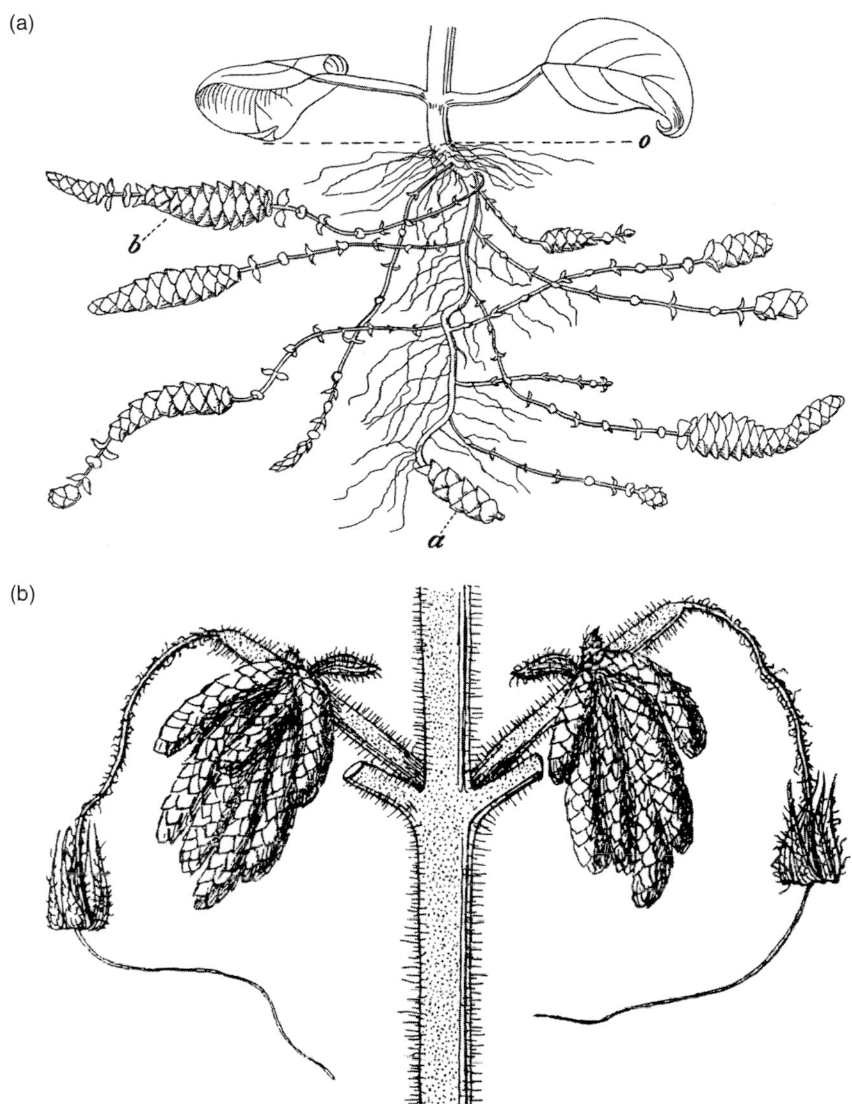

Figure 8.8 (a) *Tydaea lindeniana*. Bulb geophyte with catkin-like bulbs at the end of stolons. *a* is the mother bulb; the terminal bud has grown out into an aerial shoot above the soil surface (*o*), while the underground portion of the shoot bears scale leaves with axillary stolons which terminate in bulbs (*b*). The bulb leaves are thick and fleshy and act as storage organs. (b) *Achimenes cordata*. Bulb geophyte in which the catkin-like bulbs develop from meristems in the axils of prophylls on the flower stalk. The illustration shows two flowers after senescence of their corollas and withering of the upper portion of the flower stalk. Reproduced from Raunkiaer (1934).

(Hu *et al.*, 2003); yet it is this very persistence that is the problem in weedy grasses like *S. halepense* and in leafy spurge. Horvath and colleagues are focused on leafy spurge (*Euphorbia esula*) because it is a pernicious, deep-rooted weed of the Great Plains (Horvath *et al.*, 2002). Perenniality is, in this case, a consequence of the production of adventitious buds by the roots. These buds remain dormant until the aerial portion of the plant dies or is removed. In this case the connection to flowering might seem rather distant, but it is just more subtle than in phanerophytes or chamaephytes; dormancy regulates the release of leafy spurge adventitious buds, and therefore perenniality.

The control over dormancy and flowering by cold signals are so similar as to suggest a common regulatory basis (see Metzger, 1996; Battey, 2000; Horvath *et al.*, 2003). It will be fascinating to see whether *VRN*-type regulatory networks (see Henderson and Dean, 2004) underpin both dormancy release and the control of flowering, as suggested in Figure 8.7, with *Fragaria vesca* in mind. The case of wheat vernalisation, in which *VRN1* and *VRN2* are unrelated to genes of the same name in *Arabidopsis*, yet can be considered to have analogous functions, suggests that convergent evolution may be at work (see Amasino, 2004, for discussion).

8.5 Conclusion

In this brief review we have used Raunkiaer's classification of the life forms of plants as a basis to explore the regulation of flowering in the context of perennial development. We have emphasized the importance of the conservation of meristems in a vegetative state during flowering episodes, in those perennials that maintain an above ground structure. In those that die back to a perennating rhizome or root system, meristem dormancy is critical and a strong mechanistic connection to flowering seems probable. We have not described the behaviour of bulb-forming species (geocryptophytes; see Rees, 1989), bulbil formation or other approaches to asexual propagation that impinge both on perenniality and perennial flowering. These examples show the diversity of reproductive processes within the plant kingdom, and suggest the enormous potential for future research and discovery. In this context, we conclude with two examples from Raunkiaer that emphasize the freedom of plants in combining developmental processes (Figure 8.8). *Tydaea lindeniana* and *Achimenes cordata* are bulb geophytes from the Gesneriaceae. *Tydaea* produces catkin-like bulbs at the end of underground stolons, while *Achimenes* produces similar structures aerially from prophylls on the flower stalk, and buries them in the ground.

Acknowledgements

We are greatly indebted to Sid Thomas (IGER, Aberystwyth) for distilling Raunkiaer's system into the form presented in Table 8.1, for allowing us to reproduce that table here and for helpful comments on the manuscript. We also thank Maarten Koorneef (MPI, Cologne) and Toshi Foster (HortResearch,

Palmerston North) for supplying Figures 8.2 and 8.5. TT is grateful to the UK NERC for a CASE award, and MA to the Max Planck Society for a Fellowship.

References

Aitken, Y. (1985) Temperate forage grasses and legumes, in *Handbook of Flowering*, Vol. I (ed. A. H. Halevy), CRC Press, Florida, pp. 185–202.

Albani, M., Taylor, S., Rodriguez Lopez, C. *et al.* (2001) *Fragaria vesca* ... one way to understand flowering in perennials, *Flowering Newsletter*, **31**, 44–48.

Albani, M. C., Battey, N. H. and Wilkinson, M. J. (2004) The development of ISSR-derived SCAR markers around the *SEASONAL FLOWERING LOCUS* (*SFL*) in *Fragaria vesca*, *Theoretical and Applied Genetics*, **109**, 571–579.

Amasino, R. (2004) Vernalization, competence and the epigenetic memory of winter, *The Plant Cell* , **16**, 2553–2559.

Battey, N. H. (2000) Aspects of seasonality, *Journal of Experimental Botany*, **51**, 1769–1780.

Battey, N. H. (2005) Applications of plant architecture: haute cuisine for plant developmental biologists, in *Plant Architecture and its Manipulation* (ed. C. G. N. Turnbull), Blackwells, UK, pp. 288–314.

Battey, N. H. and Tooke, F. (2002) Molecular control and variation in the floral transition, *Current Opinion in Plant Biology*, **5**, 62–68.

Battey, N. H., Le Miere, P., Tehranifar, A. *et al.* (1998) Genetic and environmental control of flowering in strawberry, in *Genetic and Environmental Manipulation of Horticultural Crops* (eds K. E. Cockshull, D. Gray, G. B. Seymour and B. Thomas), CABI, Wallingford, pp. 111–131.

Beveridge, C. A., Weller, J. L., Singer, S. R. and Hofer, J. M. I. (2002) Axillary meristem development. Budding relationships between networks controlling flowering, branching, and photoperiod responsiveness, *Plant Physiology*, **131**, 927–934.

Brown, T. and Wareing, P. F. (1965) The genetical control of the everbearing habit and three other characters in varieties of *Fragaria vesca*, *Euphytica*, **14**, 97–112.

Buban, T. and Faust, M. (1982) Flower bud induction in apple trees: internal control and differentiation, *Horticultural Reviews*, **4**, 174–203.

Cekic, C., Battey, N. H. and Wilkinson, M. J. (2001) The potential of ISSR-PCR primer-pair combinations for genetic linkage analysis using the seasonal flowering locus in *Fragaria* as a model, *Theoretical and Applied Genetics*, **103**, 540–546.

Cox, T.S., Bender, M., Picone, C. *et al.* (2002) Breeding perennial grain crops, *Critical Reviews in Plant Sciences*, **21**, 59–91.

Dennis, F. Jr (2003) Flowering, pollination and fruit set and development, in *Apples: Botany, Production and Uses* (eds D. C. Ferree and I. J. Warrington), CABI, Wallingford, pp. 153–166.

Diomaiuto, J. (1988) Periodic flowering or continual flowering as a function of temperature in a perennial species: the Ravenelle wallflower (*Cheiranthus cheiri* L.), *Phytomorphology*, **38**, 163–171.

Duke, S. O., Scheffler, B. E., Dayan, F. E. and Dyer, W.E. (2002) Genetic engineering crops for improved weed management traits, *Crop Biotechnology ACS Symposium Series*, **829**, 52–66.

Foster, T., Johnston, R. and Seleznyova, A. (2003) A morphological and quantitative characterization of early floral development in apple (*Malus* x *domestica* Borkh.), *Annals of Botany*, **92**, 199–206.

Gendall, A. R., Levy, Y. Y., Wilson, A. and Dean, C. (2001) The *VERNALIZATION 2* gene mediates the epigenetic regulation of vernalization in *Arabidopsis*, *Cell*, **107**, 525–535.

Grbić, V. and Bleecker, A. B. (1996) An altered body plan is conferred on *Arabidopsis* plants carrying dominant alleles of two genes, *Development*, **122**, 2395–2403.

Gur, A. (1985) Rosaceae – deciduous fruit trees, in *Handbook of Flowering*, Vol. I (ed. A. H. Halevy), CRC Press, Florida, pp. 355–389.

Guttridge, C. G. (1985) *Fragaria* x *ananassa*, in *Handbook of Flowering*, Volume III (ed. A. H. Halevy), CRC Press, Florida, pp. 16–33.

Hancock, J. F. (1999) *Strawberries*, CABI, Wallingford.

Heide, O. M. (1994) Control of flowering and reproduction in temperate grasses, *New Phytologist*, **128**, 347–362.

Henderson, I. R. and Dean, C. (2004) Control of Arabidopsis flowering: the chill before the bloom, *Development*, **131**, 3829–3838.

Horvath, D. P., Anderson, J. V., Chao, W. S. and Foley, M. E. (2003) Knowing when to grow: signals regulating bud dormancy, *Trends in Plant Science*, **8**, 534–540.

Horvath, D. P., Chao, W. S. and Anderson, J. V. (2002) Molecular analysis of signals controlling dormancy and growth in underground adventitious buds of leafy spurge, *Plant Physiology*, **128**, 1439–1446.

Hu, F. Y., Tao, D. Y., Sacks, E. *et al.* (2003) Convergent evolution of perenniality in rice and sorghum, *Proceedings of the National Academy of Sciences of the United States of America*, **100**, 4050–4054.

Jackson, D. I. and Sweet, G. B. (1972) Flower initiation in temperate woody plants, *Horticultural Abstracts*, **42**, 9–25.

Jensen, C. S., Salchert, K. and Nielsen, K. K. (2001) A *Terminal Flower1*-like gene from perennial ryegrass involved in floral transition and axillary meristem identity, *Plant Physiology*, **125**, 1517–1528.

Kirchner, D. E. (2002) Phylo-geography von *Arabis alpina* L. (Brassicaceae), *Botanische Jahrbücher für Systematik, Pflanzengeschichte und Pflanzengeographie*, **124**, 183–210.

Koorneef, M., Blankestijn-de Vries, H., Hanhart, C., Soppe, W. and Peters, T. (1994) The phenotype of some late-flowering mutants is enhanced by a locus on chromosome 5 that is not effective in the Landsberg *erecta* wild-type, *The Plant Journal*, **6**, 911–919.

Kotoda, N. and Wada, M. (2004) *MdTFL1*, a *TFL1*-like gene of apple, retards the transition from the vegetative to reproductive phase in transgenic *Arabidopsis*, *Plant Science*, **168**, 95–104.

Ladizinsky, G. (1998) *Plant Evolution under Domestication*, Kluwer, The Netherlands.

Lefroy, E. C., Hobbs, R. J., O'Conner, M. H. and Pate, J. S. (1999) What can agriculture learn from natural ecosystems?, *Agroforestry Systems*, **45**, 423–436.

Levy, Y. Y., Mesnage, S., Mylne, J. S., Gendall, A. R. and Dean, C. (2002) Multiple roles of *Arabidopsis VRN1* in vernalization and flowering time control, *Science*, **297**, 243–246.

Main, C. L., Mueller, T. C., Hayes, R. M. and Wilkerson, J. B. (2004) Response of selected horseweed (*Conyza canadensis* (L.) Cronq.) populations to glyphosate, *Journal of Agricultural and Food Chemistry*, **52**, 879–883.

Metzger, J. D. (1996) A physiological comparison of vernalization and dormancy chilling requirement, in *Plant Dormancy* (ed. G. A. Lang), CABI, Wallingford, pp. 147–156.

Michaels, S. D. and Amasino, R. M. (2000) Memories of winter: vernalisation and the competence to flower, *Plant, Cell and Environment*, **23**, 1145–1153.

Mouradov, A., Cremer, F. and Coupland, G. (2002) Control of flowering time: interacting pathways as a basis for diversity, *The Plant Cell*, **14**, S111–S130.

Ng, C. H., Ratnam, W., Surif, S. and Ismail, B. S. (2004) Inheritance of glyphosate resistance in goosegrass (*Eleusine indica*), *Weed Science*, **52**, 564–570.

Paterson, A. H., Schertz, K. F., Lin, Y.-R., Liu, S.-C. and Chang, Y.-L. (1995) The weediness of wild plants: molecular analysis of genes influencing dispersal and persistence of johnsongrass, *Sorghum halepense* (L.) Pers, *Proceedings of the National Academy of Sciences of the United States of America*, **92**, 6127–6131.

Poduska, B., Humphrey, T., Redweik, A. and Grbic, V. (2003) The synergistic activation of *FLOWERING LOCUS C* by *FRIGIDA* and a new flowering gene *AERIAL ROSETTE 1* underlies a novel morphology in Arabidopsis, *Genetics*, **163**, 1457–1465.

Pollard, H., Cruzan, M. and Pigliucci, M. (2001) Comparative studies of reaction norms in *Arabidopsis*. I. Evolution of response to daylength, *Evolutionary Ecology Research*, **3**, 129–155.

Raunkiaer, C. (1934) *The Life Forms of Plants and Statistical Plant Geography*, Oxford University Press, Oxford, UK.

Rees, A. R. (1989) Evolution of the geophytic habit and its physiological advantages, *Herbertia*, **45**, 104–110.

Scheinost, P. L., Lammer, D. L., Cai, X., Murray, T. D. and Jones, S. S. (2001) Perennial wheat: the development of a sustainable cropping system for the US Pacific northwest, *American Journal of Alternative Agriculture*, **16**, 147–151.

Searle, I. and Coupland, G. (2004) Induction of flowering by seasonal changes in photoperiod, *EMBO Journal*, **23**, 1217–1222.

Sedgley, M. (1990) Flowering of deciduous perennial fruit crops, *Horticultural Reviews*, **12**, 223–264.

Soule, J. and Piper, J. K. (1992) *Farming in Nature's Image: An Ecological Approach to Agriculture*, Island Press, USA.

Sreekantan, L., McKenzie, M. J., Jameson, P. E. and Clemens, J. (2001) Cycles of floral and vegetative development in *Metrosideros excelsa* (Myrtaceae), *International Journal of Plant Sciences*, **162**, 719–727.

Sung, S. and Amasino, R. M. (2004) Vernalization in *Arabidopsis thaliana* is mediated by the PHD finger protein VIN3, *Nature*, **427**, 159–164.

Thomas, H. (2003). Do green plants age, and if so, how?, *Topics in Current Genetics*, **3**, 145–171.

Thomas, H., Thomas, H. M. and Ougham, H. (2000) Annuality, perenniality and cell death, *Journal of Experimental Botany*, **51**, 1781–1788.

Tooke, F., Ordidge, M., Chiurugwi, T. and Battey, N. H. (2005) Mechanisms and function of flower and inflorescence reversion, *Journal of Experimental Botany*, **56**, 2587–2599.

Wagoner, P. (1990) Perennial grain-development: past efforts and potential for the future, *Critical Reviews in Plant Sciences*, **9**, 381–409.

Part III A developmental genetic model for the origin of the flower

9 Flower colour

Yoshikazu Tanaka and Filippa Brugliera

9.1 Introduction

Coloured flowers have evolved mainly in response to selection pressure from pollinators. However, largely due to their aesthetic appeal, horticulturalists and garden enthusiasts have worked with flowers for centuries. Literature and art are replete with references to the beauty of flowers, illustrating that their significance to humans goes beyond their traded value. It is in more recent times that flowers have become of global commercial value.

I wandered lonely as a cloud
That floats on high o'er vales and hills,
When all at once I saw a crowd,
A host, of golden daffodils.

William Wordsworth (1770–1850)

Moreover, flower colour has provided an indelible marker for the elucidation of what has proved to be a most complex and fascinating metabolic pathway, so art, science and commerce have been joined.

The world floriculture industry recorded retail sales in Japan, USA and Europe of around €27 billion in 2002 (Chandler, 2003). The ornamental horticulture industry strives to develop new and different varieties of flowering plants to meet demand and maintain market share for particular species. An effective way to create such novel varieties is through the manipulation of flower colour and, to this end, classical breeding techniques have been used with considerable success, resulting in a wide range of colours for most commercial varieties of flowers. Traditional breeding techniques are limited, however, by the constraints of a particular species' gene pool and for this reason it is rare for a single species to have a full spectrum of coloured varieties. Significantly, until 1996, none of the three top-selling cut-flowers – roses, chrysanthemums and carnations – which between them account for more than 50% of sales, have violet to blue varieties. Yellow flowering varieties are missing from top-selling pot plants such as pelargonium, begonia and cyclamen, whilst orange-coloured flowers are absent from chrysanthemum, cymbidium and petunia.

There are a number of recent comprehensive reviews on manipulation of flower colour (Tanaka *et al.*, 1998, 2005; Mol *et al.*, 1999; Tanaka and Mason, 2003; Davies, 2004). Instead of reviewing all the material previously covered, this chapter will focus on recent developments in understanding the biochemistry and molecular biology of flower colour and the manipulation of flower colour via genetic engineering.

9.2 Flower colour

Flavonoids, carotenoids and betalains are the three main pigment classes that contribute to the diverse range of flower colours seen in nature (Plate 9.1). Betalains, nitrogen containing water-soluble compounds, are the least abundant of these three classes. They contribute to various hues of ivory, yellow, orange, red and violet (Strack *et al.*, 2003). Carotenoids are C-40 tetraterpenoids that are lipid soluble and are located in plastids or derivatives thereof. They contribute to the majority of yellow to orange hues in a number of flowers. Carotenoids, along with red or magenta anthocyanins, also contribute to the orange/red, bronze and brown colours seen in flowers such as roses (de Vries *et al.*, 1974; Eugster and Marki-Fischer, 1991) and chrysanthemums (Jordan and Reimann-Philipp, 1983). Of the three pigment classes, the water-soluble flavonoids are the most common and are responsible for a range of colours from yellow to red to violet to blue. An extensive study by Kay *et al.* (1981) has shown that flavonoids are almost exclusively located in the epidermal cells. Flavonoids are derivatives of the phenylpropanoid pathway and include aurones, chalcones, flavones, flavonols, flavanones, isoflavones and anthocyanins. Aurones contribute to the yellow colours seen in *Antirrhinum*, *Cosmos*, *Dahlia* (Nakayama *et al.*, 2000a). Chalcones are paler yellow and are found in carnation and peony. Flavones and flavonols are generally colourless but can act as copigments to enhance flower colour and bluing (Goto, 1989). However some flavonols can also contribute to yellow colours seen in *Tagetes* sp. and *Chrysosplenium americanum* (Anzelloti *et al.*, 2000; Halbwirth *et al.*, 2004). Anthocyanins are *O*-glycosides and are coloured flavonoids that are the main determinants of flower colour. They are predominantely found in the vacuoles of petal epidermal cells (Koes *et al.*, 1990). Anthocyanins can also be found in other organs such as leaves, seeds, fruit, pollen and styles.

9.3 The pigment classes

Although betalains and carotenoids do contribute to flower colour, the flavonoids are the most understood and studied. This chapter will therefore give a brief overview of the status of betalains and carotenoids with respect to flower colour but the main focus will be on the flavonoids.

9.3.1 Betalains

Betalains are water-soluble vacuolar pigments containing nitrogen, which are classified into the red–violet betacyanins and the yellow betaxanthins. The chemical structures, distribution and biosynthesis of betalains have been recently reviewed by Strack *et al.* (2003). Betalains are conjugates of betalamic acid (chromophore) with *cyclo*-dihydroxyphenylalanine (*cyclo*-DOPA) and amino acids or amines. They are found in flowers, fruits and vegetative tissues of Caryophyllales plants (such as cacti, red beet and *Portulaca*) with the exception of plants of the Caryophyllaceae and Molluginaceae that generally accumulate anthocyanins. Interestingly some mushrooms also accumulate betalains.

The biosynthetic pathway of betalains (Figure 9.1) is less understood than that of flavonoids and there is an argument about the step at which glucosylation occurs. Some important progresses have recently been achieved in understanding betalain biosynthesis. The first enzyme in the pathway is tyrosinase, which hydroxylates tyrosine to DOPA and oxidises DOPA to dopaquinine which is then spontaneously cyclised to *cyclo*-DOPA (Strack *et al.*, 2003). Tyrosinase, a polyphenol oxidase, has been partially purified from *Portulaca* and characterised (Steiner *et al.*, 1999). A *PPO* gene has been cloned from *Phytolacca americana* (Joy *et al.*, 1995); however, it is unclear that the gene is actually involved in betalain biosynthesis.

Dopa-4,5-dioxygenase (DODA) catalyzes the synthesis of seco-DOPA that is spontaneously converted to betalamic acid. The gene encoding DODA has been cloned from *Portulaca grandiflora* (Christinet *et al.*, 2004). Its transient expression following particle bombardment of *Portulaca* petals yielded yellow and violet spots of betacyanins and betaxanthins, respectively. The enzyme is homologous to bacterial extradiol 4,5-dioxygenases that require ferric ions for activity. A *DODA* gene has also been obtained from a fungus (*Amanita* sp.; Hinz *et al.*, 1997) and shown to be functional in a transient assay expression system leading to the production of betalains in *Portulaca* petals (Mueller *et al.*, 1997). The deduced amino acid sequences of plant and fungal DOPAs do not appear related. The condensation of betalamic acid with cyclo-DOPA, amino acids or amines proceeds spontaneously to form betanidin and betaxanthins, respectively (Schliemann *et al.*, 1999).

Betanidin 5-*O*- and 6-*O*-glucosyltransferases (B5GT and B6GT) catalyze the glucosylation of betanidin. Genes encoding B5GT and B6GT have been cloned from *Dorotheanthus bellidiformis* (Vogt *et al.*, 1999; Vogt, 2002). The B5GT and B6GT have broad substrate specificity and in an *in vitro* assay were shown to also glucosylate the C-4′ hydroxyl group of the flavonol, quercetin and the C-3 hydroxyl group of the anthocyanidin, cyanidin. Not surprisingly, the predicted amino acid sequence of B5GT appears closely related to the flavonoid 7/3′ GT family. Tobacco glucosyltransferases, NtGT1 and NtGT3, that also have broad substrate specificity (Taguchi *et al.*, 2001, 2003) belong to the same phylogenetic group as B6GT.

An alternative biosynthetic pathway has also been proposed recently. UDP-glucose: *cyclo*-DOPA 5-*O*-glucosyltransferase (cDOPA5GT) activity was demonstrated in *Mirabilis jalapa* (four o'clock). The activity was correlated with that of betanin accumulation during petal development and was detected in another five species of Centrospermae. It has been suggested that the glucose moiety of betanin is introduced at the *cyclo*-DOPA step rather than at the betanidin aglycon

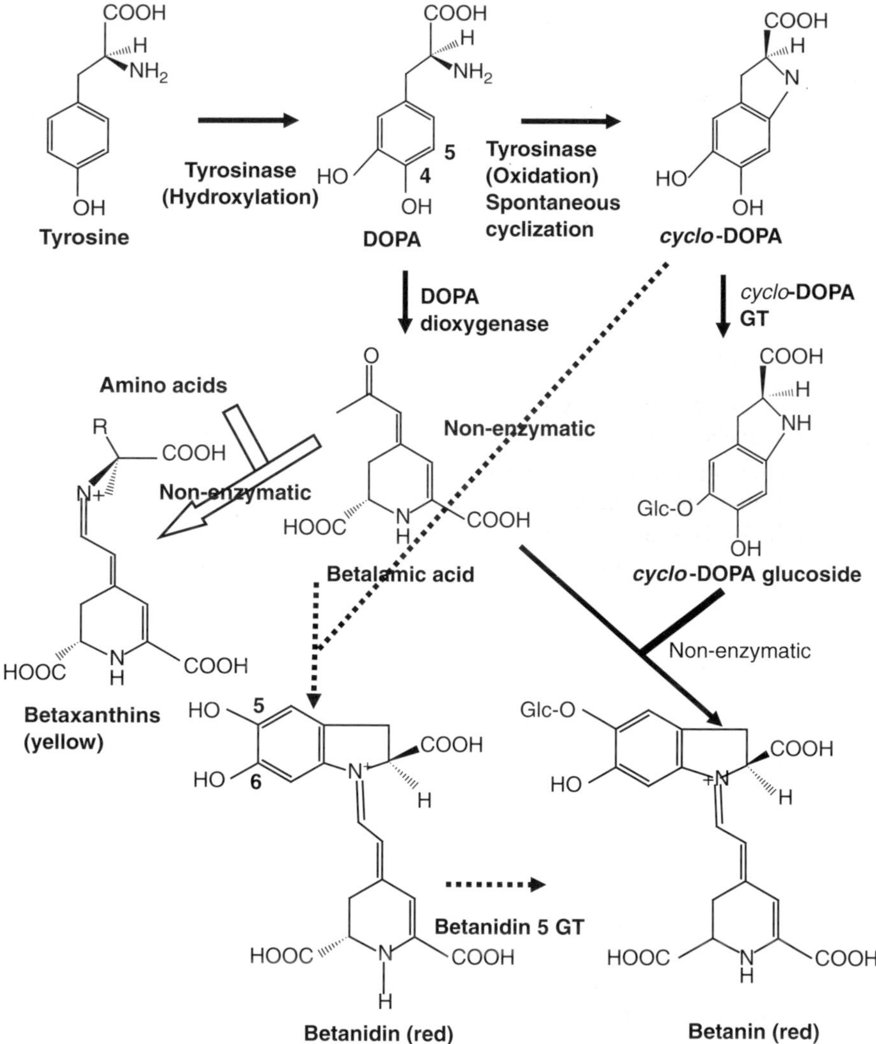

Figure 9.1 Major betalain biosynthetic pathway (adopted and simplified from Strack *et al.*, 2003; Sasaki *et al.*, 2004). Glucosylation may occur at the step of DOPA (solid arrow; Sasaki *et al.*, 2004) or betanidin (Strack *et al.* 2003).

step (Figure 9.1) (Sasaki *et al.*, 2004). The gene encoding a cDOPA5GT has been cloned from *M. jalapa* and *Celpsia cristata* (feather cockscomb). The glucosylated *cyclo*-DOPA spontaneously reacted with betalamic acid *in vitro*. Unlike betanidin 5GT described earlier, the DOPA5GT from *M. jalapa* appeared to be a specific *cyclo*-DOPA as it was unable to catalyze glucose transfer to DOPA, quercetin, cyanidin or betanidin. DOPA5GTs form a separate family from other flavonoid/betalain GT families (Sasaki *et al.*, 2005).

B5GT with its broad substrate specificity *in vitro* may possibly not be involved specifically with betanin biosynthesis. Even *Arabidopsis*, having the smallest genome, contains about 120 glucosyltransferase sequences. Out of 90 recombinant *Arabidopsis* glucosyltransferases 48 showed activity towards hydroxycoumarins as acceptors (Lim *et al.*, 2003), confirming that it is notoriously difficult to elucidate *in vivo* function from *in vitro* data.

Betanidin 5-glucoside is further modified by hydroxycinnamoyltransferase that utilises hydroxycinnamoylglucose as an acyl donor (Bokern *et al.*, 1992) while anthocyanin acyltransferases utilise hydroxycinnamoyl-CoA as acyl donors.

9.3.2 Carotenoids

Carotenoids are tetraterpene (C40) molecules derived from eight isoprene units and have a critical photoprotective role in chloroplasts. They are classified into carotenes, containing a conjugated double bond between carbon hydrogen and xanthophylls, having terminal rings (Figure 9.2; Fraser and Bramley, 2004). Carotenoids are responsible for the orange/yellow colour of flowers and fruits of higher plants and accumulate in the chromoplasts and related organella. The presence or absence of carotenoids altered the flower colour of *Mimulus* and was shown to affect the preference of pollinators. Bumblebees preferred dark pink flower colours whilst hummingbirds showed a preference for the yellow-orange colours (Bradshaw and Schemske, 2003). Chrysanthemum flower colour is due to both flavonoid and carotenoid pigments. To date, the carotenoids isolated from chrysanthemum, except for (9Z)-vioalaxanthin, are β-, ε-carotene (α-carotene) derivatives (Kishimoto *et al.*, 2004).

The carotenoid biosynthetic pathway (Figure 9.2) in plants has been elucidated using several biochemical and molecular approaches (Hirschberg, 2001; Fraser and Bramley, 2004) with the isolation of key genes from various species (Fraser and Bramley, 2004). A brief overview of the carotenoid pathway in higher plants is described later and in Figure 9.2. For a detailed review refer to Fraser and Bramley (2004). Colourless phytoene is synthesised from two molecules of geranylgeranyl pyrophosphate by the catalysis of phytoene synthase. Phytoene desaturase and ζ-carotene desaturase introduce two double bonds each to yield lycopene. Lycopene is converted to β-carotene via the catalysis of β-cyclase and then to zeaxanthin by β-hydroxylase. Vioalaxanthin epoxidase catalyzes the synthesis of vioalaxanthin from zeaxanthin. This reaction is reversible and de-epoxidation by zeaxanthin epoxidase can convert vioalaxanthin back to zeaxanthin. Vioalaxanthin is modified to neoxanthin by the catalysis of neoxanthin synthase. Lycopene can alternatively be converted to α-carotene by lycopene β-cyclase and ε-cyclase, and α-carotene can then be hydroxylated to form lutein by β- and ε-hydroxylase.

It is perhaps the pharmaceutical or nutraceutical benefits of carotenoids that have received the most attention with respect to metabolic engineering of the pathway (Giuliano *et al.*, 2000; Fraser and Bramley, 2004). Most of the genetic engineering experiments regarding carotenoids have been aimed at an improvement in nutritional quality of crops such as carrot, potato, tomato, canola and rice

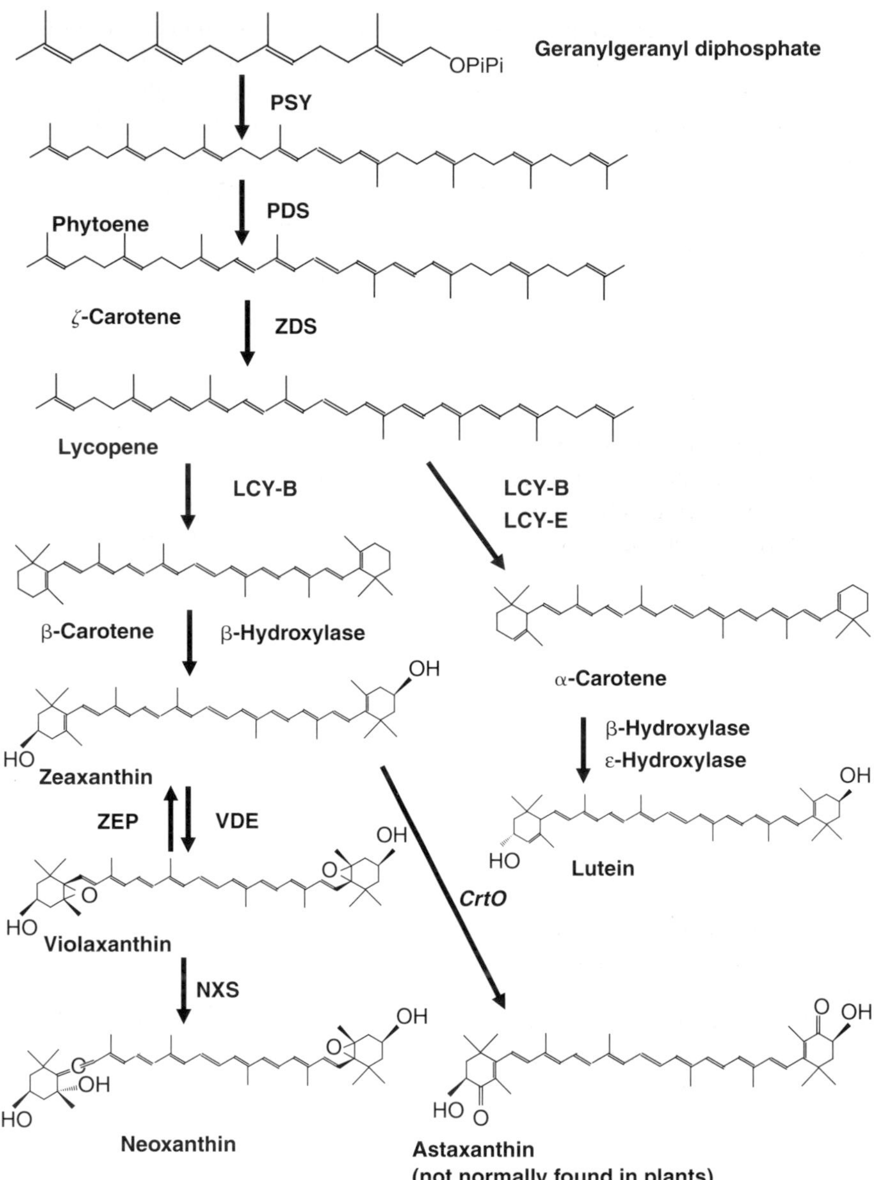

Figure 9.2 A snapshot of the carotenoid biosynthetic pathway in plants (adopted and simplified from Fraser and Bramley, 2004). The biosynthesis starts from geranylgeranyl diphospate, a diterpene. Bacteria and algae have similar but different pathways due to differing substrate specificities of the enzymes involved. Astaxanthin is not normally produced in plants but have been engineered by the introduction of the *CrtO* gene from algae. Abbreviations include: PSY, phytoene synthase; PDS, phytoene desaturase; ZDS, ζ-carotene desaturase; LCY-B, lycopene β-cyclase; LCY-E, lycopene ε-cyclase; VDE, vioalaxanthin epoxidase; ZEP, zeaxanthin epoxidase; NXS, neoxanthin synthase.

(reviewed in Fraser and Bramley, 2004). To date, reports on modification of flower colour by engineering the carotenoid pathway have used tobacco as a model system. The nectaries of tobacco flowers contain chromoplasts that accumulate β-carotene and vioalaxanthin and are normally yellow in colour. Mann *et al.* (2000) introduced the coding region of the *CrtO* gene encoding β-carotene ketolase from the alga *Haematococcus pluvialis* under the control of carotenoid pathway specific promoter into *Nicotiana tabacum*. Only the nectaries of the flowers of the transgenic plants changed colour from yellow to red and accumulated astaxanthin (not normally found in plants) and other ketocarotenoids. The total carotenoid concentration in the transgenic nectaries was also increased by 43%.

9.3.3 Flavonoids

The flavonoids are secondary metabolites having C6–C3–C6 structures and are derived from the phenylpropanoid pathway. Flavonoids are virtually ubiquitous in green plants. They absorb UV-B and thus protect plants from UV-damage (Ryan *et al.*, 2001). Plants respond to UV-light exposure by producing flavonoid pigments in organs that do not normally produce them. They are antioxidant and scavenge radicals that are formed from various biotic and abiotic processes. Their antioxidant activity promotes human and domestic animal health. Some are inhibitory to insect feeding and play roles in interactions with microorganisms. Attack by plant pathogens can also lead to the production in plants of small molecules (e.g. phytoalexins), which are flavonoid derivatives. *Phaseolus vulgaris* (common bean) and *Petroselinum crispum* (parsley) produce flavonoid phytoalexins in response to attacks by pathogens such as fungal elicitors (reviewed by Harborne and Williams, 2000).

Flavonoids, especially anthocyanins, provide pigmentation to plant parts such as flowers and fruits and therefore act as attractants to pollinators and fruit-eating animals. The number of pollinator visits received by *Ipomoea purpurea* (the common morning glory) flowers appears related to their colour. Pollinators of *I. purpurea* flowers appear to be most attracted to intensely coloured flowers and least attracted to white flowers (Ennos and Clegg, 1983). While *Ipomoea* species having blue flowers (*I. nil* – the Japanese morning glory, *I. purpurea, I. tricolor*) are bee pollinated, red flowering *Ipomoea* species such as *I. quamoclit* are hummingbird pollinated (Zufall and Rausher, 2004).

9.4 Anthocyanin structure and flower colour

The structures of hundreds of anthocyanins have been determined (Harborne and Williams, 2000; Honda and Saito, 2002). However, in spite of this, almost all of them share anthocyanidin 3-glycoside (usually glucosides) as a common structure. Further modification by glycosylation, acylation and methylation is dependent on the plant species and varieties. Glucosylation, acylation and methylation are catalyzed by specific glucosyltransferases, acyltransferases and methyltransferases, respectively. In general, when the modifications on the anthocyanin molecule are equivalent, delphinidin-based anthocyanins (including the methylated derivatives, petunidin and malvidin) are bluer than cyanidin-based anthocyanins (including its methylated derivative, peonidin), which are bluer than pelargonidin-based anthocyanins. Some plant species, e.g. sorghum, contain minor anthocyanidins such as 3-deoxyanthocyanidins (Kang *et al.*, 2003).

Plants belonging to the same family usually have structurally similar anthocyanins. Examples of anthocyanins found in rose (Rosaceae), carnation (Caryophyllaceae), chrysanthemum (Compositae), cineraria (Compositae), salvia (Laviatae), petunia (Solanaceae), gentian (Gentianaceae), butterfly pea (Leguminosae) and morning glory (Convolvulaceae) are shown in Figure 9.3. Blue flowers tend to contain delphinidin-type anthocyanins often modified with multiple aromatic acyl groups (polyacylated anthocyanins). The bluing effect and stabilisation of flower colours depend on the number of aromatic acyl groups (Honda and Saito, 2002). Aromatic acyl moieties of anthocyanins contribute to intramolecular stacking of anthocyanins and thus bluing and stabilisation of the pigment. It has been shown that the 3′-aromatic moiety of gentiodelphin (Figure 9.3), an anthocyanin found in gentian flowers, plays a major role in the intramolecular stacking (Yoshida *et al.*, 2000). Kazuma *et al.* (2003) revealed that glucosylation at the 3′- and 5′-positions of ternatins (butterfly pea anthocyanins, Figure 9.3) was critical in producing blue petals because the missing glucosylations prevented polyacylation. A butterfly pea variety containing anthocyanins without polylacylation produced mauve flowers. The 'Heavenly blue' anthocyanin of *I. tricolor* and *I. nil* is a cyanidin-based anthocyanin with a high degree of modification (three aromatic acyl groups and six glucosyl groups) and together with a high petal vacuolar pH (about 7.5) make the colour blue. This illustrates that a number of factors impact on the final colour observed in a flower. Although the predominant contribution is from the type of anthocyanidin, factors such as the modification(s) on the anthocyanidin molecule along with the vacuolar pH and the presence of copigments and metal ions all contribute to flower colour.

9.5 Complexity of blue flowers

Flowers yield blue colours via a sophisticated combination of the factors mentioned earlier.

Copigments such as flavones and flavonols are generally colourless by themselves but contribute to bathochromic shifts (blue and more intense colour) and stabilisation of anthocyanins by forming intermolecular stacking complexes (Goto, 1989). Where anthocyanin/flavone complexes exist in blue flowers there is usually only one specific flavone constituent that acts as a copigment and the flavone to anthocyanin ratios are generally high unless a metal cation is involved in the complex (Harborne and Williams, 2000).

Plate 9.1 (a) Cineraria flowers containing anthocyanins. (b) Chrysanthemum flowers containing carotenoids. (c) Four o'clock flowers containing betalains.

Plate 9.2 (Continued).

(d)

Plate 9.2 (Continued).

Plate 9.2 Colour modified flowers by genetic engineering. (a) Antisense *CHS* expression in *Petunia hybrida* cv Old Glory Blue. The host and three transgenic flowers are shown (Florigene Ltd., unpublished results). (b) *Petunia cultivar* Surfinia Purple Mini (left), transgenic petunia derived from Surfinia Purple Mini by down regulation of $F3'5'H$, *AR-AT* and *FLS*. (c) The colour varies depending on transgenic lines or even branches of one plant. (d) *Torenia hybrida* Summerwave Blue (left), a white transgenic plant obtained using RNAi of *ANS* gene (center), a pink transgenic plant (right) obtained using RNAi of $F3'5'H$and $F3'H$ genes and expression of pelargonium *DFR* gene. (e) Transgenic carnations accumulating delphinidin that are commercially available, Florigene Moondust^{TM}, Florigene Moonshadow^{TM}, Florigene Moonaqua^{TM}, Florigene Moonlite^{TM}, Florigene Moonshade^{TM}, Florigene Moonvista^{TM}. (f) Transgenic roses accumulating delphinidin have blue hue that conventionally bred roses have never achieved.

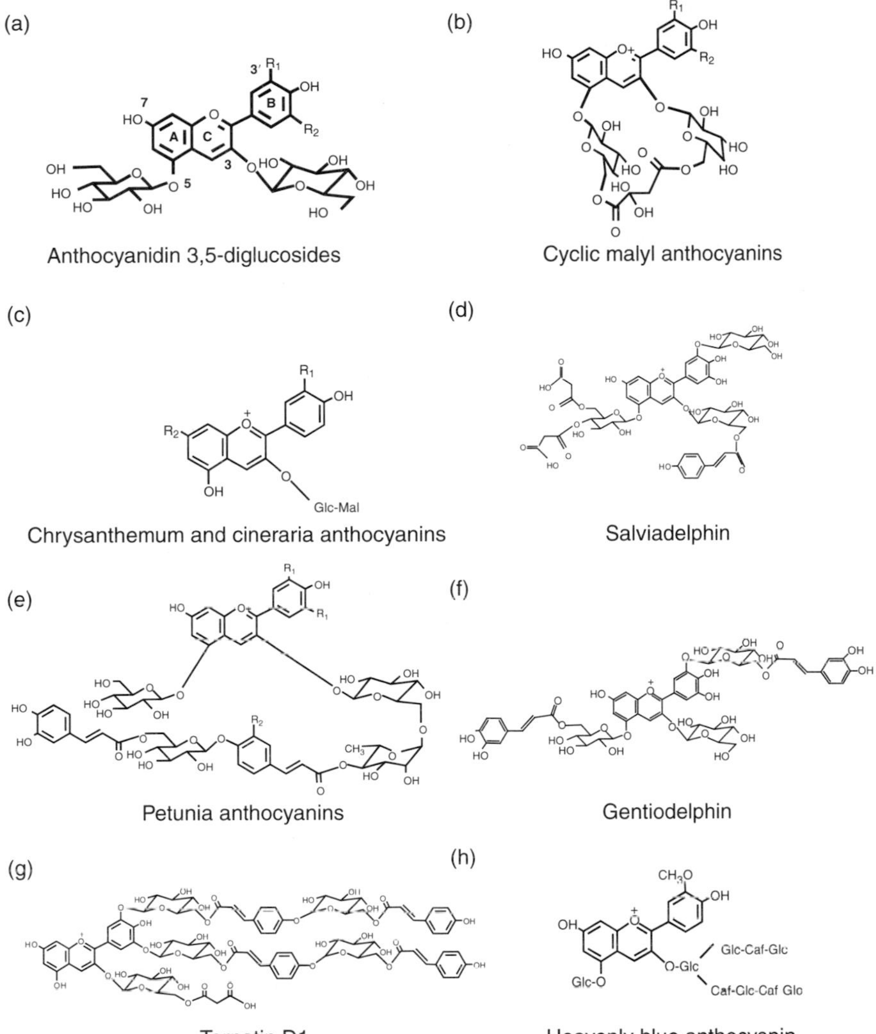

Figure 9.3 Examples of anthocyanin structures that accumulate in the vacuoles of petal epidermal cells. (a) Pelargonidin (R1 = R2 = H) or cyanidin (R1 = H, R2 = OH) 3, 5-diglucosides predominate in rose petals. Transgenic roses expressing a heterologous flavonoid 3′, 5′-hydroxylase genes (*F3′5′H*) produce delphinidin 3, 5-diglucoside (R1 = R2 = OH). (b) Carnations petals accumulate bridge anthocyanins. Native carnations contain pelargonidin-based (R1 = R2 = H) or cyanidin-based (R1 = H, R2 = OH) anthocyanins. Transgenic carnation expressing a heterologous *F3′5′H* genes accumulate delphinidin-based anthocyanins (R1 = R2 = OH). (c) Examples of anthocyanins found in chrysanthemum (R1 = R2 = H) and cineraria (cinerarin, R1 = caffeoyl-glucosyl, R2 = caffeoyl-glucosyl-caffeoyl-glucosyl). (d) Salviadelphin, a blue anthocyanin found in *Salvia* petals. (e) Anthocyanins that accumulate in petunia flowers R_1 can be H or OCH_3. R2 = H or Caffeoyl-glucosyl or coumaroyl-glucosyl. (f) A gentian anthocyanin, Gentiodelphin. (g) Ternatin, an anthocyanin found in butterfly pea flowers. (h) A morning glory anthocyanin, Heavenly blue anthocyanin.

Metal ions complexed with anthocyanins have been described from several plant species (reviewed by Harborne and Williams, 2000). Commelinin, a blue complex of *Commelina communis*, consists of six molecules of delphinidin glycoside (manonylawobanin), six molecules of flavone (flavocommelinin) and two metal ions (Fe^{3+} and Mg^{2+}). The anthocyanin complex found in corn flower (*Centaurea cyanus*) has a similar composition and consists of six succinylcyanin molecules, six flavone molecules, Fe^{3+}, and two Ca^{2+} (Shiono *et al.*, 2005). However, the anthocyanidin is based on cyanidin rather than delphinidin. The blue flower (sepal) colour observed in hydrangeas (*Hydrangea macrophylla*) is a result of a complex of delphinidin 3-glucoside, quinic acid and Al^{3+}. It is commonly known to most gardeners that the acidity of the soil and content of Al^{3+} influences the colour of hydrangea flowers. Acidic soil increases the availability of Al^{3+} in the soil and leads to a change of flower colour from pink to blue.

The impact of the pH of its surroundings also has a dramatic impact on the colour of anthocyanins. Anthocyanins predominate as their stable flavylium ions and appear orange or red in acidic environments of pH less than 2.5. The violet anhydrobase is formed in pH solutions of 4–6 but this is rapidly converted to the colourless pseudobase. Whilst in basic aqueous solutions the blue ionised quinoidal anhydrobase predominates (Goto and Kondo, 1991). The estimated pH of pigmented cell sap is generally within the range of 4.5–5.5 (Haslam, 1998). However the estimated vacuolar pH of pigmented epidermal cells ranges from 2.5 in begonia (Stewart *et al.*, 1975) to 7.7 in morning glory (Yoshida *et al.*, 1995). Yoshida *et al.* (2003) separated blue and red cells of hydrangea sepals and measured the vacuolar pH with a proton-selective microelectrode. They concluded that the vacuolar pH of blue cells (4.1) was higher than red cells (3.3). They also measured the vacuolar pH of morning glory petal epidermal cells with the microelectrode and revealed the pH increased from 6.6 to 7.7 during flower opening (Yoshida *et al.*, 1995). The flower colour changes from purple to blue during the same period.

9.6 Biosynthesis of flavonoids

9.6.1 Phenylpropanoid pathway

The basic structure of the flavonoid molecule consists of two aromatic rings (A and B) joined by a C3 unit with oxygen (C ring) (Figure 9.4). The flavonoids can be classified into several classes depending on the oxidation level of the basic structure.

The phenylpropanoid and flavonoid biosynthetic pathways have been well characterised (Haslam, 1998) and are summarised in Figure 9.4. Phenylalanine ammonia lyase (PAL) catalyzes the first step in the general phenylpropanoid pathway in which phenylalanine is converted to *trans*-cinnamic acid. The *trans*-cinnamic acid produced by the PAL reaction is converted to 4-coumaric acid by the action of cinnamate 4-hydroxylase (C4H). The activation of hydrocinnamic acids resulting in thioesters of coenzyme A is catalyzed by 4-coumarate: CoA ligase (4CL).

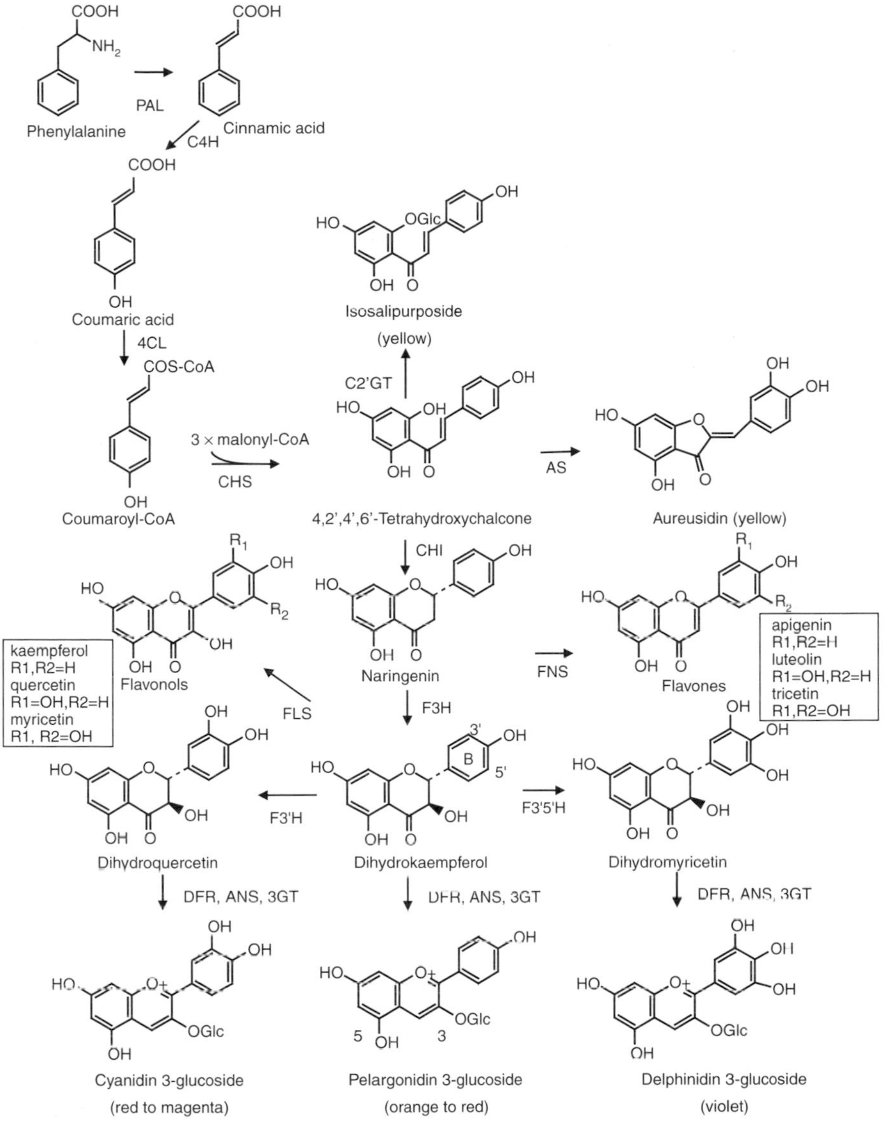

Figure 9.4 The flavonoid biosynthetic pathway (a derivative of the phenylpropanoid pathway) leading to the production of the first coloured anthocyanins, anthocyanidin 3-*O*-glucosides. Anthocyanidin 3-*O*-glucosides can be further modified with glycosyl, acyl or methyl groups in a species-specific manner. Abbreviations include: PAL, phenylalanine ammonia lyase; C4H, cinnamic acid 4-hydroxylase; 4CL, 4-coumarate Co-A ligase; C2′GT, UDP-glucose: tetrahydroxychalcone 2′-*O*-glucosyltransferase; CHS, chalcone synthase; CHI, chalcone isomerase; AS, aureusidin synthase; F3H, flavanone 3β-hydroxylase; F3′H, flavonoid 3′-hydroxylase; F3′5′H, flavonoid 3′, 5′-hydroxylase; DFR, dihydroflavonol 4-reductase; ANS, anthocyanidin synthase; FNS, flavone synthase; FLS, flavonol synthase; 3GT, UDP-glucose: anthocyanidin 3-*O*-glucosyltransferase.

The phenylpropanoid pathway has many branches at this point. These lead to the production of flavonoids, lignins, stilbenes and coumarins.

9.6.2 *Main pathway leading to anthocyanidin 3-glucoside*

The flavonoid biosynthetic pathway, leading to anthocyanidin 3-glucosides (Figure 9.4), is generally conserved among plant species and well understood. Studies into the biochemistry, genetics and molecular biology of the flavonoid pathway and its manipulation have been extensively reviewed (Winkel-Shirley, 2001a,b; Springob *et al.*, 2003; Davies, 2004; Schijlen *et al.*, 2004; Tanaka *et al.*, 2005) and the structural and regulatory genes in the pathway have been cloned from many plant species. The first committed step of flavonoid biosynthesis is the stepwise condensation of three acetate units of malonyl-CoA with *p*-coumaroyl-CoA to yield 4, 2′, 4′, 6′-tetrahydroxychalcone (THC). THC is stereospecifically isomerised to (2*S*)-naringenin (a flavanone) by the catalysis of chalcone isomerase (CHI). (2*S*)-Flavanones are subsequently converted to the corresponding (2*R*, 3*R*)-dihydroflavonols (e.g. naringenin to dihydrokaempferol) by a stereospecific hydroxylation at position 3 catalyzed by flavanone 3β-hydroxylase (F3H). (2*R*, 3*R*)-Dihydroflavonols are reduced to (*2R, 3S, 4S*)-leucoanthocyanidins by dihydroflavonol 4-reductase (DFR). Anthocyanidin synthase (ANS) then catalyzes the synthesis of coloured but unstable anthocyanidins from colourless leucoanthocyanidins. The anthocyanidins are then stabilised by glucosylation at the 3-*O*-position by UDP-glucose: anthocyanidin/flavonoid 3-*O*-glucosyltransferase (3GT) to produce anthocyanins, which are subsequently transported to and stored in the vacuole.

9.6.3 *Flavanones to flavones*

Flavones, a class of flavonoids, are synthesised from flavanones by the catalytic activity of flavone synthase (FNS). Interestingly there are two kinds of FNS; one is a 2-oxo-glutarate-dependent dioxygenase (FNSI) and the other a cytochrome P450-type mono-oxygenase (FNSII). FNSI has only been found in parsley and the gene encoding FNSI has been cloned recently (Martens *et al.*, 2003a). FNSII is more ubiquitous and genes encoding FNSII have been obtained in snapdragon (Akashi *et al.*, 1999), torenia (Akashi *et al.*, 1999), perilla (Kitada *et al.*, 2001), gerbera (Martens and Forkmann, 1999) and China aster (Martens and Forkmann, Genbank accession number AAF04115). In legumes, flavones are synthesised by a two-step reaction involving first flavanone 2-hydroxylase and then a dehydration step (Akashi *et al.*, 1998).

9.6.4 *Dihydroflavonols to flavonols*

Flavonols, another class of flavonoids, are synthesised from the corresponding dihydroflavonols by the action of flavonol synthase (FLS). FLS catalyzes the introduction of a double bond between 2 and 3 of the C-ring. FLS is a soluble dioxygenase which requires 2-oxoglutarate, Fe^{2+} and ascorbate as co-factors (Forkmann *et al.*, 1986; Stich *et al.*, 1992; Holton *et al.*, 1993b). Flavonols are glycosylated by glycosyltransferases (Griesbach and Asen, 1990). The UDP glucose: flavonoid 3-glucosyltransferase (3GT) isolated from *Petunia hybrida* has broad substrate specificity and can act on both anthocyanidins and flavonols (Kho *et al.*, 1978; Gerats *et al.*, 1983; Yamazaki *et al.*, 2002). In contrast a 3GT isolated from *Brassica oleracea* cv. Red Danish is specific for flavonol glucoside biosynthesis (Sun and Hrazdina, 1991). The glycosylated flavonols are generally colourless but are important contributors to flower colour as they act as copigments with the anthocyanins to enhance flower colour (Scott-Moncrieff, 1936; Goto, 1989).

An important point to note here is that both DFR and FLS utilise the same substrate and therefore effectively compete for substrate. FLS activity generally appears at an earlier developmental stage than DFR and so flavonols are produced at an earlier stage of development than the coloured anthocyanins (Tanaka and Mason, 2003).

9.6.5 *Hydroxylation of B-ring*

The number of hydroxyl groups on the B-ring of the flavonoid molecule is a key factor in the final colour observed. Pelargonidin-based pigments (one hydroxyl group on the B-ring) tend to result in the production of brick red/red flower colours. Cyanidin-based pigments (two hydroxyl groups) tend to give red/magenta flower colours. Delphinidin-based pigments (three hydroxyl groups) generally result in violet flower colours (Figure 9.4). Hydroxylation of B-ring at the 3′ position is catalyzed by flavonoid 3′-hydroxylase (F3′H) and at both the 3′ and 5′ positions by flavonoid 3′, 5′-hydroxylase (F3′5′H). Therefore, F3′H and F3′5′H activities are key determinants of flower colour. Both F3′H and F3′5′H are microsomal cytochrome P450 enzymes. These enzymes generally have broad substrate specificities and are able to catalyze hydroxylation of flavanones, dihydroflavonols, flavonols and flavones. Recent evidence lends support to the idea that the flavonoid pathway enzymes form multi-enzyme complexes that are anchored to the ER by the membrane bound cytochromes P450 (reviewed in Winkel, 2004). Saslowsky and Winkel-Shirley (2001) showed that CHS and CHI co-localised in electron dense particles in *Arabidopsis* root cells which were not observed in a *tt7* mutant lacking F3′H (the proposed membrane anchor). The absence of the F3′5′H activity has been correlated with the absence of violet/blue flowers in roses, carnations, chrysanthemums and gerberas, all of which are important floricultural crops.

9.6.6 *Modification of anthocyanins*

Anthocyanidin 3-glucosides are further modified by glycosylation and aromatic/aliphatic acylation in a species-specific manner. Glycosyltransferases and acyltransferases catalyze these reactions, respectively. Generally the enzymes are specific to the position of modification but less specific to substrate. Recent advances are described in this section focusing on well-characterised species.

P. hybrida is the best-characterised species as far as flower colour is concerned (Sink, 1984) (Figure 9.3e). The structural genes involved in the pathway have been cloned (3GT; Yamazaki *et al.*, 2002), UDP-rhamnose: anthocyanidin 3-glucoside rhamnosyltransferase (3RT; Brugliera *et al.*, 1994), hydroxycinnamoyl CoA: anthocyanidin 3-rutinoside acyltransferase (AR-AT; Brugliera and Koes, 2001), UDP-glucose: anthocyanin 5-glucosyltransferase (5GT; Yamazaki *et al.*, 2002), anthocyanin methyltransferase (MT; Brugliera *et al.*, 2003). However, further chemical studies identified new diacyl anthocyanins (Fukui *et al.*, 1998) and anthocyanidin 3-sophorosides (Ando *et al.*, 2000) accumulating in some petunia varieties.

Members of the Laviatae family (lavender, salvia, perilla) contain anthocyanidin 3-(hydroxycinnamoyl) glucoside 5-(mono or dimalonyl) glucosides (Figure 9.3d). Complementary DNAs (cDNAs) encoding hydroxycinnamoyl-CoA: anthocyanin 3-*O*-glucoside-6″-*O*-acyltransferase which catalyzes the transfer of a coumaroyl or caffeoyl group to the 3-*O*-glucose of anthocyanins have been isolated from perilla (Yonekura-Sakakibara *et al.*, 2000), lavender (Tanaka *et al.*, unpublished results), scarlet sage (*Salvia splendens*; Suzuki *et al.*, 2004b) and blue salvia (*Salvia guaranitica*; Florigene Ltd., unpublished results). *In vitro* assays reveal that these enzymes utilise pelargonidin, cyanidin and delphinidin 3-glucosides and 3, 5-diglucosides as acyl acceptors and the aromatic caffeoyl Co-A and coumaroyl Co-A as acyl donors but not aliphatic acyl Co-A groups. A cDNA encoding malonyl CoA: anthocyanidin 5-*O*-glucoside-6″-*O*-malonyltransferase (5MaT1) was recently cloned from scarlet sage. The enzyme was shown to require an aromatic acyl moiety on the 3-glucoside and did not accept pelargonidin 3, 5-diglucoside or delphinidin 3-glucoside as substrates (Suzuki *et al.*, 2001). *Salvia splendens* contains an alternative malonyl CoA: anthocyanin 5-*O* -glucoside-4‴-*O*-malonyltransferase (5MaT2) and the corresponding cDNA has also been cloned. The enzyme 5MaT2 did not show any 5MaT1 activity in an *in vitro* assay (Suzuki *et al.* 2004b).

Chrysanthemum, dahlia and cineraria belong to the Compositae family and their anthocyanins share a common structure but vary in the degree of modification (Figure 9.3c). Complementary DNA clones encoding malonyl CoA: anthocyanidin 3-*O*-glucoside-6″-*O*-malonyltransferase were recently isolated from dahlia petals (*Dahlia variabilis*; Suzuki *et al.*, 2002), cineraria (*Senecio cruentus*; Suzuki *et al.*, 2003a) and chrysanthemum (Dm3MaT1; Suzuki *et al.*, 2003b). The cDNA of malonyl CoA: anthocyanidin 3-*O*-glucoside-3″,6″-*O*-dimalonyltransferase (Dm3MaT2) that can catalyze consecutive malonyl transfers has been also cloned recently (Suzuki *et al.*, 2003b). DmMaT1 and DmMaT2 utilise pelargonidin, cyanidin, delphinidin and quercetin 3-glucosides as acyl acceptors but not pelargonidin 3, 5-diglucosides. DmMaT1 utilises succinyl-CoA as an acyl donor but DmMaT2 is more specific to malonyl-CoA. UDP-glucuronic acid: anthocyanidin 3-*O*-glucoside 2″-*O*-*β*-glucuronosyltransferase (BpUGAT) has been purified from red daisy (*Bellis perennis*) and the gene encoding the enzyme has been cloned

(Sawada *et al.*, 2005). This enzyme is highly specific to cyanidin 3-glucoside and cyanidin 3-(malonyl) glucoside and UDP-glucuronate (Sawada *et al.*, 2005).

Blue gentian (*Gentiana triflora*) flowers contain gentiodelphin (Figure 9.3f). Hydroxycinnamoyl-CoA: anthocyanin 5-*O*-glucoside-6″-*O*-acyltransferase (5AT) and UDP-glucose: anthocyanin 3′-*O*-glucosyltransferase (3′GT) cDNAs have been isolated and characterised from gentian (Fujiwara *et al.*, 1998; Fukuchi-Mizutani *et al.*, 2003). The 5AT utilises anthocyanidin 3, 5-diglucoside and anthocyanidin 3-(aliphatic acyl)-glucoside, 5-diglucoside but not anthocyanidin 3-(aromatic acyl)-glucoside 5-diglucoside and caffeoyl Co-A and coumaroyl Co-A but not aliphatic acyl Co-A. Gentian 3′GT appears to be specific for delphinidin 3, 5-diglucoside.

Butterfly pea (*Clitoria ternatea*) accumulates ternatins (highly modified anthocyanins; Figure 9.3g). A gene encoding UDP-glucose: anthocyanin 3′, 5′-*O*-glucosyltransferase has recently been cloned from butterfly pea. The enzyme catalyzes sequential glucosylations at 3′- and 5′-hydroxyl groups and is specific to delphinidin 3-(malonyl) glucoside (Noda *et al.*, 2004). Morning glories also contain highly modified anthocyanins (Figure 9.3h). Complementary DNA clones encoding UDP-glucose: anthocyanidin 3-glucoside glucosyltransferase (3GGT), 3GT and 5GT have been recently been cloned from *I. nil* (Morita *et al.*, 2005) but to date no functionally identified genes encoding anthocyanin acyltransferases have been cloned from morning glories.

Analysis of T-DNA knockout lines of *Arabidopsis* identified genes encoding UDP-rhamnose: flavonol 3-*O*-rhamnosyltransferase and UDP-glucose: flavonol 3-*O*-glycoside-7-*O*-glucosyltransferase (Johns *et al.*, 2003). The genes encoding anthocyanin glucosyltransferase and acyltransferase have been putatively identified making use of gene chip analysis of transgenic *Arabidopsis thaliana* over-expressing *Arabidopsis PAP1* (a positive transcriptional regulator of anthocyanin biosynthesis, belonging to the *Myb* family). Many of the flavonoid biosynthetic genes were shown to be up-regulated in the plant (Tohge *et al.*, 2005). Such a comprehensive analysis will be useful in cloning all structural genes in one pathway at the same time.

Monocotyledonous species are less understood in terms of molecular biology although non-ornamental maize, barley and rice systems have been well studied. Imayama *et al.* (2004) and Yoshihara *et al.* (2005) recently isolated a *5GT* and *3GT* genes from Dutch Iris (*Iris hollandica*).

9.6.7 *Transcriptional regulation*

Flavonoid biosynthesis is transcriptionally regulated and its structural genes are developmentally regulated in a well-coordinated manner in petal development.

Myb and *Myc* (basic helix-loop-helix) type transcriptional factors generally regulate the expression of the flavonoid biosynthetic structural genes (Mol *et al.*, 1998; Springob *et al.*, 2003). Involvement of a WD40 protein in the regulatory pathway may also be universal. These are well characterised in snapdragon, petunia, *Arabidopsis*, maize and perilla for example (Springob *et al.*, 2003). Regulatory genes associated with the anthocyanin pathway are functionally conserved among plant species but they have distinct sets of target genes, which explains some of the species-specific diversity.

An increase in anthocyanin levels has been achieved via over-expression of genes encoding such transcription factors. For example, over-expression of the maize *Lc* gene (*Myc*) led to an increased amount of anthocyanins in tobacco flowers (Lloyd *et al*., 1992) and that of *PAP1* (a Myb gene) in *Arabidopsis* caused ectopic anthocyanin accumulation in various tissues (Borevitz *et al*., 2000).

9.6.8 Regulation of vacuolar pH

There has been some progress towards understanding petal epidermal vacuolar pH and its regulation.

As mentioned earlier, the colour and stability of anthocyanins are impacted by vacuolar pH, which is usually maintained as weakly acidic. Although higher (neutral) pH generally yields bluer flower colours, anthocyanins are less stable at higher pH and must be stabilised with more than one glycosyl and aromatic acyl group (Goto and Kondo, 1991; Honda and Saito, 2002). Genetic control of petal vacuolar pH is known in petunia and morning glory. The only structural gene shown to regulate vacuolar pH to date encodes a Na^+/H^+ antiporter (*Purple*) in morning glory (Fukada-Tanaka *et al*., 2000). The gene is highly expressed just before flower opening, which elevates pH from 6.5 to 7.5 and changes the colour from purple to blue. Homologues have been isolated from petunia, torenia and *Nierembergia* (Yamaguchi *et al*., 2001) but their function *in vivo* is unclear. Higher vacuolar pH has been shown to be specific to epidermal cells where anthocyanins accumulate (Yoshida *et al*., 1995). Elevation of vacuolar pH using an antiporter gene in transgenic plants is yet to be reported.

P. hybrida has 7 *Ph* loci thought to be involved in regulation of petal pH. If any of the *Ph* loci are in a homozygous recessive state then the pH of the corolla homogenate is around pH 6.2 and the flowers have a blue hue (Wiering and de Vlaming, 1984; van Houwelingen *et al*., 1998). If all the *Ph* genes are in the dominant form, then the pH of the corolla homogenate is around 5.3. The *Ph6* gene from *P. hybrida* was cloned by transposon tagging. The maize transposable element *Activator* (*Ac*) was introduced into *P. hybrida* and it transposed into the *Ph6* gene. The tagged *Ph6* mutation led to a pH change from 5.6 to 5.9 and a bluing of some petal segments. Revertant sectors were reddish (Chuck *et al*., 1993). Spelt *et al*. (2002) further showed that *Ph6* actually represented an allele of *An1* (a regulatory gene that controls the anthocyanin pathway in petunia). In fact *AN1*, *AN2* and *AN11* not only regulated the anthocyanin pathway but also impacted on the vacuolar pH.

9.7 Flower colour and mutations

Knowledge of the flavonoid biosynthesis pathway has been gained from tracer experiments with radioactive substrates, enzymological studies and genetic mutants in flavonoid biosynthesis. These mutations were often selected because of a visible change in flower colour. Included in these are well-characterised mutations that affect an early step of the pathway and lead to acyanic flowers or sectors. Mutations in the later steps of anthocyanin modification tend to result in altered shades of a given colour. Also, mutations not directly affecting the flavonoid pathway, but altering the final colour due to alterations in pH of the vacuole or the level of copigments present, have also been described (reviewed by Schram *et al.*, 1984; Wiering and de Vlaming, 1984). As flower colours are obvious, many mutants involved in flower colour have been recorded in various plants. Snapdragon, petunia and morning glory are well studied and the latter two are described here. These mutants have been useful tools in the study of the flavonoid pathway and the subsequent isolation of flavonoid biosynthetic genes.

9.7.1 Petunia

The *Petunia* genus was described in 1803 by Jusseau (Sink, 1984) and belongs to the Solanaceae family. Any disruption to flower colour is easily observed and many mutants, especially in the flavonoid pathway, are readily available. The first genes to be identified were those involved in floral pigmentation and growth habit of which 32 loci have been implicated in the control of flower colour (de Vlaming *et al.*, 1984). Various methods have been used to assign genes to particular chromosomes (reviewed by Maizonnier, 1984 and Cornu, 1984) including the use of trisomics in crosses, chromosomal rearrangement, deletions of complete arms of chromosomes and the use of satellites as chromosome markers. A telocentric translocation where one arm of chromosome I was transferred to the end of the long arm of chromosome II allowed the first accurate gene localisation in petunia – that of *Hf1* to chromosome I. The first loci mapped were the main anthocyanin genes that control flower colour namely *An2*, *An4*, *Fl*, *Hf1*, *Po* and *Rt*. A number of other loci associated with flower colour, leaf colour and plant morphology have also been mapped (Cornu, 1984; Cornu *et al.*, 1990).

Petunia mutants have been used in the elucidation of the flavonoid pathway in petunia (reviewed in Martin and Gerats, 1993), in the isolation of key structural flavonoid pathway genes from petunia (Holton *et al.*, 1993b; Brugliera *et al.*, 1994; Kroon *et al.*, 1994) and in the corroboration of identity of newly isolated genes (Holton *et al.*, 1993a,b; Huits, 1993; Brugliera *et al.*, 1994; 1999). Groups at INRA, France and The Vrije Universiteit, Amsterdam, developed two large populations of petunia mutants. A variety of methods have been used to categorise these mutants.

9.7.2 Ipomoea

I. nil has been a popular floricultural plant in Japan since around the seventeenth century (Iida *et al.*, 2004). Many spontaneous mutants relating to flower colour have been isolated and the majority of the mutations appear to be caused by insertion of DNA transposable elements into anthocyanin biosynthetic genes. A mutable allele, *flecked*, that causes variegation in the flower contains a DNA transposable element (*Tpn1*) in the DFR gene (*DFR-B*; Inagaki *et al.*, 1996). The mutable *speckled* allele of *I. nil* that confers pale yellowish flowers with fine and round coloured spots distributed over the corolla is the CHI gene containing *Tpn2*, a *Tpn1*-related transposable element (Iida *et al.*, 2004). *I. nil* blue flower colour is mainly controlled by two genetic loci, *Magenta* and *Purple*. Double mutants exhibit red flowers and recessive *magenta* and *purple* mutants yield magenta and purple flowers, respectively (Hoshino *et al.*, 2003). The *magenta* allele is a mutation of the flavonoid 3′-hydroxylase (F3′H) gene (Iida *et al.*, 2004). The mutable *purple* allele is caused by a non-autonomous transposable element and the *Purple* gene was shown to encode a vacuolar Na^+/H^+ exchanger (Fukuda-Tanaka *et al.*, 2000; Yamaguchi *et al.*, 2001).

9.8 Evolutionary aspects of structural genes in the flavonoid pathway

Most of the enzymes encoded by structural genes involved in flavonoid biosynthesis belong to common enzyme families such as cytochrome P450 monooxygenases, 2-oxoglutarate dependent dioxygenases, glucosyltransferases, acyltransferases and polyketide synthases. It appears that these genes may have been recruited from pre-existing 'old' pathways in plants as is the case for other secondary metabolic pathways such as the alkaloid biosynthetic pathway, which also utilises enzymes from the same families.

The cytochrome P450 monooxygenases, which usually catalyze hydroxylation reactions, are NADPH-dependant and contain a protoheme as a prosthetic group and an invariant cysteine residue (Schuler and Werck-Reichhart, 2003). One plant species may contain around 300–500 cytochromes P450 (Schuler and Werck-Reichhart, 2003). Cytochromes P450 are generally classified into a family if they have greater than 40% amino acid identity and then further into subfamilies if the amino acid identity is greater than 55% (Schuler and Werck-Reichhart, 2003). C4H belongs to the CYP73A family and was one of the first plant cytochromes P450 enzymes characterised. Nucleic acid sequences of C4H, F3′H (CYP75B) and F3′5′H (CYP75A) have been cloned from many species (Figure 9.5a). The development of a P450 phylogenetic tree reveals that *F3′H* and *F3′5′H* genes may have diverged before speciation of flowering plants, because all of the *F3′H* and *F3′5′H* sequences isolated to date belong to separate subfamilies (CYP75B and CYP75A, respectively) (Figure 9.5a). Although many *Arabidopsis* and rice genes are annotated as *F3′H* or *F3′5′H* homologues in the Genbank/EMBL/DDBJ DNA databases, only a few of them actually segregate into the CYP75 A or B families. FNSII (flavone synthase) belongs to the CYP93B family and CYP93C contains isoflavone synthase. CYP73, CYP75 and CYP93 are not closely related although they are involved in the same pathway.

F3H, ANS and FLS are all 2-oxoglutarate dependent dioxygenases that require ferric ions, 2-oxoglutarate and ascorbic acid for full activity (Turnbull *et al.*, 2004).

The family contains various important plant enzymes such as ACC oxidase and gibberellin oxidase. Analysis of a 2-oxoglutarate dependent dioxygenase phylogenetic tree indicates that these enzymes had diverged before flowering plants speciated and that ANS and FLS are closely related. The three-dimensional structure of *Arabidopsis* ANS has been determined (Wilmouth *et al.*, 2002). Parsley contains a 2-oxoglutarate dependent dioxygenase-type FNSI and its sequence has high similarity to F3H (Figure 9.5b) leading to the suggestion of divergent evolution of F3H and FNSI in parsley (Martens *et al.*, 2003a).

Glucosyltransferases are highly divergent and polyphyletic and have been classified into 65 families (Coutinho *et al.*, 2003). Among them, Family 1 consists of GTs that utilise UDP-sugars as the donor and as such flavonoid/anthocyanin GTs have been grouped to Family 1. Analysis of the *Arabidopsis* genome sequence has revealed the presence of 110 glucosyltransferase homologues (Lim *et al.*, 2003; Paquette *et al.*, 2003). A phylogenetic analysis of GTs (Figure 9.5c) indicates that 3GT and 5GT form separate families indicating they had diverged before the speciation of higher plants. Butterfly pea 3′5′GT is very close to butterfly pea 3GT (Noda *et al.*, 2004) and would likely have diverged after speciation (the sequence has not been published yet). 7GT and 3′GT form another family to which the betalain 5GT also belongs. As mentioned previously, the betalain 5GT also catalyzes glucosylation of 7 and 3′ position of flavonoids (Vogt *et al.*, 1999). Taken together there are

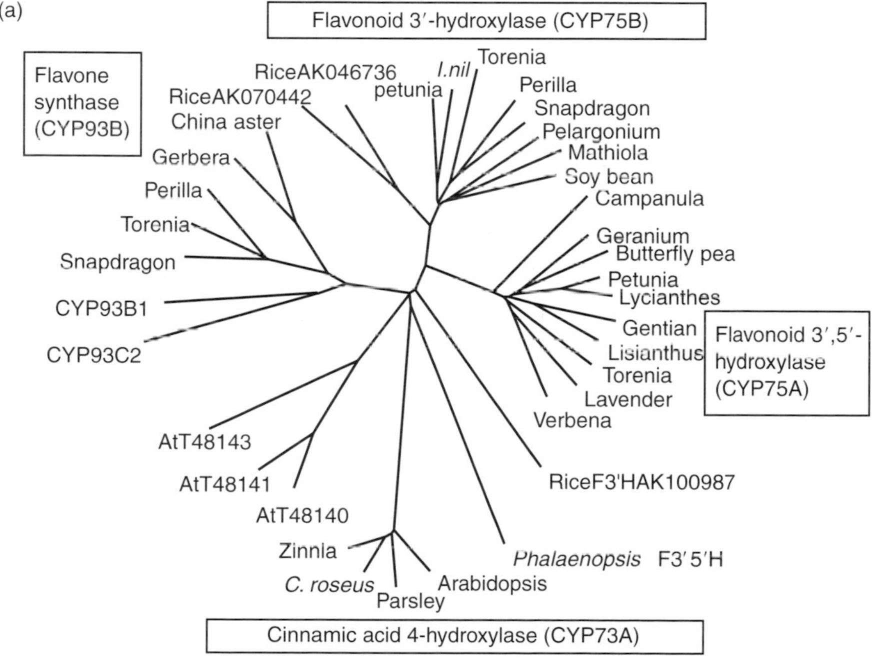

Figure 9.5 (Continued.)

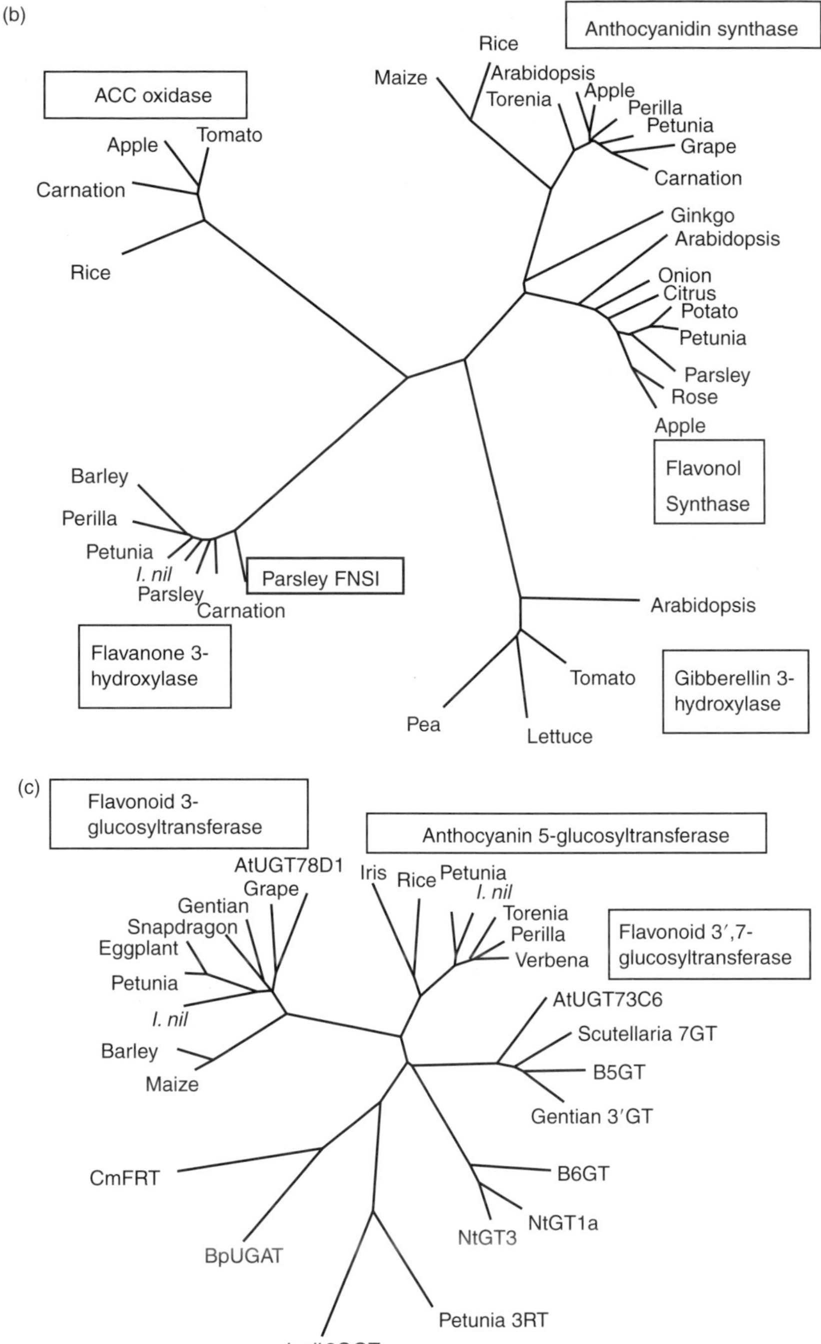

Figure 9.5 (Continued.)

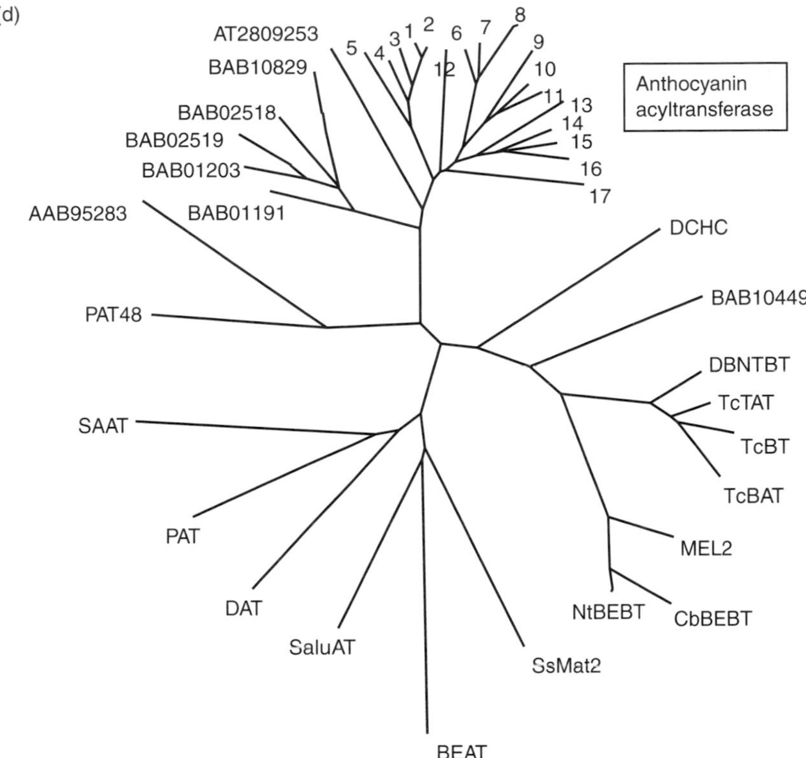

Figure 9.5 Phylogenetic relationships of (a) cytochromes P450, (b) 2-oxoglutarate dependent dioxygenases, (c) glucosyltransferases and (d) acyltransferases. Homologues from various plant species were subjected to CLUSTALW alignment and the trees were made by Treeview program (Page, 1996). See text for details and abbreviations. (a) Grouping on the bases of amino acid sequences agrees with functional groups such as F3′H, F3′5′H, C4H and FNSII. Rice and *Arabidposis* (At) homologues annotated as F3′H or F3′5′H are not always classified into the corresponding groups. CYP93B1, licorice flavanone 2-hydroxylase; CYP93C2, licorice isoflavone synthase. (b) Parsley *FNSI* gene may have evolved from the *F3H* gene (Martens et al. 2003a). ACC, 1-aminocyclopropane-1-carboxylic acid. (c) Flavonoid 3′,7-glucosyltransferase group may contain enzymes with broad substrate specificities. Scutellaria 7GT (AB031274). (d) Anthocyanin/flavonoid acyltransferases are described in the text. Compositae anthocyanin acyltransferases, 1; Chysanthemum 3MaT, 2; chrysanthemum 3MaT2, 3; cineraria 3MaT, 4; dahlia 3MaT, 5; dahliaAT929 (functionally unidentified). Laviatae anthocyanin acyltransferase, 6; perilla 3AT, 7; lavender 3AT, 8; blue salvia 3AT, 9; salvia AT201, 10; perilla5MaT, 11; salvia 5MaT, 12; Gentian 5AT, 13; torenia 5AT, 14; verbena flavonol 3MaT2, 15; verbena flavonol 3MaT1, 16. *Lamium purpureum* flavonol 3MaT, 17. GAT106. The sequences shown with accession numbers are from *Arabidopsis*. Gentian AT106 and PAT48 are functionally unidentified gentian and petunia homologues (Yonekura-Sakakibara *et al.* 1999); DCHC, carnation hydroxycinnamoyl/benzoyl-CoA: anthranilate *N*-hydroxycinnamoyl/benzyltransferase (CAB06430); DBNTBT, *T. cuspidata* 3′-*N*-debenzoyltaxol *N*-benzoyltransferase (AAM75818); TcTAT, *T. cuspidata* taxadienol acetyltransferase (AAF34254); TcBT, *T. cuspidata* 2-benzoyl-7, 13-diacetylbaccatin III-2-*O*-benzoyltransferase (AAG38049); TcBAT, *T. cuspidata* 10-deacetylbaccatin III-10-*O*-acetyltransferase (AAF27621); Mel2, a melon homologue (Z70521); CbBEAT and NtBEAT, tobacco benzyl-CoA: benzyl alcohol benzoyltransferase (AF500200, T03242); BEAT, *C. breweri* acetyl CoA: benzylalcohol acetyltransferase (AAC18062); SaluAT, poppy salutaridiniol 7-*O*-acetyltransferase (AAK73661); DAT, *Catharanthus roseus* deacetylvindoline 4-*O*-acetyltransferase (AAC99311); SAAT, strawberry alcohol acyltransferases (AAG13130).

some questions as to what the physiological role of betalain 5GT is (Sasaki *et al.*, 2005). Petunia 3RT and morning glory 3GGT, both of which catalyse the transfer of a glycosyl moiety to anthocyanidin 3-glucosides, seem to be related. *Bellis perennis* anthoyanin glucuronosyltransferase (BpUGT; Sawada *et al.*, 2005) and *Citrus maxima* flavonoid 1–2 rhamnosyltransferase (CmFRT; Frydman *et al.*, 2004) may also be related to this group (Figure 9.5c).

Anthocyanin acyltransferases belong to the BAHT acyltransferase family that includes various acyl Co-A dependent acyltransferases (Figure 9.5d) (St-Pierre and De Luca, 2000). All but AR-AT and SsMaT2 form a separate family (anthocyanin acyltransferase family). Laviate and Compositae acyltransferases seem to be closer, respectively, which implies that they emerged after speciation of plant families. BAHT acyltransferase family contains acyltransferases involved in various plant secondary metabolite biosynthesis such as morphine and taxol and fragrance of flowers and berries. About 60 homologues have been found in the *Arabidopsis* genome although most of them have not been functionally identified.

Chalcone synthase belongs to the polyketide synthase family and has been reviewed previously (Schroder, 1999).

9.9 Genetic engineering of flower colour

Although flowers are generally appreciated for their colour, other characteristics such as flower form, vase-life, number of flowers per plant and disease resistance are considered to be of importance to flower growers and breeders. Classic breeding techniques have allowed the production of a large variety of flower colours. However, classical flower breeding is limited by the size of the gene pool for any given species. Also, it is often not generally possible to alter one characteristic without altering another favourable characteristic. Molecular flower breeding therefore is an additional tool that may be used by the flower breeder to alter flower colour without affecting other desirable characteristics.

Flower colour may be genetically engineered by using a number of strategies. One is to introduce a novel gene or genes into a plant. This may in effect create a novel pathway or branch point in the plant (Meyer *et al.*, 1987) or lead to increased levels of flavonoids (Lloyd *et al.*, 1992). Another strategy is to block the flavonoid pathway at a specific step so that accumulation of a different set of precursors occurs, thereby creating a novel flower colour. This has been achieved using antisense (van der Krol *et al.*, 1988) or sense suppression (co-suppression) (Napoli *et al.*, 1990) and more recently using RNAi (RNA interference) (Fukuchi-Mizutani *et al.*, 2004) an innovative combination of the antisense and sense technologies described by Waterhouse *et al.* (1998) that relies on the production of double-stranded RNA (dsRNA). Transcription of dsRNA has been reported as being more efficient in down regulation of target gene expression (Wang and Waterhouse, 2001; Wesley *et al.*, 2001; Fukuchi-Mizutani *et al.*, 2004). A combination of the different strategies can also be used to block one branch of the pathway and introduce a novel pathway in the species or variety of plant (Tsuda *et al.*, 2004). Whichever strategy is employed it is becoming obvious that an understanding of the target variety is required at the chemical, biochemical and molecular level if the aim of the experiment is to produce an expected outcome or novel colour (Holton 1996; Martens *et al.*, 2003b).

In the case of plant transformation, it has not been possible to insert a transgene into a specific position of the genome or to regulate the number of copies inserted or regulate any rearrangements/deletions of the transgenes and thus the degree of expression or suppression by the transgene(s) can vary dramatically between individual transgenic events. For research purposes this may not be critical as any transgenic event produced provides a source for further study. However, for a commercial outcome, it is necessary to generate large numbers (50–100 or more) of independent transgenics and then to select the best coloured lines. This is followed by further elimination based upon productivity and the molecular analysis required for regulatory approvals, resulting in a few elite lines for commercialisation. In order to achieve relatively large numbers of transgenic events, an efficient transformation system for the species and, in some cases, for the specific variety, are essential. Laborious optimisation of plant transformation systems is necessary and transformation is often species- and cultivar-dependent. Floricultural plant transformation has been reviewed recently (Deroles *et al.*, 1997; Tanaka *et al.*, 2005) and this chapter will focus on flower colour modification in different species that has been achieved via genetic engineering.

9.9.1 Petunia

P. hybrida is generally used as a bedding plant in gardens or as a pot plant. Petunia species are not only grown commercially but are also used as experimental plant material and have been studied with regard to taxonomic, anatomical and morphological relationships within the genus. Petunia has been used as an experimental model plant species mainly due to the fact that it can be readily grown from seed to flower within a short time frame, the growing requirements are simple and abundant seed harvest occurs around 4 weeks after pollination (Sink, 1984). Furthermore, petunia is generally easy to transform and regenerate (Horsch *et al.*, 1985), grows quickly and flowers readily under a controlled environment. It has been an excellent model for flower colour modification by genetic engineering because its flavonoid biosynthetic pathway has been elucidated and the structural and regulatory genes involved have been isolated.

Van der Krol *et al.* (1988) first demonstrated that constitutive transcription of an antisense petunia chalcone synthase gene altered petunia flower colour. A range of flower pigmentation patterns was obtained including completely white flowers, flowers with reduced pigmentation, flowers with white sectors or white rings around the corolla limb and flowers with the same colour as the wild-type control. An attempt to over-express petunia *CHS* in pigmented petunia petals unexpectedly yielded transgenic petunia plants with completely white flowers or flowers with some sectors of reduced pigmentation (Napoli *et al.*, 1990). Van der Krol *et al.* (1990) also introduced additional copies of petunia *DFR* and petunia *CHS* genes into petunia, and observed a reduction in pigmentation in up to 25% of the transgenic flowers. Tsuda *et al.* (2004) have also demonstrated that down-regulation of one of either *CHS*, *DFR* or *F3H* flavonoid biosynthetic genes leads to white or paler flower colour, occasionally some patterns or white spots. Petunia phenotypes obtained with antisense suppression of CHS gene are shown in Plate 9.2 (Florigene Ltd., unpublished results). However, down-regulation of flavonoid biosynthesis may not be an ideal tactic when considering the production of commercially viable flowers as flavonoids play an important role in protection from various stresses. Investigation of a white petunia line where flavonoid biosynthesis was blocked with the introduction of an antisense version of *CHS* revealed that the plant was prone to mite attack (compared to the coloured control) and the flower colour phenotype was not stable (Tanaka *et al.*, 1998; Tsuda *et al.*, 2004).

The substrate specificity shown by DFR can regulate the anthocyanidin structure that a plant species accumulates. Petunia and cymbidium DFRs do not reduce dihydrokaempferol and thus they do not accumulate pelargonidin-based pigments and so do not produce brick red/orange flower colours (Forkamnn and Ruhnau, 1987; Johnson *et al.*, 1999). On the other hand, DFRs from many plants such as rose, gerbera and maize are able to utilise dihydrokaempferol as substrate. Transgenic brick-red petunias accumulating pelargonidin-type anthocyanins were obtained by introduction of a maize DFR into a mutant petunia line that accumulated dihydrokaempferol due to the deficiency of F3′5′H, F3′H and FLS (Meyer *et al.*, 1987). The flower colour was pale and unstable but after incorporating the transgenic petunia into a hybridisation breeding program, a stable bright-orange petunia variety was obtained (Oud *et al.*, 1995). Similar flower colour changes have been obtained by the expression of gerbera (Helariutta *et al.*, 1993) or rose DFRs (Tanaka *et al.*, 1995). An important point to note is that this strategy required the use of a petunia line that accumulated dihydrokaempherol (DHK). When a petunia variety that was white due to a transposon in the *DFR* gene, but contained F3′5′H activity, was transformed with a rose *DFR* sequence, the introduced rose DFR was unable to compete effectively with the endogenous F3′5′H for DHK although it was able to utilise dihydromyricetin (DHM) to produce delphinidin-based pigments (Florigene Ltd., unpublished results).

Identification of a dihydrokaempferol-accumulating line in commercially important species or cultivars can be difficult. Tsuda *et al.* (2004) were able to engineer a red petunia line that normally accumulates cyanidin-based pigments to produce pelargonidin-based pigments (orange) by down-regulation of the endogenous petunia *F3′H* gene along with expression of a rose *DFR* gene. Many important floricultural species including cyclamen, delphinium, iris, gentian, *Cymbidium* are presumed not to accumulate pelargonidin due to the substrate specificity of their endogenous DFRs. Similar strategies might be employed to generate orange coloured flowers of these species.

Tsuda *et al.* (2004) have also reported on the simultaneous down-regulation of three genes. The *P. hybrida* cultivar Surfinia Purple Mini (having F3′5′H, FLS and AR-AT activities) is a commercial petunia line that produces purple flowers that accumulate malvidin and petunidin-type anthocyanins as well as flavonols. It is normally vegetatively propagated from tissue culture plants for commercial propagation. When the expression of *F3′5′H*, *FLS* and *AR-AT* genes was simultaneously suppressed, the flower colour changed to red (Plate 9.2b). Although a few lines stably exhibited the red flower colour, they eventually lost the altered colour and the host colour was partly regained (Plate 9.2c). Conversely, when the same lines that had been maintained in tissue culture were transferred to soil after 4 years, they exhibited the altered flower colour. Although phenotypic instability is an obstacle for commercialisation, maintaining these transgenic plants in tissue culture appears to be a practical way to avoid such an obstacle in this case at least.

Constitutive expression of the *Lc* gene resulted in increased anthocyanin levels in floral and vegetative tissues, including the leaves of transgenic petunias. These leaves were purple due to accumulation of anthocyanins, and may represent a novel ornamental plant of commercial value (Bradley *et al.*, 1998).

9.9.2 Torenia

Since white flowering plant varieties are rather easily made by conventional breeding, it is difficult to justify the cost of development of white varieties by genetic engineering. Nevertheless, genetic engineering of a white variety can be commercially viable as only the flower colour is modified presumably without sacrificing any other desirable characteristics. The *Torenia hybrida* cultivar Summerwave is male and female sterile because it is an interspecies hybrid. Its sterility prevents further breeding to increase flower colour varieties, although it exhibits superior characters such as more flowers and a longer flowering period.

Transgenic white and blue/white Summerwave were generated by sense suppression of *CHS* and *DFR*, respectively (Suzuki *et al.*, 2000). However, field trial analysis revealed that the flower colour was unstable and the flowers occasionally regained the blue colour. The transgenic plants, especially Summerwave with sense *CHS* suppression, lost vigour compared to the host plant (unpublished results Suntory Ltd.). As mentioned earlier, blockage of *CHS* biosynthesis can result in flavonoid-free transgenic plants and flavonoids have been found to play an important role in UV protection, general plant defence and signalling (Winkel-Shirley, 2002). Therefore, instead of down-regulating *CHS* or *DFR* genes, the *ANS* gene was down-regulated in torenia using dsRNA (RNAi) technology. Approximately half of the transgenic torenia plants produced white flowers (Nakamura *et al.*, 2006), from which stable white lines were selected and are currently under evaluation (Plate 9.2d). Summerwave expresses two different *CHS* genes in petals. It has been shown that each of them could be down-regulated by transcribing dsRNA specific to the 3′-non-coding region of the respective gene to yield white or paler flower colours (Fukuzaki *et al.*, 2004).

Fukuchi-Mizutani *et al.* (2004) were also able to switch the metabolic flux from delphinidin to pelargonidin in torenia flowers. Down-regulation of *F3′5′H* and *F3′H* using dsRNA (RNAi) along with expression of a heterologous *DFR* resulted in torenia lines producing bright pink flowers that accumulated up to 78% pelargonidin (Plate 9.2d).

Aida *et al.* (2000a) obtained transgenic *T. frunieri* plants with paler flower colour using sense and antisense constructs containing *DFR* and *CHS*. They observed that the flowers of torenia harbouring the antisense *DFR* gene were bluer than those harbouring the antisense *CHS* gene because down-regulation of DFR lead to an increase in flavones and the resulting copigmentation effect with the anthocyanins shifted the flower colour towards blue (Aida *et al.*, 2000b).

9.9.3 Carnations

Carnations are one of the top ten selling cut-flowers world-wide. There are basically three forms of carnations available. These are the midi, spray and standard types; most sales are confined to spray and standard types. Red and pink carnations generally accumulate anthocyanins based upon cyanidin and/or pelargonidin pigments (Bloor, 1998; Gonnet and Fenet, 2000) and flavonols, whilst yellow carnations generally accumulate the chalcone, chalconaringenin 2′-*O*-glucoside (Forkmann and Dangelmayr, 1980; Yoshida *et al.*, 2004). Detailed analysis has revealed that the anthocyanins are generally macrocyclic anthocyanins 3, 5 di-*O*-glucoside (6″, 6‴-malyl diester) of cyanidin or pelargonidin (Bloor, 1998; Gonnet and Fenet, 2000; Nakayama *et al.*, 2000b; Figure 9.3b).

The first report of a modification of carnation flower colour by genetic engineering was using sense carnation *CHS* expression in a pink-flowering cultivar called Manon. Around 10% of the transgenic plants produced had reduced floral pigmentation (Gutterson, 1995). Zuker *et al.* (2002) down-regulated the carnation *F3H* thereby altering the metabolic flux from anthocyanins to methylbenzoate. The resultant carnations were both paler in colour and more fragrant.

The world's first genetically modified cut-flower to be sold commercially was a violet carnation of the midi-type called Florigene Moondust™. The specific activity of petunia DFR to preferentially act on DHM over dihydroquercetin and not to utilise DHK as a substrate was used to advantage in the production of the violet carnations. A white flowering carnation that lacked both DFR and F3′H activities and thus accumulated DHK was transformed with petunia genes for F3′5′H and DFR. In these flowers DHK was converted to DHM by the introduced F3′5′H and the introduced petunia DFR was then able to convert the DHM to leucodelphinidin. The endogenous carnation ANS and 3GT, previously shown to be able to convert leucodelphinidin into coloured delphinidin 3-glucoside by precursor feeding experiments, allowed conversion of the product into delphinidin-based pigments (Holton, 1996; Mol *et al.*, 1999). This strategy has been extended to increase the range of violet carnations available on the market today. There are currently six lines of transgenic carnations being sold world-wide (Plate 9.2e). These include Florigene Moondust™ described earlier, Florigene Moonshadow™, a purple coloured carnation from a similar background – both midi-types, and four standard varieties from the same parental background called Florigene Moonacqua™, Florigene Moonlite™, Florigene

Moonshade™ and Florigene Moonvista™. Detailed analysis of the accumulating anthocyanins in Florigene Moondust™ and Florigene Moonshadow™ revealed that the delphinidin derivatives were analogous to natural anthocyanins, the predominant anthocyanin being delphinidin 3, 5 diglucoside-6″-*O*-4, 6‴-*O*-1-cyclic malyl diester confirming that the endogenous anthocyanin biosynthetic enzymes are able to accept delphinidin-based molecules as substrates (Fukui *et al.*, 2003). Flavonol and flavone glycosides also accumulated in these flowers with the flavone apigenin 6-*C*-glucosyl-7-*O*-glucoside-6‴-malyl ester having the strongest copigment effect and thereby contributing to the bluish hue of the carnation flowers (Fukui *et al.*, 2003). The carnations are vegetatively propagated with several hundreds of thousands plants grown and tens of millions of flowers produced per year. The flower colour has remained stable through several generations of vegetative propagation.

9.9.4 Roses

Roses are the top selling flowers in the world and make up around 27% of the world's cut flower market. They are often called the 'Queen' of flowers and their cultivation abounds with legend. The anthocyanins generally found in red and pink roses are simple 3, 5 diglucosides of cyanidin and/or pelargonidin (Biolley and Jay, 1993; Figure 9.3a). However, the methylated cyanidin pigment, peonidin, has been found in some wild roses (Mikanagi *et al.*, 2000) along with some aromatic acylation. The copigments in rose flowers are generally flavonols (Biolley and Jay, 1993) and roses do not produce flavones.

Although mutation breeding is used in some plants such as chrysanthemums to produce colour variations without altering other plant characteristics, the technique has not been completely successful in rose. Generation of a variety of colours with unaltered plant characteristics has not been possible. Therefore, in order to produce a range of pink-flowering and white-flowering lines of an elite rose line (Royalty), a sense version of a rose *CHS* cDNA clone was fused to a constitutive promoter and introduced into Royalty. Royalty normally produces deep red flowers on stems that give good performance through storage and good consumer vase-life. Although white lines were not obtained, pink-flowering transgenic plants that showed a reduction in anthocyanin levels from 50 to 90% were obtained (Guttersen, 1995).

Roses, as with carnations and chrysanthemums, do not accumulate delphinidin-based pigments primarily because they lack F3′5′H activity. Some so-called blue roses such as Rhapsody in Blue have been studied and found to accumulate cyanidin 3, 5 diglucosides (cyanin) and the flavonols quercetin and kaempferol glycosides at a copigment to pigment ratio of 3 to 1. The 'blue' colour was attributed to the 'trapping' of the cyanin in anthocyanin inclusions (AVIs) allowing a high concentration of the cyanin in a quinonoidal form than would normally be possible in the vacuole (Gonnet, 2003).

Rose breeders have been working to generate blue roses and many rose cultivars with names that contain the word 'blue' have been marketed. However, their colour is mauve at best. It has been expected that roses should be able to produce delphinidin and become blue using genetic engineering techniques (Holton *et al.*, 1993a). Recently, Suntory Ltd and Florigene Ltd announced the production of "blue" roses accumulating almost 100% delphinidin-based anthocyanins, mainly delphinidin 3, 5-diglucoside (Plate 9.2f). Delphinidin production was achieved by expressing a pansy (*Viola* spp.) *F3′5′H* gene. The flower colour is blue–violet and distinct from the roses currently marketed. The regulatory procedures for trialling and selling this new genetically modified rose are currently underway. Another significance of these roses is that production of delphinidin can be inherited by subsequent generations. The introduction of the *F3′5′H* gene into rose lines also allows for classical breeding techniques to be used now to introduce delphinidin production into other rose lines. Classical rose breeding will therefore be influenced in the future, at least as far as colour is concerned, by developments in genetic engineering. Further efforts to make bluer roses are in progress by utilising the many genes impacting on flower colour.

9.9.5 Chrysanthemums

Chrysanthemums are top selling cut-flowers and make up around 20% of the world's cut-flower market. They are also sold as potted plants and garden plants. Chrysanthemum flower colour can be attributed to both anthocyanins and carotenoids. The anthocyanins generally found in chrysanthemum flowers are cyanidin 3-malonylglucosides (Saito *et al.*, 1988). Delphinidin is not found and pelargonidin is rarely found in chrysanthemum petals. Schwinn *et al.* (1994) found that the absence of pelargonidin pigments was due to the presence of F3′H activity and not due to DFR specificity as in *P. hybrida*. When F3′H activity was blocked by a cytochrome P450 inhibitor, pelargonidin pigments were produced in the chrysanthemum petals. This suggests that down-regulation of chrysanthemum F3′H should allow the production of pelargonidin pigments in chrysanthemum petals.

A white-flowering relative of the pink-flowering chrysanthemum cultivar Moneymaker was genetically engineered by introducing a sense or antisense version of the chrysanthemum *CHS* coding sequence, as no pure white lines of the Moneymaker cultivar are available. White flowering transgenic plants were obtained with both the sense and antisense chrysanthemum *CHS* constructs. The white flowering transgenic plants in which *CHS* was suppressed using sense constructs showed a dramatic reduction in endogenous *CHS* message (Courtney-Gutterson *et al.*, 1994). However, the costs associated with bringing genetically modified flowers to market have prevented flowers such as white Moneymaker chrysanthemums from successful commercialisation. A premium price would be necessary to recover costs associated with the production of a genetically modified flower and its subsequent release to market. This would be difficult to achieve when a consumer sees no obvious difference between a white genetically modified chrysanthemum flower and a white 'traditionally bred' flower.

9.9.6 *Other plant species*

Besides petunia, carnation, chrysanthemum and rose, down-regulation of anthocyanin biosynthesis structural genes has been achieved in many plant species including gerbera (Elomaa *et al.*, 1993), lisianthus (Deroles *et al.*, 1998) and gentian (Nishihara *et al.*, 2003). Lobelia has been proposed as a model plant for engineering flower colour due to its short period from gene introduction to flowering (4 months) and red to blue flower colour changes have been achieved by expression of lisianthus *F3′5′H* (Kanno *et al.*, 2003).

Forsythia x intermedia, an ornamental shrub, generally produces yellow-coloured flowers that accumulate carotenoids and the flavonol rutin (quercetin 3-rutinoside). Conversely, the vegetative organs such as the leaves and stems are coloured and accumulate anthocyanins. Molecular and biochemical analysis have revealed that most of the anthocyanin biosynthesis enzymes are active in the petals except for ANS (Rosati *et al.*, 1999). Bronze–orange *Forsythia x intermedia* flowers were engineered by introduction of an *Antirrhinum majus* DFR and a *Matthiola incana* ANS (Rosati *et al.*, 2003). The petals accumulated cyanidin pigments in a background of yellow carotenoid pigments to produce the bronze–orange colour. Presumably, carotenoid production in the petals could be down-regulated to allow for true red colours in *Forsythia*.

9.9.7 *Challenges to generate yellow flowers*

Appealing yellow colours are missing in major pot plants such as pelargonium, cyclamen and begonia. Although the transgenic approach has not yet been successful, increased knowledge in the biosynthesis of yellow pigments and the respective molecular tools for their manipulation are becoming available. As mentioned previously, yellow colours can be achieved using carotenoids and betalains. Chalcones can also contribute to the colours observed in yellow flowered varieties of carnation, peony and periwinkle. The most common chalcone, THC, is yellow but is spontaneously isomerised to naringenin *in vitro* and rapidly isomerised *in vivo* by CHI. Yellow flowers of carnation, peony and periwinkle accumulate THC as a 2′-glucoside (*iso*salipurposide). Accumulation of THC 2′-glucoside is attributed to a deficiency in CHI activity. Itoh *et al.* (2002) showed that both *CHI* and *DFR* genes are disrupted by a transposon in yellow carnation. Pale yellow cyclamen has also been shown to be deficient in CHI activity. Therefore, it appears that a lack of CHI activity and presence of a UDP-glucose: THC 2′-glucosyltransferase (C2′GT) activity are required for the accumulation of the yellow coloured *iso*salipurposide. The two groups isolated various glucosyltransferase genes from yellow carnation petals and each group identified glucosyltransferase genes encoding C2′GT activity *in vitro* (Ogata *et al.*, 2004; Okuhara *et al.*, 2004). Curiously, the two groups identified different glucosyltransferase genes as *C2′GT* genes and these recombinant enzymes are not specific to THC but have broad substrate specificity including 3-glucosylation of anthocyanidins. This suggests that non-specific glucosyltransferases catalyze glucosylation of 2′-THC in carnation and perhaps in other yellow flowers accumulating *iso*salipurposide. Furthermore, Okuhara *et al.* (2004) produced transgenic petunias expressing two of the putative *C2′GT* genes and revealed that the petals produced small amounts of *iso*salipurposide although no colour changes were observed. It seems that a reverse genetics approach in carnations may be useful to clarify which glucosyltransferase(s) is specific for THC *in vivo*.

In contrast to the pale yellow colour of chalcones, aurones are bright yellow flavonoids and therefore provide a very tempting goal for metabolic engineering of yellow pigments. Aurones are found in yellow flowers of distantly related species including snapdragon, dahlia, limonium, zinnia and morning glory. Aurone synthase, more specifically aureusidin synthase (AS), was purified from yellow snapdragon petals and the cDNA encoding the enzyme has been cloned (Nakayama *et al.*, 2000a). AS catalyzes the 3-hydroxylation and the oxidative cyclisation of THC and 2′, 4′, 6′, 3, 4-pentahydroxychalcones to synthesise aureusidin and aureusidin/bracteatin (Sato *et al.*, 2001). However, expression of the gene in petunia did not result in flower colour change or aurone accumulation (unpublished results Suntory Ltd.). AS is a polyphenol oxidase homologue and is localised in the vacuole unlike other polyphenol oxidases which are generally localised in plastids (Ono *et al.*, in press).

Yellow cosmos and dahlia accumulate 6′-deoxychalcones as the predominant pigment (Davies and Schwinn, 1997). Polyketide reductase (PKR; formally called chalcone reductase) catalyzes deoxylation at the 6′-position of THC to stabilise the chalcone and results in a yellow colour. Davies *et al.* (1998) expressed a *PKR* cDNA from *Medicago sativa* in a white petunia line and obtained pale yellow flowers that accumulated chalcones identified as butein 3-*O*-glucoside and butein 4-*O*-glucoside. Similar results were obtained by transforming petunia with a licorice *PKR* gene (Tanaka *et al.*, 2005). However, in both cases, the colour was only visible in flower buds and was not intense enough to represent a new yellow variety with commercial value.

The flavonol 6-hydroxyquercetin owes its yellow colour to the presence of an additional hydroxyl group in position of the A-ring in addition to 5, 7-hydroxylation. The 6-hydroxylation has been shown to be catalyzed by a cytochrome P450-dependent monooxygenase in *Tagetes* sp. (Halbwirth *et al.*, 2004) and 2-oxoglutarate dependent dioxygenase in *C. americanum* (Anzelloti *et al.*, 2000). The genes encoding these flavonoid 6-hydroxylases may therefore be useful molecular tools to engineer yellow flower colour.

9.10 Concluding remarks

Progress in the elucidation of diverse pigment structures, the biochemical and molecular biology of their biosynthesis and plant transformation capabilities have enabled the molecular breeding of flower colour as a feasible approach to the production of novel flower colours. In addition to a better understanding of floral pigment biosynthetic pathways, improvements in the regulation of heterologous gene expression in plants should lead to further developments in flower colour. However, although a number of novel-coloured flowers have been developed to date, only transgenic mauve/violet carnations have successfully passed the numerous production and commercialisation hurdles to be sold as cut-flowers on the world flower market. The production of these novel coloured carnations combines knowledge gleaned from a range of scientific disciplines through field-trialling, product development, intellectual property, regulatory approvals, marketing and distribution world-wide. The cost of taking a genetically modified plant to market is high and can only be offset if a premium is obtained for the novel product. Thus, genetic engineering of flower colour is ultimately driven by the consumer.

Acknowledgements

We thank John Mason for his critical reading of the manuscript. We also thank Yoshihiro Ozeki, Yasumasa Morita and Toru Nakayama for providing unpublished results or preprints.

References

Aida, R., Kishimoto, S., Tanaka, Y. and Shibata, M. (2000a) Modification of flower color in torenia (*Torenia fournieri* Lind.) by genetic transformation, *Plant Science*, **153**, 33–42.

Aida, R., Yoshida, K., Kondo, T., Kishimoto, S. and Shibata, M. (2000b) Copigmentation gives bluer flowers on transgenic torenia plants with the antisense dihydroflavonol-4-reductase gene, *Plant Science*, **160**, 49–56.

Akashi, T., Aoki, T. and Ayabe, S. (1998) Identification of a cytochrome P450 cDNA encoding (2S)-flavanone 2-hydroxylase of licorice (*Glycyrrhiza echinata* L.; Fabaceae) which represents licodione synthase and flavone synthase II, *FEBS Letters*, **431**, 287–290.

Akashi, T., Fukuchi-Mizutani, M., Aoki, T. *et al.* (1999) Molecular cloning and biochemical characterization of a novel cytochrome P450, flavone synthase II, that catalyzes direct conversion of flavanones to flavones, *Plant Cell Physiology*, **40**, 1182–1187.

Ando, T., Tatsuzawab, F., Saito, N. *et al.* (2000) Differences in the floral anthocyanin content of red petunias and *Petunia exserta*, *Phytochemistry*, **54**, 495–501.

Anzelloti, D. and Ibrahim, R. K. (2000) Novel flavonol 2-oxuglutarate dependent dioxygenase: affinity purification, and kinetic properties, *Archives of Biochemistry and Biophysics*, **382**, 161–172.

Biolley, J. P. and Jay, M. (1993) Anthocyanins in modern roses: chemical and colorimetric features in relation to the colour range, *Journal of Experimental Botany*, **44**(268), 1725–1734.

Bloor, S. J. (1998) A macrocyclic anthocyanin from red/mauve carnation flowers, *Phytochemistry*, **49**(1), 225–228.

Bokern, M., Heuer, S. and Strack, D. (1992) Hydroxycinnamic acid transferases in the biosynthesis of acylated betacyanins: purification and characterization from cell cultures of *Chenopodium rubrum* and occurrences in some members of the Caryophyllales, *Botanica Acta*, **105**, 146–151.

Borevitz, J. O., Xia, Y., Blount, J., Dixon, R. A. and Lamb, C. (2000) Activation tagging identifies a conserved *MYB* regulator of phenylpropanoid biosynthesis, *The Plant Cell* **12**, 2383–2393.

Bradley, J. M., Davies, K. M., Deroles, S. C., Bloor, S. J. and Lewis, D. H. (1998) The maize *Lc* regulatory gene up-regulates the flavonoid biosynthetic pathway of Petunia, *The Plant Journal*, **13**, 381–392.

Bradshaw, H. D. and Schemske, S. D. (2003) Allele substitution at a flower colour locus produces a pollinator shift in monkeyflowers, *Nature*, **426**, 176–178.

Brugliera, F. and Koes, R. (2001) Plant anthocyanidin rutinoside aromatic acyl transferases, *International Patent Publication Number WO01/72984*, Assignee: International Flower Developments.

Brugliera, F., Barri-Rewell, G., Holton, T. A. and Mason, J. G. (1999) Isolation and characterization of a flavonoid 3′-hydroxylase cDNA clone corresponding to the *Ht1* locus of *Petunia hybrida*, *Plant Journal*, **19**, 596–602.

Brugliera, F., DeMelis, L., Koes, R. and Tanaka, Y. (2003) Genetic sequences and uses there of, *International PCT Application Number PCT/AU03/00079*, Assignee: International Flower Developments.

Brugliera, F., Holton, T. A., Stevenson, T. W., Farcy, E., Lu, C-Y. and Cornish, E. C. (1994) Isolation and characterization of a cDNA clone corresponding to the *Rt* locus of *Petunia hybrida*, *The Plant Journal*, **5**, 81–92.

Chandler, S. F. (2003) Commercialization of genetically modified ornamental plants, *Journal of Plant Biotechnology*, **5**, 69–77.

Christinet, L., Burdet. F. X., Zaiko, M., Hinz, U. and Zryd, J. P. (2004) Characterization and functional identification of a novel 4, 5-extradiol dioxygenase involved in betalain pigment biosynthesis in *Portulaca grandiflora*, *Plant Physiology*, **134**, 265–274.

Chuck, G., Robbins, T. Nijjar, C. Ralston, E., Courtney-Gutterson, N. and Dooner, H. K. (1993) Tagging and cloning of a petunia flower colour gene with the maize transposable element *Activator*, *The Plant Cell*, **5**, 371–378.

Cornu, A. (1984) Genetics, in *Petunia* (ed. Sink, K.), Springer-Verlag, Berlin, pp. 34–48.

Cornu, A., Farcy, E., Maizonnier, D., Haring, M., Veerman, W. and Gerats, A. G. M. (1990) *Petunia hybrida* in *Genetic Maps – Locus Maps of Complex Genomes*, 5th edn (ed. S. J. O'Brien), Cold Spring Harbor Laboratory Press, New York, pp. 113–124.

Courtney-Gutterson, N., Napoli, C., Lemieux, C., Morgan, A., Firoozabady, E. and Robinson, K. E. P. (1994) Modification of flower color in Florist's Chrysanthemum: production of a white-flowering variety through molecular genetics, *Bio/Technology*, **12**, 268–271.

Coutinho P. M., Deleury, E., Davies, G. J. and Henrissat, B. (2003) An evolving hierarchical family classification for glycosyltransferases, *Journal of Molecular Biology* **328**, 307–317.

Davies, K., ed. (2004) *Plant Pigments and Their Manipulation*, *Annual Plant Reviews*, Vol. 14, Blackwell Publishing, Oxford.

Davies, K. M. and Schwinn, K. E. (1997) Biotechnology of ornamental plants, in *Biotechnology of Ornamental Plants* (eds R. L. Geneve, J. E. Preece and S. A. Markle), CAB International, Wallingford, pp. 259–294.

Davies, K. M., Bloor, S. J., Spiller, G. B. and Deroles, S. C. (1998) Production of yellow colour in flowers: redirection of flavonoid biosynthesis in Petunia, *Plant Journal* **13**, 259–266.

Deroles, S. G., Boase, M. R. and Konczak, I. (1997) Transformation protocols for ornamental plants, in *Biotechnology of Ornamental Plants* (eds R. L. Geneve, J. E. Preece and S. A. Markle), CAB International, Wallingford, pp. 87–119.

Deroles, S., Bradley, J. M., Schwinn, K. E. *et al.* (1998) An antisense chalcone synthase cDNA leads to novel colour patterns in lisianthus (*Eustoma grandiflorum*) flowers, *Molecular Breeding*, **4**, 59–66.

Elomaa, P., Honkanen, J., Puska, R. *et al.* (1993) *Agrobacterium*-mediated transfer of antisense chalcone synthase cDNA to *Gerbera hybrida* inhibits flower pigmentation, *Bio/Technology*, **11**, 508–511.

Ennos, R. A. and Clegg, M. T. (1983) Flower colour variation in the morning glory *Ipomoea purpurea*, *Journal of Heredity*, **74**, 247–250.

Eugster, C. H. and Marki-Fischer, E. (1991) The chemistry of rose pigments, *Angewandte Chemie International Edition in English*, **30**, 654–672.

Fraser, P. D. and Bramley, P. M. (2004) The biosynthesis and nutritional uses of carotenoids, *Progress in Lipid Research*, **43**, 228–265.

Forkmann, G. and Dangelmayr, B. (1980) Genetic control of chalcone isomerase activity in flowers of *Dianthus caryophyllus*, *Biochemical Genetics*, **18**, 519–527.

Forkmann, G. and Ruhnau, B. (1987) Distinct substrate specificity of dihydroflavonol 4-reductase from flowers of *Petunia hybrida*, *Zeitschrift fur Naturforschung*, **42c**, 1146–1148.

Forkmann, G., de Vlaming, P., Spribille, R., Wiering, H. and Schram, A. W. (1986) Genetic and biochemical studies on the conversion of dihydroflavonols to flavonols in flowers of *Petunia hybrida*, *Zeitschrift fur Naturforschung*, **41c**, 179–186.

Frydman, A., Weisshaus, O., Bar-Peled, M. *et al.* (2004) Citrus fruit bitter flavors: isolation and functional charaterization of the gene Cm1,2RhaT encoding 1, 2 rhamonosyltransferase, a key enzyme in the biosynthesis of the bitter flavonoids of citrus, *The Plant Journal*, **40**, 88–100.

Fujiwara, H., Tanaka, Y., Yonekura-Sakakibara, K. *et al.* (1998) cDNA cloning, gene expression and subcellular localization of anthocyanin 5-aromatic acyltransferase from *Gentiana triflora*, *The Plant Journal*, **16**, 421–431.

Fukuchi-Mizutani, M., Katsumoto, Y., Brugliera, F. *et al.* (2004) Flower colour modification of the pot plants torenia and Nierembergia, by down regulation of flavonoid hydroxylase genes, in *Abstracts of 7th Symposium on Cytochrome P450 Biodiversity and Biotechnology* (ed. H. Ohkawa), p. 58.

Fukuchi-Mizutani, M., Okuhara, H., Fukui, Y. *et al.* (2003) Biochemical and molecular characterization of a novel UDP-glucose:anthocyanin 3′-O-glucosyltransferase, a key enzyme for blue anthocyanin biosynthesis, from gentian, *The Plant Journal*, **132**, 1652–1663.

Fukuda-Tanaka, S., Inagaki, Y., Yamaguchi, T., Saito, N. and Iida, S. (2000) Colouring-enhancing protein in blue petals, *Nature*, **407**, 581.

Fukui, Y., Kusumi, T., Yoshida, K., Kondo, T., Matsuda, C. and Nomoto, K. (1998) Structures of two diacylated anthocyanins from *Petunia hybrida* cv. Surfinia Violet Mini, *Phytochemistry*, **47**, 1409–1416.

Fukui, Y., Tanaka, Y., Kusumi, T., Iwashita, T. and Nomoto, K. (2003) A rationale for the shift in colour towards blue in transgenic carnation flowers expressing the flavonoid 3′, 5′-hydroxylase gene, *Phytochemistry*, **63**, 15–23.

Fukuzaki, E. Kawasaki, K., Kajiyama, S. *et al.* (2004) Flower color modifications of *Torenia hybrida* by downregulation of chalcone synthase genes with RNA interference, *Journal of Biotechnology*, **111**, 229–240.

Gerats, A. G. M., Wallroth, M., Donker-Koopman, W., Groot, S. P. C. and Schram, A. W. (1983) The genetic control of the enzyme UDP-glucose: 3-O-flavonoid-glucosyltransferase in flowers of *Petunia hybrida*, *Theoritical and Applied Genetics*, **65**, 349–352.

Giuliano, G. Aquilani, R. and Dharmapuri, S. (2000) Metabolic engineering of plant carotenoids, *Trends in Plant Science*, **5**, 1360–1385.

Gonnet, J.-F. (2003) Origin of the color of cv. Rhapsody in Blue rose and some other so-called 'blue' roses, *Journal of Agricultural and Food Chemistry*, **51**, 4990–4994.

Gonnet, J.-F. and Fenet, B. (2000) 'Cyclamen red' colors based on a macrocyclic anthocyanin in carnation flowers, *Journal of Agricultural and Food Chemistry*, **48**, 22–25.

Goto, T. (1989) Structure, stability and color variation of natural anthocyanins, in *Progress in the Chemistry of Organic Natural Products* (eds G. Herz, H. Grisebach, G. G. E. Kirby and C. Tamm), Springer-Verlag, Wien, vol 52, pp. 113–158.

Goto, T. and Kondo, T. (1991) Structure and molecular stacking of anthocyanins – flower color variation, *Angewante Chemie International Edition in English*, **30**, 17–33.

Griesbach, R. J. and Asen, S. (1990) Characterization of the flavonol glycosides in *Petunia*, *Plant Science*, **70**, 49–56.

Gutterson, N. (1995) Anthocyanin biosynthetic genes and their application to flower color modification through sense suppression, *HorticultureScience*, **30**, 964–966.

Halbwirth, H., Forkmann, G. and Stich, K. (2004) The A-ring specific hydroxylation of flavonols in position 6 in *Tagetes* sp. is catalyzed by a cytochrome P450 dependent monooxygenase, *Plant Science*, **167**, 129–135.

Harborne, J. B. and Williams, C. A. (2000) Advances in flavonoid research since 1992, *Phytochemistry*, **55**, 481–504.

Haslam, E., ed. (1998) *Practical Polyphenolics, from Structure to Molecular Recognition and Physiological Action*, Cambridge University Press, Cambridge, UK.

Helariutta, Y., Elomaa, P., Kotilainen, M., Seppänen, P. and Teeri, T. H. (1993) Cloning of cDNA coding for dihyrdoflavonol-4-reductase (DFR) and characterization of expression in the corollas of *Gerbera hybrida* var. Regina (Compositae), *Plant Molecular Biology*, **22**, 183–193.

Hinz, U. G., Fivaz, J., Girod, P.-A. and Zryd, J.-P. (1997) The gene coding for the DOPA dioxygenase involved in betalain biosynthesis in *Amanita muscaria* and its regulation, *Molecular and General Genetics*, **256**, 1–6.

Hirschberg, J. (2001) Carotenoid biosynthesis in flowering plants, *Current Opinion in Plant Biology*, **4**, 210–218.

Holton, T. A. (1996) Transgenic plants exhibiting altered flower colour and methods for producing same, *International Patent Publication Number WO 96/36716*, Assignee: International Flower Developments.

Holton, T. A., Brugliera, F., Lester, D. R. *et al.* (1993a) Cloning and expression of cytochrome P450 genes controlling flower colour, *Nature* , **366**, 276–279.

Holton, T. A., Brugliera, F. and Tanaka, Y. (1993b) Cloning and expression of flavonol synthase from *Petunia hybrida*, *The Plant Journal*, **4**, 1003–1010.

Honda, T. and Saito, N. (2002) Recent progress in the chemistry of polyacylated anthocyanins as flower colour pigment, *Heterocycles*, **56**, 633–692.

Horsch, R. B., Fry, J. E., Hoffmann, N. L., Eicholtz, D., Rogers, S. G. and Fraley, R. T. (1985) A simple and general method for transferring genes into plants, *Science*, **227**, 1229–1231.

Hoshino, A., Morita, Y., Choi, J. D. *et al.* (2003) Spontaneous mutations of the flavonoid 3′-hydroxylase gene conferring reddish flowers in the three morning glory species, *Plant Cell Physiology*, **44**, 990–1001.

van Houwelingen, A., Souer, E., Spelt, C., Kloos, D., Mol, J. and Koes, R. (1998) Analysis of flower pigmentation mutants generated by random transposon mutagenesis in *Petunia hybrida*, *The Plant Journal*, **13**, 39–50.

Huits, H. (1993) Mutable genes and control of flower pigmentation in *Petunia hybrida*, Ph.D. thesis, Vrije Universiteit te Amsterdam.

Iida, S., Morita, Y., Choi, J. D., Park, K. I. and Hoshino, A. (2004) Genetics and epigenetics in flower pigmentation associated with transposable elements in morning glories, *Advances in Biophysics*, **38**,141–159.

Imayama, T., Yoshihara, N., Fukuchi-Mizutani, M., Tanaka, Y., Ino, I. and Yabuya, T. (2004) Isolation and characterization of a cDNA clone of UDP-glucose: anthocyanin 5-*O*-glucosyltransferase in *Iris hollandica*, *Plant Science*, **167**, 1243–1248.

Inagaki, Y., Hisatomi, Y. and Iida, S. (1996) Somatic excision of the transposable element, Tpn1, from the DFR gene for pigmentation in subepidermal layer of periclinally chimeric flowers and sexual transmission of empty donor sequences in their progeny of Japanese morning glory bearing variegated flowers, *Theoretical and Applied Genetics*, **92**, 449–504.

Itoh, Y., Higeta, D., Suzuki, A., Yoshida, H. and Ozeki, Y. (2002). Excision of transposable elements from the chalcone isomerase and dihydroflavonol 4-reductase genes may contribute to the variegation of the yellow-flowered carnation (*Dianthus caryophyllus*), *Plant and Cell Physiology*, **43**, 578–585.

Johns, P., Messner, B., Nakajima, J., Schaffner, A. R. and Saito, K. (2003) UGT53C6 and UGT78D1, glycosyltransferases in involved in flavonol glucoside biosynthesis in *Arabidopsis thaliana*, *Journal of Biology and Chemistry*, **278**, 43910–43918.

Johnson, E. T., Yi, H., Shin, B., Oh, B. J., Cheong, H. and Choi, G. (1999) *Cymbidium hybrida* dihydroflavonol 4-reductase does not efficiently reduce dihydrokaempferol to produce orange pelargonidin-type anthocyanins, *The Plant Journal*, **19**, 81–85.

Jordan, C. and Reimann-Philipp, R. (1983) Untersuchurgen über typ and grad der polyploidie von *Chrysanthemum morifolium* Ramat. durch erbanalysen von zwei blütenfarbmerkmalen, *Zeitschrift für Pflanzenzuchtung*, **91**, 111–122.

Joy, R. W. 4th, Sugiyama, M., Fukuda, H. and Komamine, A. (1995) Cloning and characterization of polyphenol oxidase cDNAs of *Phytolacca americana*, *Plant Physiology*, **107**, 1083–1089.

Kang, J.-M., Chia, L.-S., Goh, N.-K., Chia, T.-F. and Brouillard, R. (2003) Analysis and biological activities of anthocyanins, *Phytochemistry*, **64**, 923–933.

Kanno, Y., Noda, N., Kazuma, K., Tsugawa, H. and Suzuki, M. (2003) Transformation of *Lobelia erinus*, in *Abstracts of 21st Annual Meeting of Japanese Society on Plant Cell Molecular Biology*, Nara, pp. 121.

Kay, Q. O. N., Daoud, H. S. and Stirton, C. H. (1981) Pigment distribution, light reflection and cell structure in petals, *Botanical Journal of Linnean Society*, **83**, 57–84.

Kazuma, K., Noda, N. and Suzuki, M. (2003) Flavonoid composition related to petal color in different lines of *Clitoria ternatea*, *Phytochemistry*, **64**, 1133–1139.

Kho, K. F. F., Kamsteeg, J. and van Brederode, J. (1978) Identification, properties and genetic control of UDP-glucose: cyanidin 3-*O*-glucosyltransferase in *Petunia hybrida*, *Zeitschrift für Pflanzen physiologie*, **88**, 449–464.

Kishimoto, S., Maoka, T., Nakayama, M. and Ohmiya, A. (2004) Carotenoid composition in petals of chrysanthemum (*Dendranthema grandiflorim* (Ramat.) Kitamura), *Phytochemistry*, **65**, 2781–2787.

Kitada, C., Gong, Z., Tanaka Y., Yamazaki, M. and Saito, K. (2001) Differential expression of two cytochrome P450s involved in the biosynthesis of flavones and anthocyanins in chemo-varietal forms of *Perilla frutescens*, *Plant and Cell Physiology*, **42**, 1338–1344.

Koes, R. E., van Blokland, R., Quattrocchio, F., van Tunen, A. J. and Mol, J. N. M. (1990) Chalcone synthase promoters in petunia are active in pigmented and unpigmented cell types, *The Plant Cell*, **2**, 379–392.

van der Krol, A. R., Lenting, P. E., Veenstra, J. *et al.* (1988) An antisense chalcone synthase gene in transgenic plants inhibits flower pigmentation, *Nature*, **333**, 866–869.

van der Krol, A. R., Mur, L. A., Beld, M., Mol, J. N. M. and Stuitje, A. R. (1990) Flavonoid genes in petunia: addition of a limited number of gene copies may lead to suppression of gene expression, *The Plant Cell*, **2**, 291–299.

Kroon, J., Souer, E., de Graaf, A., Xue, Y., Mol, J. and Koes, R. (1994) Cloning and structural analysis of the anthocyanin pigmentation locus *Rt* of *Petunia hybrida*: characterization of insertion sequences in two mutant alleles, *The Plant Journal*, **5**, 69–80.

Lim, E. K., Baldauf, S., Li, Y. *et al.* (2003) Evolution of substrate recognition across a mutigene family of glycosyltransferases in Arabidopsis, *Glycobiology*, **13**, 139–145.

Lloyd, A. M., Walbot, V. and Davis, R. W. (1992). *Arabidopsis* and *Nicotiana* anthocyanin production activated by maize regulators *R* and *C1*, *Science*, **258**, 1773–1775.

Maizonnier, D. (1984) Cytology, in *Petunia* (ed. K. Sink), Springer-Verlag, Berlin, pp. 21–33.

Mann, V., Harker, M., Pecker, I. and Hirschberg, J. (2000) Metabolic engineering of astaxanthin production in tobacco flowers, *Nature*, **18**, 888–892.

Martens, S. and Forkmann, G. (1999) Cloning and expression of flavone synthase II from *Gerbera hybrida*, *The Plant Journal*, **20**, 611–618.

Martens, S., Forkmann, G., Britsch, L., Wellmann, F., Matern, U. and Lukacin, R. (2003a) Divergent evolution of flavonoid 2-oxoglutarate-dependent dioxygenases in parsley, *FEBS Letters*, **544**, 93–98.

Martens, S., Knott, J. Setz, C. A., Janvari, L., Yu, S.-N. and Forkmann, G. (2003b) Impact of biochemical pre-studies on specific metabolite engineering strategies of flavonoid biosyhtnesis in plant tissues, *Biochemical Engineering Journal*, **14**, 227–235.

Martin, C. and Gerats, T. (1993) The control of flower coloration, in *The Molecular Biology of Flowering* (ed. B. R. Jordan), CAB International, UK, pp. 219–255.

Meyer, P., Heidmann, I., Forkmann, G. and Saedler, H. (1987) A new petunia flower colour generated by transformation of a mutant with a maize gene, *Nature*, **330**, 677–678.

Mikanagi, Y., Saito, N., Yokoi, M. and Tatsuzawa, F. (2000) Anthocyanins in flowers of genus *Rosa*, sections *Cinnamomeae* (=*Rosa*), *Chinenses*, *Gallicanae* and some modern garden roses, *Biochemical Systematics Ecology*, **28**, 887–902.

Mol, J., Cornish, E., Mason, J. and Koes, R. (1999) Novel coloured flowers, *Current Opinion in Biotechnology*, **10**, 198–201.

Mol, J., Grotewold, E. and Koes, R. (1998) How gene paint flowers and seeds, *Trends in Plant Science*, **3**, 212–217.

Morita, Y., Hoshino, A., Kikuchi, Y., Okuhara, H., Ono, E., Tanaka, Y., Fukui, Y., Saito, N., Nitasako, E., Noguchi, H. and Ida, S. (2005) Japanese morning glory *dusky* mutants displaying reddish-brown or purplish-gray flowers are deficient in a novel glycosylation enzyme for anthocyanin biosynthesis, UDP-glucose: anthocyanidin 3-*O*-glucoside-2″-*O*-glucosyltransferase, due to 4-bp insertions in the gene, *The Plant Journal*, **42**, 353–363.

Mueller, L. A., Hinz, U., Uze, M., Sautter, C. and Zryd, J.-P. (1997) Biochemical complementation of the betalain biosynthetic pathway in *Portulaca grandiflora* by a fungal 3, 4-dihydroxyphenylalanine dioxygenase, *Planta*, **203**, 260–263.

Nakamura, N., Fukuchi-Mizutani, M., Miyazaki, K., Suzuki, K. and Tanaka, Y. (2006) RNAi suppression of the anthocyanidin synthase gene of *Torenia hybrida* yields white flowers with higher frequency and better stability than antisense and sense suppression, *Plant Biotechnology*, 23, in press.

Nakayama, T., Yonekura-Sakakibara, K., Sato, T. *et al.* (2000a) Aureusidin synthase: a polyphenol oxidase homolog responsible for flower coloration, *Science*, **290**, 1163–1166.

Nakayama, Y., Koshida, M., Yoshida, H. *et al.* (2000b) Cyclic malyl anthocyanins in *Dianthus caryophyllus*, *Phytochemistry*, **55**, 937–939.

Napoli, C., Lemieux, C. and Jorgensen, R. (1990) Introduction of a chimeric chalcone synthase gene into petunia results in reversible co-suppression of homologous genes *in trans*, *The Plant Cell*, **2**, 279–289.

Nishihara, M., Nakatsuka, T., Mishiba, K., Kikuchi, A. and Yamamura, S. (2003) Flower color modification by suppression of chalcone synthase gene in gentian, *Plant and Cell Physiology*, **44**, s159.

Noda, N., Kato, N., Kogawa, K., Kazuma, K. and Suzuki, M. (2004) Cloning and characterization of the gene encoding anthocyanin 3′, 5′-O-glucosyltransferase involved in ternatin biosynthesis from blue petals of butterfly pea (*Clitoria ternatea*), *Plant and Cell Physiology*, **45**, s132.

Ogata, J., Itoh, Y., Ishida, M., Yoshida, H. and Ozeki, Y. (2004) Cloning and heterologous expression of a cDNA encoding flavonoid glucosyltransferases from *Dianthus caryophyllus*, *Plant Biotechnology*, **21**, 367–376.

Okuhara, H., Ishiguro, K., Hirose, C. *et al.* (2004) Molecular cloning and functional expression of tetrahydroxychalcone 2′-glucosyltransferase genes, *Plant Cell and Physiology*, **45**, s133.

Ono, E., Hatayama, M., Isono, Y., Sato, T., Watanabe, R., Yonekura-Sakakibara, K., Fukuchi-Mizutani, M., Tanaka, Y., Kusumi, T., Nishino, T., Nakayama, T., (in press) Unexpected subcellular localization of a flavonoid biosynthetic polyphenol oxidase in vaculoes, *The Plant Journal*.

Oud, J. S. N., Scheniders, H. Kool, A. J. and van Grinsven M. Q. M. (1995) Breeding of transgenic orange *Petunia hybrida* varieties, *Euphytica*, **85**, 403–409.

Page, R. D. M. (1996) An application to display phylogenetic trees on personal computers, *Computer Applications Biosciences*, **12**, 357–358.

Paquette, S., Moller, B. L. and Bak, S. (2003) On the origin of family 1 plant glycosyltransferases, *Phytochemistry*, **62**, 399–413.

Rosati, C., Cadic, A., Duron, M., Ingouff, M. and Simoneau, P. (1999) Molecular characterisation of the anthocyanidin synthase gene in *Forsythia x intermedia* reveals organ-specific expression during flower development, *Plant Science*, **149**, 73–79.

Rosati, C., Simoneau, P., Treutter, D. *et al.* (2003) Engineering of flower color in forsythia by expression of two independently-transformed dihydroflavonol 4-reductase and anthocyanidin synthase genes of flavonoid pathway, *Molecular Breeding*, **12**(3), 197–208.

Ryan, K. G., Swinny, E. E., Winefiled, C. and Markham, K. R. (2001) Flavonoids and UV photoprotection in *Arabidopsis* mutants, *Zeitschrift fur Naturforschung*, **56c**, 745–754.

Saito, N., Toki, K., Honda, T. and Kawase, K. (1988) Cyanidin 3-malonylglucuronylglucoside in *Bellis* and cyanidin 3-malonlyglucoside in *Dendranthema*, *Phytochemistry*, **27**, 2963–2966.

Sasaki, N., Adachi, T., Koda, T. and Ozeki, Y. (2004) Detection of UDP-glucose: *cyclo*-DOPA 5-glucosyltransferase activity in four o'clocks (*Mirabilis jalapa* L.), *FEBS Letters*, **568**, 159–162.

Sasaki, N., Wada, K., Koda, T., Kasahara, K., Adachi, T. and Ozeki, Y. (2005) Isolation and characterization of cDNAs encoding an enzyme with glucosyltransterase activity for cyclo- DOPA from four o'clocks and feather cockscombs, *Plant and Cell Physiology*, **46**, 666–670.

Saslowsky, D. and Winkel-Shirley, B. (2001) Localization of flavonoid enzymes in *Arabidopsis* roots, *The Plant Journal*, **27**, 37–48.

Sato, T., Nakayama, T., Kikuchi, S. *et al.* (2001) Enzymatic formation of aurones in the extracts if yellow snapdragon flowers, *Plant Science*, **160**, 229–236.

Sawada, S., Suzuki, H., Ichimaida, F. *et al.* (2005) UDP-glucuronic acid: anthocyanin glucuronosyltransferase from red daisy (*Bellis perennis*) flowers, *Journal of Biological Chemistry*, **280**, 899–906.

Schram, A. W., Jonsson, L. M. V. and Bennink, G. J. H. (1984) Biochemistry of flavonoid synthesis in *Petunia hybrida*, in *Petunia* (ed. K. C. Sink), Springer-Verlag, Berlin, pp. 68–75.

Schijlen, E. G. W. M., Ric de Vos, C. H., van Tunen, A. J. and Bovy, A. G. (2004) Modification of flavonoid biosynthesis in crop plants, *Phytochemistry*, **65**, 2631–2648.

Schliemann, W., Kobayashi, N. and Strack, D. (1999) The decisive step in betaxanthin biosynthesis is a spontaneous reaction, *Plant Physiology*, **119**, 1217–1232.

Schroder, J. (1999) The chalcone/stilbene synthase-type family of condensing enzymes, in *Polyketides and Other Secondary Metabolites Including Fatty Acid and their Derivatives* (ed. U. Sankawa), Elsevier, Amsterdam, pp. 749–771.

Schuler, M. A. and Werck-Reichhart, D. (2003) Functional genomics of P450s, *Annual Review of Plant Biology*, **54**, 629–667.

Schwinn, K. E., Markham, K. R. and Giveno, N. (1993) Floral flavonoids and the potential for pelargonidin biosythesis in commercial chrysanthemum cultivars, *Phytochemistry*, **35**, 145–150.

Scott-Moncrieff, R. (1936) A biochemical survey of some mendelian factors for flower colour, *Journal of Genetics*, **32**, 117–170.

Shiono, M., Matsugaki, N. and Takeda, K. (2005) Phytochemistry: structure of the blue cornflower pigment. *Nature* **436**, 791.

Sink, K. C., ed. (1984) *Petunia*, Springer-Verlag, Berlin.

Spelt, C., Quattrocchio, F., Mol, J. and Koes, R. (2002) *ANTHOCYANIN1* of petunia controls pigment synthesis, vacuolar pH, and seed coat development by genetically distinct mechanisms, *The Plant Cell*, **14**, 2121–2135.

Springob, K., Nakajima, J., Yamazaki, M. and Saito, K. (2003) Recent advances in the bisosynthesis and accumulation of anthocyanins, *Natural Product Reports*, **20**, 288–303.

Steiner, U., Schliemann, W., Bohm, H. and Strack, D. (1999) Tyrosinase involved in betalain biosynthesis of higher plants, *Planta*, **208**, 114–124.

Stewart, R. N., Norris, K. H. and Asen, S. (1975) Microspectrophotometric measurement of pH and pH effect on colour of petal epidermal cells, *Phytochemistry*, **14**, 937–942.

Stich, K., Eidenberger, T., Wurst, F. and Forkmann, G. (1992) Flavonol synthase activity and the regulation of flavonol and anthocyanin biosynthesis during flower development in *Dianthus caryophyllus* L. (carnation), *Zeitschrift fur Naturforschung*, **47c**, 553–560.

St-Pierre, B. and De Luca, V. (2000) Evolution of acyltransferase genes: origin and diversification of the BAHD superfamily of acyltransferases involved in secondary metabolisim, in *Evolution of Metabolic Pathways* (ed. J. T. Romeo, R. Ibrahim, L. Varin and V. De Luca), Elsevier, Amsterdam, *Recent advances in phytochemistry*, **34**, pp. 285–315.

Strack, D., Vogt, T. and Schliemann, W. (2003) Recent advances in betalain research, *Phytochemistry*, **62**(3), 247–269.

Sun, Y. and Hrazdina, G. (1991) Isolation and characterization of a UDP glucose: flavonol O^3-glucosyltransferase from illuminated red cabbage (*Brassica oleracea* cv Red Danish) seedlings, *Plant Physiology*, **95**, 570–576.

Suzuki, H., Nakayama, T., Nagae, S. *et al.* (2004a) cDNA cloning and functional characterization of flavonol 3-O-glucoside-6″-O-malonyltransferases from flowers of *Verbena hybrida* and *Lamium purpureum*, *Journal of Molecular Catalysis B: Enzymatic*, **28**, 87–93.

Suzuki, H., Nakayama, T., Yamaguchi, M.-A. and Nishino, T. (2003b) cDNA cloning and characterization of two *Dendranthema x morifolium* anthocyanin malonyltransferases with different functional activities, *Plant Science*, **166**, 89–96.

Suzuki, H., Nakayama, T., Yonekura-Sakakibara, K. *et al.* (2001). Malonyl CoA:anthocyanidin 5-*O*-glucoside-6"-*O*-malonyltransferase gene from scarlet sage (*Salvia splendens*) flowers. Enzyme purification, gene cloning, expression, and characterization, *Journal of Biological Chemistry*, **276**, 49013–49019.

Suzuki, H., Nakayama, T., Yonekura-Sakakibara, K. *et al.* (2002) cDNA cloning, heterologous expressions, and functional characterization of malonyl CoA: anthocaynidin 3-*O*-glucoside-6″-*O*-malonyltransferase from dahlia flowers, *Plant Physiology*, **130**, 2142–2151.

Suzuki, H., Sawada, S., Yonekura-Sakakibara, K. *et al.* (2003a) Identification of a cDNA encoding malonyl-coenzyme A: anthocyanidin 3-O-glucoside 6″-O-malonyltransferase from cineraria (*Senecio cruentus*) flowers, *Plant Biotechnology*, **20**, 229–234.

Suzuki, H., Sawada, S., Watanabe, K. *et al.* (2004b) Identification and characterization of a novel anthocyanin malonyltransferase from scarlet sage (*Salvia splendens*) flowers: an enzyme that is phylogenetically separated from other anthocyanin acyltransferases, *The Plant Journal*, **38**, 994–1003.

Suzuki, K., Zue, H., Tanaka, Y. *et al.* (2000) Flower color modifications of *Torenia hybrida* by cosuppression of anthocyanin biosynthesis genes, *Molecular Breeding*, **6**, 239–246.

Taguchi, G., Yazawa, T., Hayashida, N. and Okazaki, M. (2001) Molecular cloning and heterologous expression of novel glucosyltransferases from tobacco cultured cells that have broad substrate specificity and are induced by salicylic acid, *European Journal of Biochemistry*, **268**, 4086–4094.

Taguchi, G., Ubukata, T., Hayashida, N., Yamamoto, H. and Okazaki, M. (2003) Cloning and characterization of a glucosyltransferase that reacts on 7-hydroxyl group of flavonol and 3-hydroxyl group of coumarin from tobacco cells, *Archives of Biochemistry and Biophysics*, **420**, 95–102.

Tanaka, Y. and Mason, J. (2003) Manipulation of flower colour by genetic engineering, in *Plant Genetic Engineering* (ed. P. Singh), SCI Tech Publishing, Housten, pp. 361–385.

Tanaka, Y., Fukui, Y., Fukuchi-Mizutani, M., Holton, T. A., Higgins, E. and Kusumi, T. (1995) Molecular cloning and characterization of *Rosa hybrida* dihydroflavonol 4-reductase, *Plant Cell Physiology*, **36**, 1023–1031.

Tanaka, Y., Katsumoto, Y., Brugliera, F. and Mason, J. (2005) Genetic Engineering in Floriculture, *Plant Cell, Tissue and Organ Culture*, **80**, 1–24.

Tanaka, Y., Tsuda, S. and Kusumi, T. (1998) Metabolic engineering to modify flower colour, *Plant Cell Physiology*, **39**, 1119–1126.

Tohge, T., Nishiyama, Y., Hirai, M. Y., Yano, M., Nakajima, J., Awazuhara, M., Inoue, E., Takahashi, H., Goodenowe, D. B., Kitayama, M., Noji, M., Yamazaki, M. and Kazuki, S. (2005) Functional Genomics by integrated analysis of metabolome and transcriptome of Arabidopsis plants over-expressing an Myb transcription factor. *The Plant Journal*, **42**, 218–235.

Tsuda, S., Fukui, Y., Nakamura, N. *et al.* (2004) Flower color modification of *Petunia hybrida* commercial varieties by metabolic engineering, *Plant Biotechnology*, **21**, 377–386.

Turnbull, J. J., Nakajima, J., Welford, R. W., Yamazaki, M., Saito, K. and Schofield, C. J. (2004) Mechanistic studies on three 2-oxoglutarate-dependent oxygenases of flavonoid biosynthesis: anthocyanidin synthase, flavonol synthase, and flavanone 3 beta-hydroxylase, *Journal of Biological Chemistry*, **279**, 1206–1216.

de Vlaming, P., Gerats, A. G. M., Wiering, H., Wijsman, H. J. W., Cornu, A., Farcy, D. and Maizonnier, D. (1984) *Petunia hybrida*: a short description of the action of 91 genes, their origin and their map location, *Plant Molecular Biology Reporter*, **2**, 21–42.

de Vries, D. P., van Keulen, H. A. and de Bruyn, J. W. (1974) Breeding research on rose pigments. I. The occurence of flavonoids and carotenoids in rose petals, *Euphytica*, **23**, 447–457.

Vogt, T. (2002) Substrate specificity and sequence analysis define a polyphyletic origin of betanidin 5- and 6-*O*-glucosyltransferase from *Dorotheanthus bellidiformis*, *The Plant Journal*, **21**, 492–495.

Vogt, T., Grimm, R. and Strack, D. (1999) Cloning and expression of a cDNA encoding betanidin 5-O-glucosyltransferase, a betanidin- and flavonoid-specific enzyme with high homology to inducible glucosyltranferases from the Solanaceae, *The Plant Journal*, **19**, 509–519.

Wang, M.-B. and Waterhouse, P.M. (2001) Application of gene silencing in plants, *Current Opinion in Plant Biology*, **5**, 146–150.

Waterhouse, P. M., Graham, M. W. and Wang, M.-B. (1998) Virus resistance and gene silencing in plants can be induced by simultaneous expression of sense and antisense RNA, *Proceedings of the National Academy of Sciences of the United States of America*, **95**, 13959–13964.

Wesley, S. V., Helliwell, C. A., Smith, N. A. *et al.* (2001) Construct design for efficient, effective and high-throughput gene silencing in plants, *The Plant Journal*, **27**, 581–590.

Wiering, H. and de Vlaming, P. (1984) Inheritance and biochemistry of pigments, in *Petunia* (ed. K.C. Sink), Springer-Verlag, Berlin, pp. 49–65.

Wilmouth, R. C., Turnbull, J. J., Welford, R. W., Clifton, I. J., Prescott, A. G. and Schofield, C. J. (2002) Structure and mechanism of anthocyanin synthase from *Arabidopsis thaliana*, *Structure*, **10**, 93–103.

Winkel, B. S. J. (2004) Metabolic channelling in plants, *Annual Review of Plant Biology*, **55**, 85–107.

Winkel-Shirley, B. (2001a) It takes a garden. How work on a diverse plant species has contributed to an understanding of flavonoid metabolism, *Plant Physiology*, **127**, 1399–1404.

Winkel-Shirley, B. (2001b) Flavonoid biochemistry. A colorful model for genetics, biochemistry, cell biology, and biotechnology, *Plant Physiology*, **126**, 485–493.

Winkel-Shirley, B. (2002). Biosynthesis of flavonoids and effects of stress, *Current Opinion in Plant Biology*, **5**, 218–223.

Yamaguchi, T., Fukada-Tanaka, S., Inagaki, Y. *et al.* (2001) Genes encoding the vacuolar Na+/H+ exchanger and flower coloration, *Plant and Cell Physiology*, **42**, 451–461.

Yamazaki, M., Yamagishi, E., Gong, Z. *et al.* (2002) Two flavonoid glucosyltransferases from *Petunia hybrida*: molecular cloning, biochemical properties and developmentally regulated expression, *Plant Molecular Biology*, **48**, 401–411.

Yonekura-Sakakibara, K., Tanaka, Y. *et al.* (2000) Molecular and biochemical characterization of hydroxycinnamoyl-CoA: anthocyanin 3-O-glucoside-6″-O-hydroxycinnamoyltransferase from *Perilla frutescens*, *Plant Cell Physiology*, **132**, 1652–1663.

Yoshida, H., Itoh, Y., Ozeki, Y., Iwashina, T. and Yamaguchi, M. (2004) Variation in chalcononaringenin 2′-*O*-glucoside content in the petals of carnations (*Dianthus caryophyllus*) bearing yellow flowers, *Scienta Horticulturae*, **99**, 175–186.

Yoshida, K., Kondo, T., Okazaki, Y. and Katou, K. (1995) Cause of blue petal colour, *Nature*, **373**, 291.

Yoshida, K., Toyama, Y., Kameda, K. and Kondo, T. (2000) Contribution of each caffeoyl residue of the pigment molecules of gentiodelphin to blue color development, *Phytochemistry*, **54**, 85–92.

Yoshida, K., Toyama-Kato, Y., Kameda, K. and Kondo, T. (2003) Sepal color variation of *Hydrangea macrophylla* and vacuolar pH measurement with proton-selective microelectrode, *Plant Cell Physiology*, **44**, 262–268.

Yoshihara, N., Imayama, T., Fukuchi-Mizutani, M., Okuhara, H., Tanaka, Y., Ino, I. and Yabuya. (2005) cDNA cloning and characterization of UDP-glucose: anthocyanidin 3-*O*-glucosyltransferase in Iris hollandica, *Plant Science*, **169**, 496–501.

Zufall, R. A. and Rausher, M. D. (2004) Genetic changes associated with floral adaptation restrict future evolutionary potential, *Nature*, **428**, 847–850.

Zuker, A., Tzfira, T., Ben-meir, H. *et al.* (2002) Modification of flower colour and fragrance by antisense suppression of the flavanone 3-hydroxylase gene, *Molecular Breeding*, **9**, 33–41.

10 Floral scent: biosynthesis, regulation and genetic modifications

Jennifer Schnepp and Natalia Dudareva

10.1 Introduction

Floral scent is a diverse blend of low molecular weight (under 300 Da) compounds, mostly lipophilic, emitted from flowers into the surrounding atmosphere. Alone, or with visual cues, floral scents provide chemical signals to pollinators thereby ensuring plant reproductive and evolutionary success (Buchmann and Nabhan, 1996; Dudareva and Pichersky, 2000). The chemical composition of floral fragrances varies widely among species in terms of the number, identity and relative amounts of constituent volatile compounds (Knudsen and Tollsten, 1993; Knudsen *et al.*, 1993). Within a species, the level of emission changes in response to endogenous diurnal rhythms, flower age, pollination status and environmental conditions. Floral scent research witnessed significant progress in the last decade, including the identification and isolation of genes encoding enzymes catalyzing the synthesis of scent compounds and advances in our understanding of the regulation of scent biosynthesis. Here we will present the biochemical, physiological, molecular and regulatory aspects of floral scent, as well as potential future directions for metabolic engineering of the odorant trait.

10.2 Biological functions of floral scents

The relative abundance of different volatile compounds results in a flower's distinctive fragrance. Some volatiles emitted from flowers function as both long- and short-distance attractants and play a prominent role for insects in the location and selection of flowers, especially in moth-pollinated flowers which are detected and visited at night (Dobson, 1994). To date, very little is known about how insects respond to individual components found in floral scents, but it is clear that they are able to distinguish between complex floral scent mixtures. This discriminatory visitation, based on floral scent, has important implications for plant reproductive success. In addition to attracting insects to flowers and guiding them to food resources within the flower, floral volatiles are essential in allowing insects to discriminate among plant species and even among individual flowers of a single species. Closely related plant species that rely on different types of insects for pollination produce different odors, reflecting the olfactory sensitivities or preferences of the pollinators

(Henderson, 1986; Raguso and Pichersky, 1995). By providing species-specific signals, flower fragrances facilitate an insect's ability to learn particular food sources, thereby increasing its foraging efficiency. At the same time, successful pollen transfer and thus sexual reproduction is ensured, which is beneficial to plants.

Fragrances also play an important role in the crop economy of insect-pollinated plants, including most fruit trees, berries, nuts, oilseeds and vegetables (McGregor, 1976; DeGrandi-Hoffman, 1987). The presence or absence of scent appropriate to the locally available insect pollinators has a substantial impact on the level of pollination and subsequent seed and fruit set (Traub *et al.*, 1942; Galen, 1985; Sugden, 1986; Henning *et al.*, 1992). A decrease in fragrance emission reduces the ability of flowers to attract pollinators (Williams, 1982; Jakobsen and Olsen, 1994) resulting in considerable losses for growers (Wood, 1975), whereas intensifying the flower odor can increase the pollinator's recruitment rate (Von Frisch, 1971). In addition, when some of the volatiles found in floral scent are synthesized in fruits, they contribute to fruit aromas, thereby enriching the overall flavor quality in the food industry (Berger, 1995) and attracting seed disseminators thus increasing seed dispersal (Pichersky and Gang, 2000).

Besides improving pollination, some volatile compounds found in floral scent have important functions in vegetative processes (Pichersky and Gershenzon, 2002). They may attract natural predators to herbivore-damaged plants (Pare and Tumlinson, 1997, 1999). As a consequence, a plant could reduce the number of herbivores by more than 90% by releasing volatiles (Kessler and Baldwin, 2001). They may also function as repellents against herbivores (Levin 1973; Rodriguez and Levin, 1976; Wood, 1982; Pellmyr *et al.*, 1987; Gershenzon and Croteau, 1991) or as airborne signals that activate disease resistance via the expression of defense-related genes in neighboring plants and in the healthy tissues of infected plants (Farmer and Ryan, 1990; Shulaev *et al.*, 1997; Seskar *et al.*, 1998).

Just as it is in nature that scent is a critical factor in attracting pollinators to flowers, scent is also important in attracting consumers to cut flowers, potted flowering plants and flowering herbaceous plants. While floral scent is not a necessity for consumers, the presence of floral scents plays an important role in human life from an esthetic point of view and contributes to the decision to cultivate and propagate specific plant species. Similar to insects, humans strongly associate scent with specific flowers, e.g. rose or jasmine. While there is a wide variation in human taste, most people prefer the scents of bee- and moth-pollinated flowers, which are often described as 'sweet-smelling'. For people, an attraction to scent is not only dictated by the composition of the fragrance but also by the amount of volatile emission (Burdock, 1995).

10.3 Biosynthesis of floral volatiles

Analysis of floral scents from a broad spectrum of plant species revealed that terpenoid, phenylpropanoid and benzenoid compounds are the major classes comprising floral bouquets. Fatty acid derivatives and other chemicals containing nitrogen or sulfur are also sometimes present (Croteau and Karp, 1991; Knudsen *et al.*, 1993; Knudsen and Tollsten, 1993). Although flower volatiles are synthesized via a few major biochemical pathways, their diversity originates from enzymatic modifications (e.g. hydroxylations, acetylations, methylations) that increase the volatility of compounds at the final step of their formation (Dudareva *et al.*, 2004).

Our knowledge of the occurrence and identification of floral volatiles has been significantly extended in the last 15 years, thanks to the adoption of simple, sensitive methods for headspace sampling and the availability of relatively inexpensive bench-top instruments for gas-chromatography–mass-spectrometry. In general, the investigation of the chemical composition of floral scents has been done by 'headspace' analysis (Raguso and Pellmyr, 1998). In this procedure, a flower that is still connected to the rest of the plant is placed inside a small glass or plastic chamber and its emitted volatiles are collected by continually purging the air inside the chamber through a cartridge packed with a polymer that binds these volatiles. After a fixed period of time, trapped volatiles are extracted from the cartridge with an organic solvent. A variation of this procedure, the highly sensitive solid phase microextraction method, or SPME, allows for 'instant' sampling of headspace volatiles (Matich *et al.*, 1996). The eluted solution is injected into a gas chromatograph, which separates the different volatiles, and then each volatile is identified by mass spectrometry. Presently, a total of more than 1000 floral scent compounds have been identified from more than 60 plant families, which constitute about 1% of the plant secondary metabolites known (Dudareva *et al.*, 2004).

10.3.1 Terpenes

Terpenes compose the largest class of plant secondary metabolites, which include hemiterpenes (C_5), monoterpenes (C_{10}), sesquiterpenes (C_{15}) and diterpenes (C_{20}), many of which have a high vapor pressure allowing their release into the air. Monoterpenes such as linalool, limonene, myrcene and trans-ß-ocimene, and also some sesquiterpenes, such as farnesene, nerolidol and caryophyllene, are common constituents of floral scent (Figure 10.1). Volatile terpenoids are synthesized from the universal five-carbon precursors, isopentenyl diphosphate (IPP) and dimethylallyl diphosphate (DMAPP), which are derived from two alternative pathways (Figure 10.2).

In the cytosol, IPP is synthesized via the condensation of acetyl-CoA (Qureshi and Porter, 1981; Newman and Chappell, 1999) by the classical mevalonic acid (MVA) pathway (McCaskill and Croteau, 1995). Two molecules of IPP and one molecule of DMAPP are condensed in a sequential head-to-tail addition of two IPP units to DMAPP catalyzed by the enzyme farnesyl pyrophosphate synthase (FPPS) to form FPP(C_{15}) (McGarvey and Croteau, 1995) and ultimately, the sesquiterpenes and triterpenes. FPPS is a member of a large family of homodimeric prenyltransferases called FPP synthases.

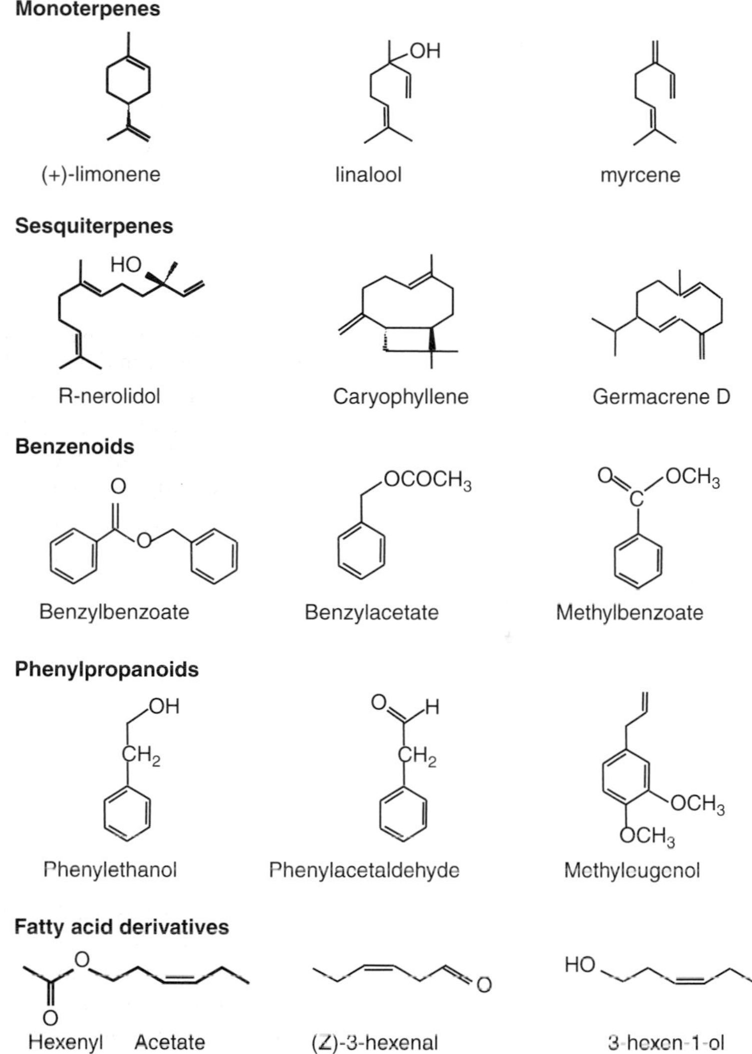

Figure 10.1 Structures of representative floral scent compounds including monoterpenes, sesquiterpenes, phenylpropanoids, benzenoids and fatty acid derivatives.

In plastids, IPP is derived from pyruvate and glyceraldehyde-3-phosphate via the methyl-erythritol-phosphate (MEP) pathway (Eisenreich *et al.*, 1998; Lichtenthaler, 1999; Rohmer, 1999) which was also referred to as the 'Rohmer' pathway (Lichtenthaler *et al.*, 1997), discovered in the past decade (for review, see Rodriguez-Concepcion and Boronat, 2002). One molecule of IPP is condensed to one molecule of DMAPP in a head-to-tail condensation reaction catalyzed by the enzyme geranyl pyrophosphate synthase (GPPS, EC 2.5.1.1) to form GPP, the

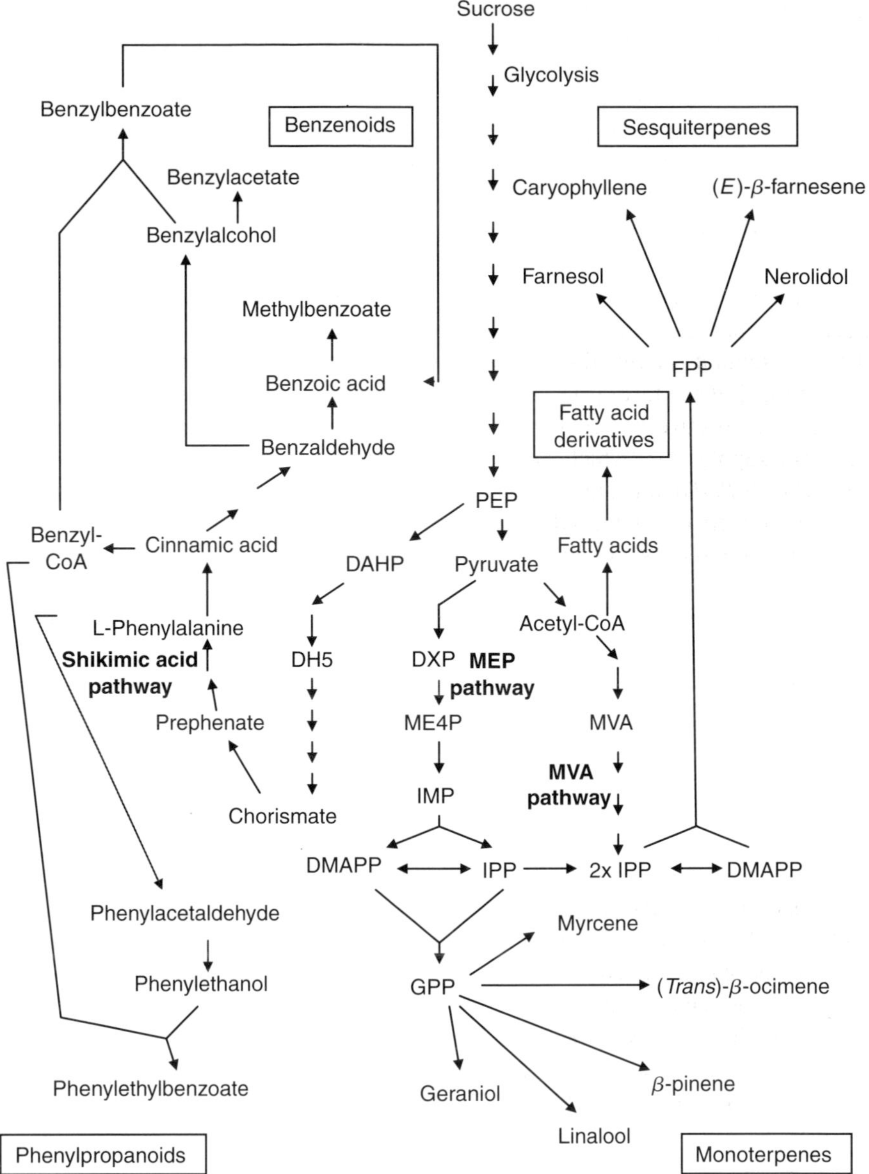

Figure 10.2 Proposed network of metabolic pathways from sucrose to branchways leading to floral volatile compounds including monoterpenes, sesquiterpenes, phenylpropanoids, benzenoids and fatty acid derivatives. Volatile compounds are shown in grey.

universal precursor of all the monoterpenes (Poulter and Rilling, 1981; Ogura and Koyama, 1998). Recent isolation of GPP synthases from different plant species revealed the existence of two fundamentally different architectural structures. While the GPP synthases of Arabidopsis (Bouvier *et al.*, 2000) and grand fir (*Abies grandis*, Burke and Croteau, 2002) are homodimers, those reported from peppermint (*Mentha* x *piperita*) leaves (Burke *et al.*, 1999) and the flowers of snapdragon (*Antirrhinum majus*) and *Clarkia breweri* (Burke *et al.*, 1999) are heterodimeric enzymes, with each subunit classified as a member of the prenyltransferase protein family. In peppermint each subunit alone is catalytically inactive and only their coexpression leads to the production of the active heterodimeric enzyme (Burke *et al.*, 1999). In contrast, in snapdragon only the small subunit is inactive when expressed alone while the large subunit is a functional GGPP synthase on its own (Tholl *et al.*, 2004).

Although the cytosolic and plastidial pool of IPP serve as the precursor of FPP and GPP, respectively, recent evidence suggested that the plastidic MEP pathway can also supply IPP (which is interconvertible with DMAPP) to the cytosol in *Arabidopsis thaliana* (Laule *et al.*, 2003).

The conversion of the GPP (C_{10}), FPP (C_{15}) and GGPP (C_{20}) to monoterpenes, sesquiterpenes and diterpenes, respectively, is carried out by a large family of enzymes known as terpene synthases (Cane, 1999; Wise and Croteau, 1999). One of the most exceptional properties of these enzymes is their tendency to make multiple products from a single substrate. Many of the terpene volatiles are direct products of terpene synthases; however, others are formed through alterations of the initial products by oxidation, dehydrogenation, acylation and other reaction types (Dudareva *et al.*, 2004).

10.3.2 Phenylpropanoids/benzenoids

Phenylpropanoids (Figure 10.1) constitute a large class of secondary metabolites in plants and are derived from phenylalanine via a complex series of branched pathways (Figure 10.2). While most of these aromatic compounds are usually non-volatile, those that are reduced at the C9 position (to either the aldehyde, alcohol or alkane/alkene) and/or contain alkyl additions to the hydroxyl groups of the phenyl ring or to the carboxyl group (i.e. ethers and esters) are volatile. Additionally, many benzenoid compounds (Figure 10.1), which lack the three-carbon chain and originate from *trans*-cinnamic acid as a side branch of the general phenylpropanoid pathway, are also volatile. These volatile phenylpropanoids/benzenoids are common constituents of floral scent (Knudsen *et al.*, 1993).

The first committed step in the biosynthesis of benzenoid and some phenylpropanoid compounds is catalyzed by the well known and widely distributed enzyme L-phenylalanine ammonia-lyase (PAL; EC 4.3.1.5). PAL catalyzes the deamination of L-phenylalanine (Phe) to produce *trans*-cinnamic acid. Formation of benzenoids from cinnamic acid requires the shortening of the side chain by a C_2 unit, for which several routes have been proposed. The side-chain shortening could occur via a CoA-dependent β-oxidative pathway, CoA-independent non-β-oxidative pathway or via the combination of these two mechanisms. The CoA-dependent β-oxidative pathway is analogous to that underlying β-oxidation of fatty acids and proceeds through the formation of four CoA-ester intermediates. The CoA-independent non-β-oxidative pathway involves hydration of the free *trans*-cinnamic acid to 3-hydroxy-3-phenylpropionic acid and side-chain degradation via a reverse aldol reaction with formation of benzaldehyde, which is then oxidized to benzoic acid by an $NADP^+$-dependent aldehyde dehydrogenase.

Recent *in vivo* stable isotope labeling and computer-assisted metabolic flux analysis revealed that both the CoA-dependent β-oxidative and CoA-independent non-β-oxidative pathways are involved in the formation of benzenoid compounds in *Petunia hybrida* (Boatright *et al.*, 2004). While there is no information available about enzymes and genes responsible for these metabolic steps, significant progress has been made in the discovery of common modifications, such as hydroxylation, acetylation and methylation, of downstream products (Dudareva *et al.*, 2004).

10.3.3 Fatty acid derivatives

Volatile fatty acid derivatives including saturated and unsaturated short-chain alcohols, aldehydes and esters originate from membrane lipids. The metabolism of these lipids through the formation of oxidized polyenoic fatty acids and subsequent reactions compose the lipoxygenase pathway (Feussner and Wasternack, 1998). In the past several years, many genes involved in the lipoxygenase pathway have been isolated and characterized (Feussner and Wasternack, 2002), however, to date none of these genes have been cloned from floral tissue.

10.4 Genes responsible for scent production

In the last eight years, investigations into floral scent in many laboratories have resulted in the characterization of a large number of genes encoding enzymes responsible for the synthesis of scent compounds. The initial breakthrough began in 1996 when the (*S*)-linalool synthase (LIS) gene encoding an enzyme responsible for the formation of the acyclic monoterpene linalool, was isolated from *C. breweri* flowers using a classical biochemical approach through enzyme purification from petal tissues with the highest activity (Pichersky *et al.*, 1995; Dudareva *et al.*, 1996). Thereafter, four more genes responsible for the biosynthesis of floral volatiles were isolated using the same protein-based cloning strategy. These include *S*-adenosyl-L-methionine (SAM):(iso) eugenol *O*-methyl transferase (IEMT; Wang *et al.*, 1997), acetyl-coenzyme A:benzyl alcohol acetyltransferase (BEAT; Dudareva *et al.*, 1998) and *S*-adenosyl-L-methionine:salicylic acid carboxyl methyl transferase (SAMT; Ross *et al.*, 1999), all from *C. breweri*, and *S*-adenosyl-L-methionine:benzoic acid carboxyl methyl transferase (BAMT) from *A. majus* (Dudareva *et al.*, 2000; Murfitt *et al.*, 2000), which encode the enzymes responsible for the formation of methyl(iso)eugenol, benzylacetate, methylsalicylate and methylbenzoate, respectively.

Development of functional genomic technology in recent years allowed the isolation of more genes responsible for scent production. This list includes myrcene synthase and ocimene synthase from snapdragon *A. majus* (Dudareva *et al.*, 2003), germacrene D synthase from roses *Rosa hybrida* (Guterman *et al.*, 2002), LIS and caryophyllene synthase from *A. thaliana* flowers (Chen *et al.*, 2003b), geraniol/citronellol acetyl transferase from roses *R. hybrida* (Shalit *et al.*, 2003), SAMT from Madagascar jasmine *Stephanotis floribunda* (Pott *et al.*, 2002), *S*-adenosyl-L-methionine:benzoic acid/salicylic acid carboxyl methyl transferase (BSMT) from petunia *P. hybrida*, *A. thaliana* and tobacco *Nicotiana suaveolens* (Negre *et al.*, 2003; Chen *et al.*, 2003a; Pott *et al.*, 2004), benzoyl-coenzyme A:benzyl alcohol benzoyl transferase (BEBT) from *C. breweri* (D'Auria *et al.*, 2002), benzoyl-coenzyme A:benzyl alcohol/phenylethanol benzoyl transferase (BPBT) from *P. hybrida* (Boatright *et al.*, 2004) and phloroglucinol O-methyl transferase (POMT) and orcinol O-methyl transferase (OOMT) from rose *Rosa chinensis* and *R. hybrida*, respectively (Lavid *et al.*, 2002; Wu *et al.*, 2004). This list could be extended by including genes isolated from vegetative tissues which are responsible for the biosynthesis of volatile compounds also found in floral scents. However, to date, the total number of the isolated genes does not exceed 5% of the reported volatile compounds.

10.5 Regulation of floral scent accumulation and emission

The isolation of genes responsible for the formation of floral scent volatiles has facilitated investigations into the regulation of scent emission. It has been found that volatiles are synthesized *de novo* in the tissues from which they are emitted. Of the flower organs, petals usually produce the highest amount of volatile compounds, although other floral organs may contribute as well to the total floral scent output (Dobson 1994; Dudareva *et al.*, 1999). Within the floral organs, scent genes are expressed almost exclusively in the cells of the epidermis (Figure 10.3; Dudareva *et al.*, 1996; Dudareva and Pichersky, 2000; Kolosova *et al.*, 2001b). Moreover, in snapdragon flowers, the conical cells of the inner epidermal cell layer, which face pollinators, are involved in scent production to a higher extent than the cells of the outer epidermis (Kolosova *et al.*, 2001b; Figures 10.3a and b).

During the lifespan of the flower, production and emission of scent volatiles is developmentally regulated, increasing during the early stages, peaking when the flowers are ready for pollination and decreasing thereafter (Dudareva and Pichersky, 2000). Over development, the concurrent changes in activities of enzymes responsible for the final steps of volatile formation, enzyme protein levels and the expression of corresponding structural genes suggest that the developmental biosynthesis of volatiles is regulated largely at the level of gene expression (Dudareva *et al.*, 1999, 2000). However, the level of the enzyme responsible for the final step of the

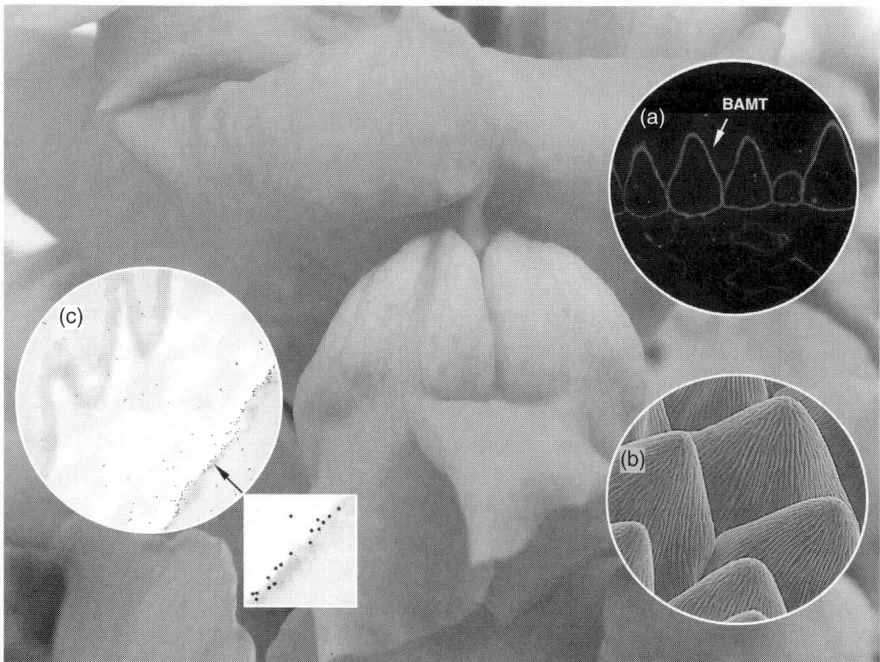

Figure 10.3 Immunolocalization of BAMT in snapdragon flowers. (a) A transverse section of a seven-day-old snapdragon petal treated with anti-BAMT antibodies and visualized by fluorescent FITC-conjugated secondary antibodies. (b) Environmental scanning electron micrograph of conical cells from the inner epidermis of a petal lobe. (c) Transmission electron microscopy of conical cells of seven-day-old snapdragon petal lobe labeled with anti-BAMT antibodies and gold conjugated goat anti-rabbit antibodies. The arrow indicates a region which was magnified three times to show gold particles.

biosynthesis of a particular volatile compound is not the only limiting factor. The amount of available substrate also contributes to the regulation of volatile product formation (Dudareva *et al.*, 2000). Moreover, in the case of enzymes that are able to use several similar substrates, such as some carboxyl methyl transferases and acytransferases, the level of supplied precursor controls the type of product that is produced (Negre *et al.*, 2003; Boatright *et al.*, 2004; Pott *et al.*, 2004).

Emission of floral volatiles from some plant species changes rhythmically during a 24-h period with a maximum at day or night, whereas other flowers continuously emit volatiles at a constant level. The rhythmic release of scent is often correlated with the corresponding temporal activity of flower pollinators and is controlled by a circadian clock or regulated by light (Jakobsen and Olsen, 1994; Helsper *et al.*, 1998; Kolosova *et al.*, 2001a; Pott *et al.*, 2003). In snapdragon, emission of volatiles (methylbenzoate, myrcene and ocimene) occurs rhythmically with a maximum emission during the day and coincides with the foraging activity of bumblebees (Kolosova *et al.*, 2001a; Dudareva *et al.*, 2003). The emission of these volatiles displays 'free-running' cycles in the absence of environmental cues (in continuous dark or continuous light) indicating the circadian nature of diurnal rhythmicity. A detailed time-course analysis during a 48-h period of the activity of BAMT, an enzyme catalyzing the final step in methylbenzoate formation, revealed that there were similar levels of activity at night as well as during the day, suggesting that the activity of this enzyme is not an oscillation-determining factor in snapdragon petals. Oscillations during the daily light/dark cycle, which were also retained in continuous dark, were found in the amount of benzoic acid, the immediate precursor of methylbenzoate, indicating the involvement of a circadian clock in the control of substrate level (Kolosova *et al.*, 2001a). A similar scenario was found for regulation of nocturnal emission of methylbenzoate in tobacco and petunia flowers. These results show that rhythmic emission of volatiles can be regulated at the cellular level of substrate for the final step of the formation of the volatile compound (Kolosova *et al.*, 2001a). Moreover, the regulation of rhythmic emission can also include transcriptional and post-translational control of the expression of a gene responsible for the final step of volatile production, as was shown for nocturnally emitting *S. floribunda* flowers (Pott *et al.*, 2003).

The scent of many flowers is markedly reduced soon after pollination. Such quantitative and/or qualitative post-pollination changes in floral bouquets, shown mostly in orchids (Arditti, 1979; Tollsten and Bergstrom, 1989; Tollsten, 1993; Schiestl *et al.*, 1997), lower the attractiveness of these flowers, as well as increase the overall reproductive success of the plant by directing pollinators to the unpollinated flowers. This is particularly important for plants with a low visitation rate where reproductive success is mostly pollinator-limited (Neiland and Wilcock, 1998). The investigation of the molecular mechanisms responsible for post-pollination changes in floral scent emission in snapdragon and petunia flowers revealed that the decrease in emission begins only after pollen tubes reach the ovary, suggesting that fertilization is a prerequisite for the reduction of floral scent after pollination (Negre *et al.*, 2003). Using methylbenzoate as an example, it has been shown that in snapdragon the decrease in ester emission after pollination is the result of down-regulation of both the methylation index (the ratio of *S*-adenosyl-L-methionine to *S*-adenosyl-L-homocysteine) and activity of the enzyme responsible for methylbenzoate formation (BAMT). In petunia the BAMT gene expression is suppressed by ethylene which is produced in response to pollination (Negre *et al.*, 2003).

In general, more than one biochemical pathway is responsible for the blend of volatile compounds released from flowers. A comparative analysis of the regulation of benzenoid and monoterpene emission in snapdragon flowers revealed that the orchestrated emission of phenylpropanoid and isoprenoid compounds is regulated upstream of the phenylpropanoid and MEP pathways and includes the coordinated expression of genes that encode enzymes involved in the final steps of scent biosynthesis (Dudareva *et al.*, 2000, 2003). However, transcription factors that regulate multiple biosynthetic pathways leading to the formation of the floral scent bouquet have not yet been discovered.

In general, emission of a particular volatile compound into the atmosphere depends on both the rate of its biosynthesis and the rate of its release. While some progress was achieved in the last eight years in understanding the biosynthesis of volatile compounds, very little is known about their release from plant tissues. The rate of release is a function of the physical properties of the compound itself (i.e. the compounds volatility) and the properties of the cellular and intracellular membranes through which the compound has to diffuse. Comparative analysis of volatile compounds emitted and present within plant tissue revealed that the emission of volatiles is not merely a function of their differential volatility but could also involve a cytologically organized excretory process (Altenburger and Matile, 1990). Virtually nothing is presently known about metabolite trafficking between various subcellular compartments, the mechanism of the release process and how these processes contribute to the regulation of volatile emission.

10.6 Genetic engineering of floral scent

Mankind's fascination with floral fragrance and the lack of scent in many modern floricultural varieties, as well as the recent availability of genes responsible for the formation of floral volatiles has stimulated interest in modifying the floral scent bouquet.

Several approaches have been taken in the last three years to alter floral scent through genetic modifications. The LIS from *C. breweri* (Dudareva *et al.*, 1996), which converts GPP to (3*S*)-linalool, was expressed under the control of the cauliflower mosaic virus (CaMV) 35S constitutive promoter in *P. hybrida* W115 (Lucker *et al.*, 2001) and carnation (*Dianthus caryophyllus*; Lavy *et al.*, 2002), both of which lack this monoterpene. Although linalool was synthesized in these transgenic plants, its emission from the flowers and vegetative tissues was undetectable by humans in both cases due to different reasons. While in carnations the amount of linalool was either below the threshold for human perception or masked by other volatiles (Lavy *et al.*, 2002), in petunia most of the linalool was sequestered as a non-volatile linalool glycoside (Lucker *et al.*, 2001). Moreover, in petunia the amount of synthesized linalool or its glycoside depended more on the availability of the substrate GPP in the tissue than on the expression of the LIS gene (Lucker *et al.*, 2001), suggesting that the availability of substrate is an important parameter in the regulation of the biosynthesis of volatile compounds.

The level of precursor for the introduced enzyme was also crucial in the case of the ectopic expression of the strawberry alcohol acyltransferase (SAAT) gene in *P. hybrida* (Beekwilder *et al.*, 2004). Although both *SAAT* expression and the corresponding enzyme activity were detected in transgenics and inherited in the T2 generation, the volatile profile was unaltered due to the lack of substrate for the introduced enzyme. An exogenous supply of isoamyl alcohol to petunia explants led to the emission of the corresponding acetyl ester (Beekwilder *et al.*, 2004).

The potential for the olfactory enhancement of flower fragrance by increasing the level of substrate was recently demonstrated in carnations (Zucker *et al.*, 2002). The increase in the substrate level was achieved by blocking the competitive anthocyanin pathway by the antisense suppression of the flavanone 3-hydroxylase and redirecting the metabolic flux toward benzoic acid, a precursor of methylbenzoate. As a result, the flowers lost their original orange/reddish color and increased methylbenzoate emission thereby altering the floral scent detected by humans (Zucker *et al.*, 2002). While the redistribution of the metabolic flux in the phenylpropanoid pathway did not have a deleterious effect on the plant, the redirection of the metabolic flux from one pathway to another could conceivably reduce essential compounds resulting in a reduction in plant survival or health (Aharoni *et al.*, 2003).

Recently, multigene engineering has been performed to modify the floral scent profile in tobacco plants. The simultaneous introduction of three *Citrus limon* monoterpene synthases [(−)-terpinene cyclase, (+)-limonene cyclase and (−)-β-pinene cyclase] into *Nicotiana tabacum* resulted in the emission of β-pinene, limonene, (−)-terpinene and a number of side products from their leaves and flowers, in addition to the regular terpenoids emitted by the parental line (Lucker *et al.*, 2004b). The total level of monoterpenes in transgenics increased by 10- to 25-fold leading to drastic changes in leaf and flower fragrance profiles, which were sufficient for detection by the human nose (El Tamer *et al.*, 2003). Although the introduced monoterpene synthases competed for GPP, the considerable level of emitted terpenoids indicated that a sufficient amount of substrate was available for the introduced enzymes in tobacco plants.

Monoterpene emission in the transgenic tobacco lines expressing all three monoterpene synthases was further modified by introducing the mint gene for limonene-3-hydroxylase, which catalyzes the hydroxylation of (+)-limonene with the formation of (+)-*trans*-isopiperitenol, under the control of the 35S promoter (Lucker *et al.*, 2004a). The resulting (+)-*trans*-isopiperitenol, an uncommon compound in the plant kingdom, was emitted as a major component of the volatile spectrum of the plants possessing all four integrated transgenes. While monoterpene synthases are localized in the plastids, the limonene-3-hydroxylase enzyme is an endoplasmic reticulum localized protein in mint and was likely localized to the ER in the transgenic tobacco as well, although this was not directly determined (Lucker *et al.*, 2004a). These results suggest that the manipulation of metabolic pathways involving multiple cellular compartments can be implemented by multiple gene transfer, thereby facilitating the production of desirable molecules in transgenic plants.

The pioneering experiments described earlier show that volatile compounds can be synthesized in heterologous systems via metabolic engineering. Availability of genes responsible for the formation of floral volatile compounds is no longer a limiting factor in metabolic engineering of floral scent. Additionally, transformation techniques have been successfully developed for several cut flower species, including roses, chrysanthemums, carnations and gerbera, although for most varieties it is still an 'art form' (summarized in a thorough review by Zuker *et al.*, 1998). Many obstacles still to overcome include a limitation in substrate availability, the undesirable metabolism of the compound of interest and/or the sequestering of the compound of interest, all of which can significantly reduce or prevent volatile emission, indicating an insufficient understanding of scent metabolic networks. The biosynthesis and accumulation of volatile compounds in cell types from which they cannot be emitted could potentially cause lethality since many of the scent compounds are cytotoxic. The redirection of metabolic flux toward a targeted compound may also have a deleterious effect on the plant as a result of depletion of the general precursors required for normal plant development. Targeting the expression of the introduced gene to the specific cell and tissue types using cell- and tissue-specific promoters will help us to avoid the deleterious effects of engineering on normal plant development and to achieve desirable emission. The experiments designed to engineer floral scent have not yet used flower-specific promoters which are available now (e.g. Cseke *et al.*, 1998).

10.7 Conclusions

Floral scent research is just emerging. Recent advances in the isolation and characterization of genes and enzymes responsible for the formation of floral volatiles has led to our understanding of the regulation of volatile biosynthetic machinery and has provided the foundation for metabolic modifications of the odorant trait. Although this information has focused primarily on end product reactions, the application of isotopic labeling methods with computer-assisted modeling will uncover entire metabolic networks leading to volatile compounds, greatly improving our ability to manipulate scent biosynthesis. These advances will not only benefit the floriculture industry but will also enable us to investigate the role of individual floral scent compounds in pollinator attraction and, subsequently, enhance plant reproductive success.

Acknowledgements

Work in the authors' laboratory is supported by the US National Science Foundation (grant numbers MCB-0212802 and MCB-0331333), the US Department of Agriculture (grant number 2003-35318-13619), the US Israel Binational Agriculture Research and Development funds (grant number US-3437-03) and Fred Gloeckner Foundation, Inc.

References

Aharoni, A., Giri, A. P., Deuerlein, S. *et al.* (2003) Terpenoid metabolism in wild-type and transgenic Arabidopsis plants, *The Plant Cell*, **15**(12), 2866–2884.

Altenburger, R. and Matile, P. (1990) Further observations on rhythmic emission of fragrance in flowers, *Planta*, **180**(2), 194–197.

Arditti, J. (1979) Aspects of the physiology of orchids, in *Advances in Botanical Research*, Vol. 7 (ed. H. W. Woolhouse), Academic Press, London, pp. 422–697.

Beekwilder, J., Alvarez-Huerta, M., Neef, E., Verstappen, F. W. A., Bouwmeester, H. J. and Aharoni, A. (2004) Substrate usage by recombinant alcohol acyltransferases from various fruit species, *Plant Physiology*, **135**(4), 1865–1878.

Berger, R. (1995) Aroma compounds in food, in *Aroma Biotechnology*, Springer-Verlag, New York, pp. 1–10.

Boatright, J., Negre, F., Chen, X. *et al.* (2004) Understanding *in vivo* benzenoid metabolism in petunia petal tissue, *Plant Physiology*, **135**(4), 1993–2011.

Bouvier, F., Suire, C., d'Harlingue, A., Backhaus, R. A. and Camara, B. (2000) Molecular cloning of geranyl diphosphate synthase and compartmentation of monoterpene synthesis in plant cells, *Plant Journal*, **24**(2), 241–252.

Buchmann, S. L. and Nabhan, G. P. (1996) The Forgotten Pollinators, Island Press, Washington D.C./Covelo, California.

Burdock, G. A. (1995) *Fenaroli's Handbook of Flavor Ingredients*, 3rd edn, CRC Press, Boca Raton, Florida.

Burke, C. C. and Croteau, R. (2002) Geranyl diphosphate synthase from *Abies grandis*: cDNA isolation, functional expression, and characterization, *Archives of Biochemistry and Biophysics*, **405**(1), 130–136.

Burke, C. C., Wildung, M. R. and Croteau, R. (1999) Geranyl diphosphate synthase: cloning, expression, and characterization of this prenyltransferase as a heterodimer, *Proceedings of the National Academy of Sciences of the United States of America*, **96**(23), 13062–13067.

Cane, D. E. (1999) Sesquiterpene biosynthesis: cyclization mechanisms, in *Comprehensive Natural Products Chemistry. Isoprenoids Including Carotenoids and Steroids*, Vol. 2. (ed. D. E. Cane), Pergamon Press, Oxford, pp. 155–200.

Chen, F., D'Auria, J. C., Tholl, D. *et al.* (2003a) An *Arabidopsis thaliana* gene for methylsalicylate biosynthesis, identified by a biochemical genomics approach, has a role in defense, *The Plant Journal*, **36**(5), 577–588.

Chen, F., Tholl, D., D'Auria, J. C., Farooq, A., Pichersky, E. and Gershenzon, J. (2003b) Biosynthesis and emission of terpenoid volatiles from Arabidopsis flowers, *The Plant Cell*, **15**(2), 481–494.

Croteau, R. and Karp, F. (1991) Origin of natural odorants, in *Perfume: Art, Science and Technology* (eds P. Muller and D. Lamparsky), Elsevier Appl. Sci. Inc., New York, pp. 101–126.

Cseke, L., Dudareva, N. and Pichersky, E. (1998) Structure and evolution of linalool synthase, *Molecular Biology and Evolution*, **15**(11), 1491–1498.

D'Auria, J. C., Chen, F. and Pichersky, E. (2002) Characterization of an acyltransferase capable of synthesizing benzylbenzoate and other volatile esters in flowers and damaged leaves of *Clarkia breweri*, *Plant Physiology*, **130**(1), 466–476.

DeGrandi-Hoffman, G. (1987) The honeybee pollination component of horticultural crop production systems, *Horticultural Reviews*, **9**, 237–272.

Dobson, H. E. M. (1994) Floral volatiles in insect biology, in *Insect–Plant Interactions*, Vol. V, (ed. E. Bernays), CRC Press, Boca Raton, Florida, pp. 47–81.

Dudareva, N. and Pichersky, E. (2000) Biochemical and molecular genetic aspects of floral scents, *Plant Physiology*, **122**(3), 627–633.

Dudareva, N., D'Auria, J. C., Nam, K. H., Raguso, R. A. and Pichersky, E. (1998) Acetyl-CoA: benzylalcohol acetyltransferase – an enzyme involved in floral scent production in *Clarkia breweri*, *The Plant Journal*, **14**(3), 297–304.

Dudareva, N., Cseke, L., Blanc, V. M. and Pichersky, E. (1996) Evolution of floral scent in *Clarkia*: novel patterns of *S*-linalool synthase gene expression in the *C. breweri* flower, *The Plant Cell*, **8**(7), 1137–1148.

Dudareva, N., Martin, D., Kish, C. M. *et al.* (2003) (E)-β-Ocimene and myrcene synthase genes of floral scent biosynthesis in snapdragon: function and expression of three terpene synthase genes of a new TPS-subfamily, *The Plant Cell*, **15**(5), 1227–1241.

Dudareva, N., Murfitt, L. M., Mann, C. J. *et al.* (2000) Developmental regulation of methyl benzoate biosynthesis and emission in snapdragon flowers, *The Plant Cell*, **12**(6), 949–961.

Dudareva, N., Pichersky, E. and Gershenzon, J. (2004) Biochemistry of plant volatiles, *Plant Physiology*, **135**(4), 1893–1902.

Dudareva, N., Piechulla, B. and Pichersky, E. (1999) Biogenesis of floral scent, *Horticultural Reviews*, **24**, 31–54.

Eisenreich, W., Schwarz, M., Cartayrade, A., Arigoni, D., Zenk, M. H. and Bacher, A. (1998) The deoxyxylulose phosphate pathway of terpenoid biosynthesis in plants and microorganisms, *Chemistry and Biology*, **5**(9), R221-R233.

El Tamer, M. K., Smeets, M., Holthuysen, N. *et al.* (2003) The influence of monoterpene synthase transformation on the odour of tobacco, *Journal of Biotechnoogy*, **106**(1), 15–21.

Farmer, E. E. and Ryan, C. A. (1990) Interplant communication: airborne methyl jasmonate induces synthesis of proteinase inhibitors in plant leaves, *Proceedings of the National Academy of Sciences of the United States of America*, **87**(19), 7713–7716.

Feussner, I. and Wasternack, C. (1998) Lipoxygenase catalyzed oxygenation of lipids, *Fett-Lipid*, **100**(4–5), 146–152.

Feussner, I. and Wasternack, C. (2002) The lipoxygenase pathway, *Annual Review of Plant Biol*ogy, **53**, 275–297.

Galen, C. (1985) Regulation of seed-set in *Polemonium viscosum*: floral scents, pollination, and resources, *Ecology*, **66**(3), 792–797.

Gershenzon, J. and Croteau, R. (1991) Terpenoids, in *Herbivores: Their Interaction with Secondary Metabolites* (eds G. A. Rosenthal and M. Berenbaum), Academic Press, New York, pp. 165–219.

Guterman, I., Shalit, M., Menda, N. *et al.* (2002) Rose scent: genomics approach to discovering novel floral fragrance-related genes, *The Plant Cell*, **14**(10), 2325–2338.

Helsper, J. P. F. G., Davies, J. A., Bouwmeester, H. J., Krol, A. F. and van Kampen, M. H. (1998) Circadian rhythmicity in emission of volatile compounds by flowers of *Rosa hybrida* L. cv. Honesty, *Planta*, **207**(1), 88–95.

Henderson, A. (1986) A review of pollination studies in the Palmae, *Botanical Rev*iew, **52**, 221–259.

Henning, J. A., Peng, Y. S., Montague, M. A. and Teuber, L. R. (1992) Honey-bee (*Hymenoptera: Apidae*) behavioral response to primary alfalfa (*Rosales: Fabaceae*) floral volatiles, *Journal of Economic Entomol*ogy, **85**(1), 233–239.

Jakobsen, H. B. and Olsen, C. E. (1994) Influence of climatic factors on rhythmic emission volatiles from *Trifolium repens* L. flowers in situ, *Planta*, **192**(3), 365–371.

Kessler, A. and Baldwin, I. T. (2001) Defensive function of herbivore-induced plant volatile emissions in nature, *Science*, **291**(5511), 2142–2143.

Knudsen, J. T. and Tollsten, L. (1993) Trends in floral scent chemistry in pollination syndromes – floral scent composition in moth-pollinated taxa, *Botanical Journal of Linnean Society*, **113**(3), 263–284.

Knudsen, J. T., Tollsten, L. and Bergstrom, G. (1993) Floral scents – a checklist of volatile compounds isolated by head-space techniques, *Phytochemistry*, **33**(2), 253–280.

Kolosova, N., Gorenstein, N., Kish, C. M. and Dudareva, N. (2001a) Regulation of circadian methylbenzoate emission in diurnally and nocturnally emitting plants, *The Plant Cell*, **13**(10), 2333–2347.

Kolosova, N., Sherman, D., Karlson, D. and Dudareva, N. (2001b) Cellular and subcellular localization of *S*-adenosyl-L-methionine:benzoic acid carboxyl methyltransferase, the enzyme responsible for biosynthesis of the volatile ester methylbenzoate in snapdragon flowers, *Plant Physiology*, **126**(3), 956–964.

Laule, O., Fürholz, A., Chang, H.-S. *et al.* (2003) Crosstalk between cytosolic and plastidial pathways of isoprenoid biosynthesis in *Arabidopsis thaliana*, *Proceedings of the National Academy of Sciences of the United States of America*, **100**(11), 6866–6871.

Lavid, N., Wang, J. H., Shalit, M. *et al.* (2002) O-methyltransferases involved in the biosynthesis of volatile phenolic derivatives in rose petals, *Plant Physiology*, **129**(4), 1899–1907.

Lavy, M., Zuker, A., Lewinsohn, E. *et al.* (2002) Linalool and linalool oxide production in transgenic carnation flowers expressing the *Clarkia breweri* linalool synthase gene, *Molecular Breeding*, **9**(2), 103–111.

Levin, D. A. (1973) The role of trichomes in plant defence, *Quarterly Review of Biology*, **48**(1), 3–15.

Lichtenthaler, H. K. (1999) The 1-deoxy-D-xylulose-5-phosphate pathway of isoprenoid biosynthesis in plants, *Annual Review of Plant Physiology and Plant Molecular Biology*, **50**, 47–66.

Lichtenthaler, H. K., Rohmer, M. and Schwender, J. (1997) Two independent biochemical pathways for isopentenyldiphosphate and isoprenoid biosynthesis in higher plants, *Physiologia Plantarum*, **101**(3), 643–652.

Lucker, J., Bouwmeester, H. J., Schwab, W., Blaas, J., Van der Plas, L. H. W. and Verhoeven, H. A. (2001) Expression of *Clarkia S*-linalool synthase in transgenic petunia plants results in the accumulation of *S*-linalyl-beta-D-glucopyranosid, *The Plant Journal*, **27**(4), 315–324.

Lucker, J., Schwab, W., Franssen, M. C. R., van der Plas, L. H. W., Bouwmeester, H. J. and Verhoeven, H. A. (2004a) Metabolic engineering of monoterpene biosynthesis: two-step production of (+)-trans-isopiperitenol by tobacco, *The Plant Journal*, **39**(1), 135–145.

Lucker, J., Schwab, W., van Hautum, B. *et al.* (2004b) Increased and altered fragrance of tobacco plants after metabolic engineering using three monoterpene synthases from lemon, *Plant Physiology*, **134**(1), 510–519.

Matich, A. J., Rowan, D. D. and Banks, N. H. (1996) Solid phase microextraction for quantitative headspace sampling of apple volatiles, *Analytical Chem*istry, **68**(23), 4114–4118.

McCaskill, D. and Croteau, R. (1995) Isoprenoid synthesis in peppermint (*Mentha* x *piperita*): development of a model system for measuring flux of intermediates through the mevalonic acid pathway in plants, *Biochemical Society Transactions*, **23**(2), 290S.

McGarvey, D. J. and Croteau, R. (1995) Terpenoid metabolism, *The Plant Cell*, **7**(7), 1015–1026.

McGregor, S. E. (1976) Economics of plant pollination, in *Insect Pollination of Cultivated Crop Plants*, USDA, Agriculture Handbook No. 496, Washington, D.C., pp. 1–7.

Murfitt, L. M., Kolosova, N., Mann, C. J. and Dudareva, N. (2000) Purification and characterization of S-adenosyl-L-methionine:benzoic acid carboxyl methyltransferase, the enzyme responsible for biosynthesis of the volatile ester methyl benzoate in flowers of *Antirrhinum majus*, *Archives of Biochemistry and Biophysics*, **382**(1),145–151.

Negre, F., Kish, C. M., Boatright, J. *et al.* (2003) Regulation of methylbenzoate emission after pollination in snapdragon and petunia flowers, *The Plant Cell*, **15**(12), 2992–3006.

Neiland, M. R. M. and Wilcock, C. C. (1998) Fruit set, nectar reward, and rarity in the Orchidaceae, *American Journal of Botany*, **85**(12), 1657–1671.

Newman, J. D. and Chappell, J. (1999) Isoprenoid biosynthesis in plants: carbon partitioning within the cytoplasmic pathway, *Critical Reviews in Biochemistry and Molecular Biology*, **34**(2), 95–106.

Ogura, K. and Koyama, T. (1998) Enzymatic aspects of isoprenoid chain elongation, *Chemical Reviews*, **98**(4), 1263–1276.

Pare, P. W. and Tumlinson, J. H. (1997) *De novo* biosynthesis of volatiles induced by insect herbivory in cotton plants, *Plant Physiology*, **114**(4), 1161–1167.

Pare, P. W. and Tumlinson, J. H. (1999) Plant volatiles as a defense against insect herbivores, *Plant Physiology*, **121**(2), 325–331.

Pellmyr, O., Bergstrom, G. and Groth, I. (1987) Floral fragrances in *Acteae*, using differential chromatograms to discern between floral and vegetative volatiles, *Phytochemistry*, **26**(6), 1603–1606.

Pichersky, E. and Gang, D. R. (2000) Genetics and biochemistry of secondary metabolites in plants: an evolutionary perspective, *Trends in Plant Science*, **5**(10), 439–445.

Pichersky, E. and Gershenzon, J. (2002) The formation and function of plant volatiles: perfumes for pollinator attraction and defense, *Current Opinion in Plant Biology*, **5**(3), 237–243.

Pichersky, E., Lewinsohn, E. and Croteau, R. (1995) Purification and characterization of S-linalool synthase, an enzyme involved in the production of floral scent in *Clarkia breweri*, *Archives of Biochemistry and Biophysics*, **316**(2), 803–807.

Pott, M. B., Effmert, U. and Piechulla, B. (2003) Transcriptional and post-translational regulation of S-adenosyl-L-methionine: salicylic acid carboxyl methyltransferase (SAMT) during Stephanotis floribunda flower development, *Journal of Plant Physiology*, **160**(6), 635–643.

Pott, M. B., Hippauf, F., Saschenbrecker, S. *et al.* (2004) Biochemical and structural characterization of benzenoid carboxyl methyltransferases involved in floral scent production in Stephanotis floribunda and Nicotiana suaveolens, *Plant Physiology*, **135**(4), 1946–1955.

Pott, M. B., Pichersky, E. and Piechulla, B. (2002) Evening specific oscillations of scent emission, SAMT enzyme activity, and SAMT mRNA in flowers of *Stephanotis floribunda*, *Journal of Plant Physiology*, **159**, 925–934.

Poulter, C. D. and Rilling, H. C. (1981) Prenyl transferases and isomerase, in *Biosynthesis of Isoprenoid Compounds* (eds J. W. Porter and S. L. Spurgeon), John Wiley & Sons, New York, pp. 161–224.

Qureshi, N. and Porter, W. (1981) Conversion of acetyl-Coenzyme A to isopentenyl pyrophosphate, in *Biosynthesis of Isoprenoid Compounds* (eds J. W. Porter and S. L. Spurgeon), John Wiley & Sons, New York, pp. 47–94.

Raguso, R. A. and Pellmyr, O. (1998) Dynamic headspace analysis of floral volatiles: a comparison of methods, *OIKOS*, **81**(2), 238–254.

Raguso, R. A. and Pichersky, E. (1995) Floral volatiles from *Clarkia breweri* and *C. concinna* (Onagraceae): recent evolution of floral scent and moth pollination, *Plant Systematics and Evolution*, **194**(1–2), 55–67.

Rodriguez, E. and Levin, D. A. (1976) Biochemical parallelisms of repellants and attractants in higher plants and arthropods, *Recent Advances in Phytochemistry*, **10**, 214–270.

Rodriguez-Concepcion, M. and Boronat, A. (2002) Elucidation of the methylerythritol phosphate pathway for isoprenoid biosynthesis in bacteria and plastids. A metabolic milestone achieved through genomics, *Plant Physiology*, **130**(3), 1079–1089.

Rohmer, M. (1999) The discovery of a mevalonate-independent pathway for isoprenoid biosynthesis in bacteria, algae and higher plants, *Natural Product Reports*, **16**(5), 565–574.

Ross, J. R., Nam, K. H., D'Auria, J. C. and Pichersky, E. (1999) *S*-adenosyl-L-methionine: salicylic acid carboxyl methyltransferase, an enzyme involved in floral scent production and plant defense, represents a new class of plant methyltransferases, *Archives of Biochemistry and Biophysics*, **367**(1), 9–16.

Schiestl, F. P., Ayasse, M., Paulus, H. F., Erdmann, D. and Francke, W. (1997) Variation of floral scent emission and post pollination changes in individual flowers of *Ophrys sphegodes* subsp. *Sphegodes*, *Journal of Chemical Ecology*, **23**(12), 2881–2895.

Seskar, M., Shulaev, V. and Raskin, I. (1998) Endogenous methyl salicylate in pathogen-inoculated tobacco plants, *Plant Physiology*, **116**(1), 387–392.

Shalit, M., Guterman, I., Volpin, H. *et al.* (2003) Volatile ester formation in roses: identification of an acetyl-CoA:geraniol acetyltransferase in developing rose petals, *The Plant Physiology*, **131**(4), 1868–1876.

Shulaev, V., Silverman, P. and Raskin, I. (1997) Airborne signalling by methyl salicylate in plant pathogen resistance, *Nature*, **385**(6618), 718–721.

Sugden, E. A. (1986) Anthecology and pollinator efficacy of *Styrax officinale* subsp. *redivivum* (Styracaceae), *American Journal of Botany*, **73**(6), 919–930.

Tholl, D., Kish, C. M., Orlova, I. *et al.* (2004) Formation of monoterpenes in *Antirrhinum majus* and *Clarkia breweri* flowers involves heterodimeric geranyl diphosphate synthases, *The Plant Cell*, **16**(4), 977–992.

Tollsten, L. (1993) A multivariate approach to post-pollination changes in the floral scent of *Platanthera bifolia* (Orchidaceae), *Nordic Journal of Botany*, **13**(5), 495–499.

Tollsten, L. and Bergstrom, G. (1989) Headspace volatiles of whole plants and macerated plant-parts of *Brassica* and *Sinapis*, *Phytochemistry*, **27**(12), 4013–4018.

Traub, H., Robinson, T. and Stevens, H. (1942) *Papaya Production in the United States*, US Dept. Agr. Cir. 633: 36 pp.

Von Frisch, K. (1971) *Bees: Their Vision, Chemical Senses, and Language*, Cornell University Press, Ithaca, New York.

Wang, J., Dudareva, N., Bhakta, S., Raguso, R. A. and Pichersky E. (1997) Floral scent production in *Clarkia breweri* (Onagraceae). II. Localization and developmental modulation of the enzyme

S-adenosyl-L-methionine: (iso)eugenol O-methyltransferase and phenylpropanoid emission, *Plant Physiology*, **114**(1), 213–221.

Williams, N. H. (1982) The biology of orchids and euglossine bees, in *Orchid Biology, Reviews and Perspectives II* (ed. J. Arditti), Cornell University Press, Ithaca, New York, pp. 119–171.

Wise, M. L. and Croteau, R. (1999) Monoterpene biosynthesis, in *Comprehensive Natural Products Chemistry. Isoprenoids Including Carotenoids and Steroids*, Vol. 2. (ed. D. E. Cane), Pergamon Press, Oxford, pp. 97–153.

Wood, D. L. (1982) The role of pheromones, kairomones and allomones in the host selection and colonization behavior of bark beetles, *Annual Review of Entomology*, **27**, 411–446.

Wood, G. J. (1975) Pollination trials on black currants, New Zealand Commercial Grower, **30**, 17.

Wu, S. Q., Watanabe, N., Mita, S. *et al.* (2004) The key role of phloroglucinol *O*-methyltransferase in the biosynthesis of *Rosa chinensis* volatile 1,3,5-trimethoxybenzene, *Plant Physiology*, **135**(1), 95–102.

Zuker, A., Tzfira T., Ben-Meir, H. *et al.* (2002) Suppression of anthocyanin synthesis by antisense *fht* enhances flower fragrance, *Molecular Breeding*, **9**(1), 33–41.

Zuker, A. T., Tzfira, T. and Vainstein, A. (1998) Genetic engineering for cut-flower improvement, *Biotechnology Advances*, **16**(1), 33–79.

Part IV Senescence

11 Flower senescence: fundamental and applied aspects

Anthony D. Stead, Wouter G. van Doorn, M. L. Jones and C. Wagstaff

11.1 Why senesce?

That flowers senesce and die or the petals abscise is obvious to all, but the questions as to when, why and how remain largely unanswered. Moreover, even if the answers to these questions were known, there would still be the desire to manipulate the process to improve either the visual appearance, the timing of flowering or the quantity of seeds produced. The interest in flower senescence therefore impinges upon ecology, physiology, horticulture and biotechnology.

From an ecological perspective, the function of the perianth in entomophilous species is to attract pollination vectors. Thus, once pollinated, the persistence of the perianth might be regarded as superfluous as, from the plant's perspective, it is important to redirect pollinators to unpollinated flowers (van Doorn, 2002a). Furthermore, persistence of the perianth may even be deleterious since the maintenance of such elaborate structures is costly in terms of water and nutrients. In the first hour after detachment from the plant, the losses from young *Helianthus* and *Hibiscus* petals were lower (6 and 13%, respectively) than from leaves (10 and 22%, respectively) but in *Hemerocallis* the situation was reversed. In the first hour, the loss from petals detached from freshly opened flowers was about 13% of the initial fresh weight; a similar rate of loss occurred from petals removed from one-day-old flowers but the loss from leaves was less than 10%. Furthermore, over longer periods the rate of water loss from old petals was maintained as they began to senesce, whereas the rate of weight loss decreased from leaves (Figure 11.1).

A further suggestion as to why it might be advantageous for petals to senesce rapidly following pollination was provided by Dole (1990) who suggested that the persistence of the flowers could increase the probability of herbivore damage by making the plants more noticeable. However, data for such a theory seems to be lacking.

Conversely, the persistence of the perianth, even on pollinated flowers, may contribute to the overall attractiveness of the inflorescence and thus attract more pollination vectors to other flowers on the inflorescence. Such a strategy is exemplified by *Lantana* and *Lupinus* species (Wainright, 1978; Stead and Reid, 1990; Weiss, 1991) where subtle changes in petal colour appear to dissuade further insect visits once the flower is pollinated but removal of these flowers may reduce the frequency of insect visits to other, unpollinated flowers on the inflorescence

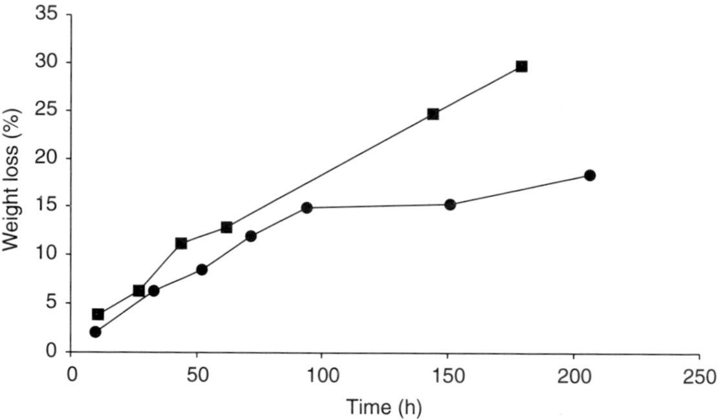

Figure 11.1 Rate of water loss (as a percentage of initial fresh weight) from detached leaves (circle) and petals from fully open flowers (square) of daylily. The continued loss of water from petals occurred as the petals began to senesce suggesting that wilting petals can represent a significant cost to the plant in terms of water lost.

(Nuttman and Willmer, 2003). Thus, the colour change acts as a 'local' signal and reduces wastage of pollen on flowers in which sufficient pollen has been deposited to ensure reasonable seed set. Indeed, in *Digitalis,* the reduction in the time to corolla abscission, another way in which the functional lifespan of the flower may be terminated, was shown to be related to the quantity of pollen loaded (Stead and Moore, 1979). Interestingly, in foxgloves and tobacco, this appears to be a response to pollination, not fertilization, and it appears that there is some form of signal passing through the stylar tissue ahead of the growing pollen tubes. Similarly, ovule maturation in some species appears to be completed only when pollen tube growth has commenced (Mol *et al.*, 2003). The chemical identity of this signal is unknown although early suggestions that it might be the ethylene precursor 1-aminocyclopropane carboxylic acid (Whitehead *et al.*, 1984; Hill *et al.*, 1987) are now discounted (Hoekstra and Weges, 1986; Hilioti *et al.*, 2000). It also appears that the signal differs between pollinated and wounded petunia flowers, although the outcome increased ethylene production and corolla wilting is the same (Woltering *et al.*, 1997). Suggestions that the signal might be electrical (Spanjers, 1978) seem not to have been investigated further.

In a few species, the petals, following pollination, turn green and become photosynthetic (reviewed in van Doorn, 2002a) and thus contribute photosynthates to the growing fruit and seeds. However, such a strategy is uncommon. This lack of reversibility of senescence contributes to the very rigid floral longevity that most species show. Indeed it was more than a century ago that the lifespan of many flowers was recorded and catalogued (Kerner von Marilaun, 1891; Molisch, 1928) but similar data for leaves could not be generated as leaf senescence, and hence longevity, is influenced considerably by the genotype and the environment (Levey and Wingler, 2005).

11.2 Defining senescence

The function of flowers is to attract pollinators and hence produce seeds. Thus, when the capacity to achieve this reduces during the normal lifespan of the flower it could be said that the flower is senescing, a process that must ultimately lead to death. Similarly, before the flower is open and likely to be visited by pollination vectors, the flower and its component parts could be thought of as immature. This definition of flower senescence, based upon the ability to attract pollination vectors, is suitable when considering flowers in their natural habitat, but for cut flowers it is the perception by the consumer that must be considered. Clearly, termination of vase life will be determined by the attractiveness of the flower(s) but sometimes a floral stem may become unattractive when the leaves yellow or in other cases if fungal infections, notably *Botrytis*, become unsightly. A further complication is that in multi-flowered stems, especially chrysanthemums and carnations, the senescence of an individual flower may or may not cause the consumer to discard the whole stem.

11.2.1 Patterns of senescence

In terms of preventing further insect visits there appear to be three or four main strategies (Stead and van Doorn, 1994) employed amongst flowering plants. In some a colour change in part or all of the perianth may discourage insect visits (Weiss, 1991; Nuttman and Willmer, 2003). Even differences in epidermal cell morphology (Glover and Martin, 1998; Comba *et al.*, 2000) influence the frequency of insect visits but whether such changes occur during the life of the flower is unclear. However, it would seem likely that wilting of the corolla, a common feature of senescence that discourages insect visits, would affect the turgidity of the epidermal cells and bring about the collapse of the often conical epidermal cells thus providing another cue to visiting insects that a particular flower should not be visited. Alternatively, it is possible that changes in either the quantity or composition of nectar might be detected and provide insects with signals as to which flowers should be visited.

Recently, another subtle pollination-induced change has been reported in *Petunia* and *Antirrhinum* whereby there is a 70–75% reduction in scent emission as a result of pollination (Negre *et al.*, 2003). However, the reduction only occurs in response to growing pollen tubes and ethylene appears to be implicated; this parallels the pollination-induced wilting response in *Petunia* where the presence of growing pollen tubes is also needed. It would be curious to know if wounding, which also induces corolla wilting in *Petunia* (Gilissen, 1976, 1977), can interfere with scent production and whether, in species such as *Digitalis* where wounding does not induce corolla abscission (Stead and Moore, 1979), scent production is modified by pollination and/or wounding.

In many flower species, the wilting and collapse or closure of the corolla reduces the attractiveness of the flower. In other species, however, abscission of the petals or corolla deters further insect visits even though significant loss of fresh weight or metabolic constituents has not occurred. The evolutionary advantages of these strategies have not been fully evaluated but in species exhibiting perianth abscission there is little chance of extensive remobilization of metabolites whereas in wilting perianths it would appear that considerable remobilization may occur. For example, where corolla wilting occurred as in petunia there was an 8-fold difference in corolla protein content between flowers just prior to opening and after wilting (Lucas, 1989), whilst in daylily the difference was 6-fold (Lay-Yee *et al.*, 1992) and in *Ipomoea*, 2.8 fold (Matile and Winkenbach, 1971). Conversely, in species where corolla abscission terminates the functional life of the flower the losses are much less with a 1.5-fold drop in *Digitalis* (Stead and Moore, 1979) and only a 1.4-fold decline in *Alstroemeria* (Wagstaff *et al.*, 2002).

The remobilization of material from older flowers is critical to the development of younger flowers in spikes of *Gladiolus* (Waithaka *et al.*, 2001) but removal of younger flower buds can enhance the size and longevity of older flowers in lily (van der Meulen-Muisers *et al.*, 2001) and *Alstroemeria* (Chanasut *et al.*, 2003). Some *Alstroemeria* growers exploit this latter fact as they remove, or trim off, the young buds of each side branch (scorpoid cyme) to promote the growth and longevity of the single remaining flower bud.

11.3 Applied means of controlling senescence in cut flowers

One of the more useful resources detailing recommendations for the post-harvest care of many cut flowers species can be found at http://www.chainoflifenetwork.org/

11.3.1 Mechanisms controlling senescence and/or abscission

As with distinctions between senescence strategies there appears to be a pattern as to which taxonomic groups show ethylene-sensitive flower senescence and which do not (Woltering and van Doorn, 1988; McKenzie and Lovell, 1992; van Doorn, 2001). Thus, in Liliaceous species, senescence usually involves ethylene-insensitive petal wilting. Whilst the wilting is ethylene-insensitive, it may be followed by petal abscission which is invariably ethylene-sensitive (van Doorn, 2002b; Wagstaff *et al.*, 2005). The only exception reported so far is one cultivar of *Tulipa* in which tepal abscission has been reported to be ethylene-insensitive (Sexton *et al.*, 2000). In Solanaceous species, ethylene appears to be involved in either corolla wilting (e.g. *Petunia*) or abscission (e.g. tobacco). Whilst species exhibiting colour changes are also ethylene-sensitive (van Doorn, 2002a). In species showing pollination-induced responses, exposure to ethylene usually mimics the pollination response although not all species that show ethylene-sensitivity show pollination-induced responses (van Doorn, 1997, 2002a). The importance of ethylene in the control of floral senescence has resulted in the development of many commercial treatments or practices to reduce ethylene production or sensitivity. A great deal of effort is invested by commercial growers to minimize the sources of ethylene and thus avoid accelerated flower senescence. Examples are the use of electric carts around the glasshouses rather than those with internal combustion engines and the removal of any dead or decaying tissues (from the growing environment to ensure cleanliness) (Bishop, 2002).

11.3.2 Water

Once removed from the plant, the flower cannot obtain water or nutrients via the roots. Although water is easily supplied, the ability of the stem to translocate water may be impaired by either embolisms or the growth of bacteria (see later). The formation of embolisms or air bubbles can effectively block the ability of tracheids and vessels to conduct water and so the recommendation to recut stems under water has a scientific basis; it is the intention to remove any tissues in which the xylem is blocked with encroaching air/water interfaces that have been pulled into the tissue as the transpiring tissues lose water during transport. The idea of plunging cut stems, particularly roses, in deep water before placing in the vase can also be explained in terms of the hydraulic conductivity of the stem (Mensink and van Doorn 2001; van Ieperen *et al.*, 2002). The ability to take up water, compared to the rate of water loss via transpiration, of course, determines if flower fresh weight increases or decreases but, in at least one species, there appears to be a circadian rhythm associated with theses processes akin to that which might be expected in intact plants (Doi *et al.*, 1999).

11.3.2.1 Vascular blockage

Vascular blockage can occur as a result of occlusions caused by oxidative reactions that occur during both dry storage and in stems held in water (van Doorn and Cruz, 2000); this can be reduced by the use of antioxidants or ensuring that stems are initially placed in water of low pH. The production of lipid-phenolic compounds such as suberin may be responsible for this physiological blockage (Williamson *et al.*, 2002) and may explain why inhibitors such as S-carvone (Williamson *et al.*, 2002) 4-hexylresorcinol (van Doorn and Vaslier, 2002) are effective at delaying flower wilting in some species.

A more common problem occurs when vascular occlusion occurs due to bacterial growth in the vase water (Put and Jansen, 1989; van Doorn *et al.*, 1995). This is especially likely if a floral preservative is used as the sugar supplied creates a wonderful environment for bacterial and fungal growth. It also explains why sanitation is so important and thus there are a range of commercially available formulations for cleaning flower buckets, dipping cut stems, and so on, all of which aim to reduce the bacterial count reaching the consumer's vase. Despite this, surveys have shown that the level of bacterial contamination increases as flowers progresses through the supply chain with low numbers at the grower to higher numbers with retailers (Hoogerwerf and van Doorn, 1992). To prevent vascular blockage by bacteria the inclusion of a bactericide in floral preservatives is common; these can range from simple bleaches to more complex molecules (van Doorn *et al.*, 1990; Florack *et al.*, 1996; Knee, 2000). The benefits of using electrolyzed anode water or electrolyzed neutral water in the vase solution may be related to their ability to sterilize the tissue and thus reduce bacterial numbers (Ohta and Harada, 2000; Izumi, 2001) although increased decomposition of ethylene has also been reported (Harada and Yasui, 2003). However, the breakdown of ethylene, once it has been released from the tissue, is unlikely to be important as the reaction only occurs at very low pH (pH 2.5). It seems improbable that the reaction could proceed within the plant tissues where the pH will be above this optimum level.

Despite the possible precautions available to reduce the growth of bacterial populations, there remain concerns over the possibility that the vase water of cut flowers could harbour harmful bacteria, especially the MRSA (Methicillin-resistant *Staphylococcus aureas*) of such concern in hospitals. Unfortunately, there are no data to support or refute such claims at present although *Staphylococcus* is not one of the bacteria usually isolated from vase water (Florack *et al.*, 1996). Nevertheless more attention needs to be given to the ability of floral preservatives to kill different types of bacteria and fungi, especially potentially pathogenic strains.

11.3.3 Inorganic nutrients

Inorganic nutrients are unlikely to become limiting during the vase life of a cut flower but, if plants are grown under poor nutrient conditions, it might be anticipated that the quality and potential longevity of the cut flower would be adversely affected. However, most studies suggest that vase life is largely unaffected by the nutrient levels applied to the plant, although fertilizer levels most certainly affect flower production rates. For example, varying levels of nitrogen application did not affect vase life in *Sandersonia* (Clark and Burge, 1999), carnation (Huett, 1994) or some varieties of *Chamelaucium*, and even when vase life was affected the effects were inconsistent between seasons (Maier *et al.*, 1996). Application of calcium, pre-harvest, did not delay flower senescence of *Lupinus havadii* although addition of calcium to the post-harvest vase solution was beneficial (Picchioni *et al.*, 2002). In potted roses, flower longevity was increased when calcium levels, relative to potassium, were increased (Mortensen *et al.*, 2001). In evolutionary terms, the stability of floral longevity is perhaps understandable since, for a plant growing under poor nutrient conditions, it may be sensible to produce fewer flowers but ensure that they have the maximum probability of being successfully pollinated, thus ensuring some seed production, as seeds offer a much greater opportunity to withstand adverse climactic or edaphic conditions.

11.3.4 Carbohydrates

The reduced photosynthetic capacity, due either to the removal of the leaves or because cut flowers are usually held at relatively low light levels, is a critical consideration, although whether sugar starvation induces petal senescence is debated (van Doorn, 2004). However, in some studies the levels of soluble sugars were deemed to be responsible for the reduced vase life of 'Sonia' roses (Ichimura *et al.*, 2003) and for the wilting of *Delphinium* flowers when, due to the use of silver thiosulfate (STS), ethylene did not cause petal abscission (Kikuchi *et al.*, 2003). However, increasing photosynthetic photon flux immediately prior to harvest of *Eustoma* flowers had only a minimal effect on floral longevity, far less than the supply of sucrose or other preservatives in the vase solution (Islam *et al.*, 2003). In other crops, notably lettuce, the time of harvest influences the post-harvest performance with leaves harvested late in the day performing better than those harvested at the traditional time of early morning (Clarkson *et al.*, 2005) possibly because they have a higher sugar content having just completed a photosynthetic cycle. To the best of our knowledge nobody has investigated if the time of harvest during the day influences floral longevity.

One of the main ingredients in floral preservatives is, therefore, a carbohydrate source. One of the main effects on petal senescence, it is suggested, is that sugars reduce the sensitivity to ethylene. The effects of sugars on petal senescence is definitely greater in species where petal senescence is regulated by ethylene (van Doorn, 2004). Verlinden and Garcia (2004) found evidence that supports this contention; the addition of sugar, either continuously in the vase solution, or as a short pulse immediately after harvest, increased vase life of carnations. This was in part due to decreased ethylene production but also as a consequence of decreased sensitivity to applied ethylene.

Commercially available preservatives usually aim to provide a final concentration of 1–2% sugar, even though higher concentrations may improve floral longevity, flower size and pigment levels in some species. The optimum concentration varies from species to species, or even between cultivars of the same species. For example, in *Alstroemeria,* increases in sucrose concentration beyond 1% can improve floral longevity as well as pigment development but can be deleterious to leaves, causing accelerated yellowing (Chanasut *et al.*, 2003). Although 5% sucrose enhanced floral longevity and the opening of subsequent floral buds in cv. Calgary, in cv. Modena it caused accelerated dehydration of the tepals of the first flower although it did not accelerate tepal abscission (Figure 11.2).

The addition of sugar had little effect on floral longevity in Stargazer lilies but, as in *Alstroemeria*, significantly enhanced pigment development (Han, 2003) and thus improved floral quality. Some authors have advocated the use of brief (up to 24 h) pulsing with a high concentration (up to 25%) of sucrose in preference to the use of sugars in the vase solution as this avoids depending on the consumer to provide the correct vase solution. However, such treatments are less effective in the case of *Liatrus* (Han, 1992), *Zinnia* (Carneiro *et al.*, 2002) and a number of other species (Ichimura, 1998). Glucose or sucrose were only effective in *Dendrobium* when aminoxyacetic acid (AOA) was added to the vase solution (Rattanawisalanon *et al.*, 2003), although the effect of AOA may be related to the reduced solution pH which facilitated lower bacterial numbers and improved water (and carbohydrate) uptake rather than any effect on ethylene biosynthesis.

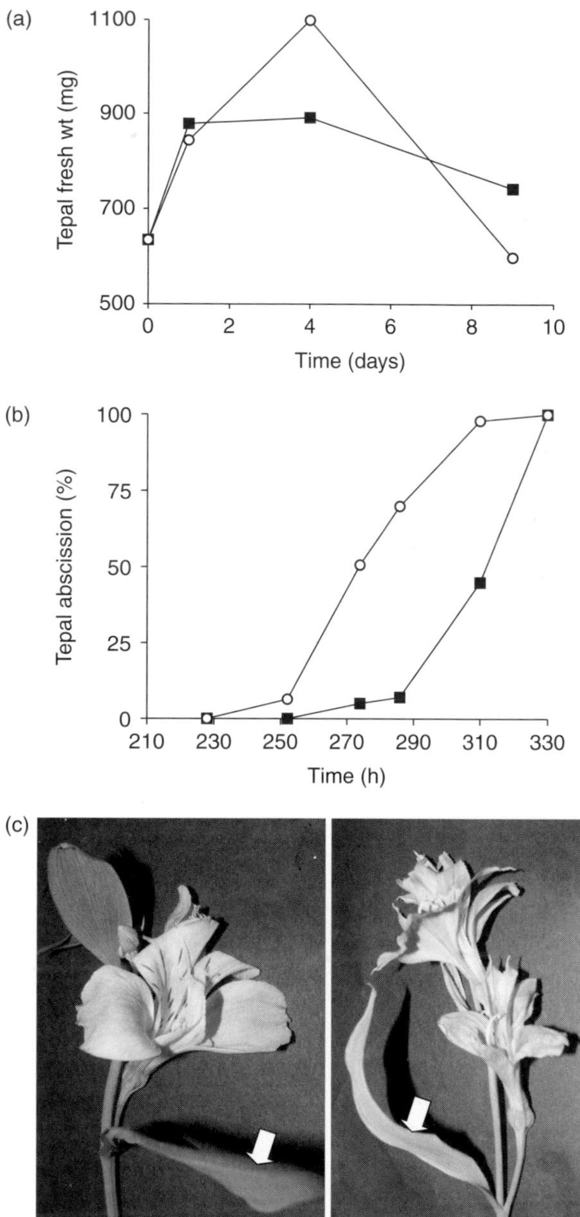

Figure 11.2 (a) Fresh weight of *Alstroemeria* cv. Modena tepals held in either water (closed squares) or 5% sucrose (open circles), $n = 10$. (b) Tepal abscission (%) from isolated cymes of *Alstroemeria* cv. Modena flowers in either water (closed squares) or 5% sucrose (open circles), $n = 50$. (c) Appearance of *Alstroemeria* cv. Modena flowers held in water (left) or 5% sucrose (right); pictures taken 10 d after flower opening. Whilst the second flower has fully developed on those cymes held in sucrose the first flower has dehydrated and there is extensive degradation of chlorophyll in the bracts (arrowed).

Relatively little attention has been paid to which carbohydrate is best, although in *Gentiana*, sucrose, glucose and fructose were all effective at prolonging vase life whereas sorbitol was not (Zhang and Leung, 2001). There have been few studies with other sugars, possibly because the cost of carbohydrates other than glucose or sucrose would be prohibitive. In *Gladiolus*, however, trehalose was the most effective sugar (Otsubo and Iwaya-Inoue, 2000). In tulips trehalose feeding caused delayed tepal abscission and improved tepal pigmentation (Wada *et al.*, 2005; Iwaya-Inoue and Takata, 2001), although it is unclear, however, if trehalose was more effective than other sugars.

11.3.5 Plant growth regulators

11.3.5.1 Ethylene

As more has been learned about the control of senescence, particularly foliar senescence, there have been many studies on the ability of plant growth regulators to delay floral senescence. In particular, ethylene is clearly implicated in the control of senescence of flowers and leaves and the effects of ethylene can be reduced by the use of silver thiosulphate (STS) (Veen and van der Geijn, 1978) which binds to the ethylene receptor thus reducing sensitivity to ethylene. The optimum duration of the STS pulse varies slightly from species to species but for commercial purposes a 1 h pulse with 4 mM STS is usually sufficient. Longer exposure or higher concentrations can cause blackening to the leaves of many species. More recently, the use of cyclic olefins such as 1-methylcyclopropene (1-MCP) has been encouraged as there are environmental concerns regarding the handling of the waste silver remaining after treating flowers with STS. 1-MCP has been shown to be very effective in extending the lifespan of many cut flowers and potted plants although often not quite as effectively as STS (Reid *et al.*, 1999; Serek and Sisler, 2001; Jones *et al.*, 2001). Unlike STS, which is administered immediately after harvest as a pulse treatment in water, 1-MCP is gaseous and harvested material can be treated either before or even during transportation. Currently 1-MCP is licensed for use on flowers and selected vegetables in the United States and it is anticipated that it will be registered for use with cut flowers in 2005 or 2006 in Europe. Other cyclic olefins appear to offer even greater efficacy than 1-MCP (Kebenei *et al.*, 2003a,b) but although patented, there seems no move to develop these molecules commercially. Other means of reducing damage due to ethylene by inhibiting ethylene biosynthesis have not been adopted commercially, in part because the most effective inhibitor 1-aminoethoxyvinylglycine (AVG) is too expensive and the only analogue AOA, is not as effective as STS or 1-MCP (Bichara and Van Staden, 1993; Serek and Anderson, 1993; Shimamura *et al.*, 1997; Finger *et al.*, 2004).

11.3.5.2 Cytokinin and gibberellin

Cytokinins, whilst very effective at delaying or even reversing leaf senescence, are not widely used to increase floral longevity. Commercial preparations that do contain cytokinins are used to prevent leaf yellowing, e.g. with *Solidago* or lilies (Philosoph-Hadas *et al.*, 1996; Han, 2001). However, transgenic petunia plants overexpressing the genes responsible for cytokinin biosynthesis, whilst showing delayed leaf senescence, also showed delayed flower senescence (see Section 11.5). Similarly, activation tagging has shown a correlation between delayed petal senescence and increased cytokinin production (Zubko *et al.*, 2002). This implies that cytokinins can increase floral longevity but, as yet, the benefits of adding them to cut flowers have not been demonstrated, possibly because it is difficult for the molecule to be taken up or to be translocated to the active site (Philosoph-Hadas *et al.*, 1996) or perhaps because other factors are more often limiting. Where cytokinins, particularly 6-benzylaminopurine, have shown a beneficial effect on vase life (Setyadit *et al.*, 2004), these have been applied as a dip since uptake from solution is poor, but even then the effect varies between species and cultivars (Paull and Chantrachit, 2001). In *Oncidium* species, cytokinin as a pulse treatment, was more effective if followed by sucrose treatment (Chen *et al.*, 2001). Application of cytokinins extracted from, and then reapplied to, 'Sonia' roses did increase longevity significantly, a fact that may suggest that it is critical to apply the correct, naturally occurring, cytokinin (Lukaszewska *et al.*, 1994). This would also explain why transgenic plants overexpressing gene(s) that otherwise limit cytokinin biosynthesis show extended floral longevity.

Gibberellins are rarely thought of as being involved in senescence but, in combination with sucrose, gibberellins have helped overcome the effects of harvesting *Gentiania* flowers prematurely and prolonged storage of harvested stems (Eason *et al.*, 2004). In *Sandersonia*, gibberellin delayed both the rise in proteolytic activity associated with senescence and the visual symptoms of sepal fading and wilting (Eason, 2002). In both *Narcissus* (Ichimra and Goto, 2000) and *Alstroemeria* (Hicklenton, 1991), GA_3 has been shown to delay leaf yellowing which in turn means that stems remain attractive to consumers for longer. Indeed, commercial floral preservatives designed specifically for Liliaceous species usually contain either a gibberellin and/or a cytokinin (Leonard and Nell, 2004). The study of Hicklenton (1991) showed that both cytokinin and gibberellin were equally effective but that there was no synergistic effect of applying both.

11.3.5.3 Other plant growth regulators

There are even less data to imply that other Plant Growth Regulators (PGRs) might play a pivotal role in controlling floral senescence. In daylilies (Panavas *et al.*, 1998), carnation (Serrano *et al.*, 2001) and other flowers (Zhang *et al.*, 1991) it has been suggested that increases in abscisic acid may be critical but with few, if any, effective inhibitors of ABA biosynthesis, this theory has not been thoroughly tested. Certainly, in *Narcissus,* treatment with ABA hastens flower senescence but putative ABA biosynthesis inhibitors were ineffective although gibberellin counteracted the effects of applied ABA as it does in other biological systems (Hunter *et al.*, 2004). Although not a cut flower, abscission of cocoa flowers was inhibited by fluidone, an ABA biosynthesis inhibitor, and levels of ABA increased 20 times during the effective pollination period of these flowers (Aneja *et al.*, 1999). This is one the very few reports where ABA, rather than ethylene, seems to control an abscission event.

There is some evidence to suggest that jasmonates can hasten senescence but this may be due their ability to stimulate ethylene production (Porat and Halevy, 1993; Porat *et al.*, 1995). Strangely, one report actually suggests that application of methyl jasmonates can improve flower quality after storage (Gast, 1999) but, given the effects on leaves where jasmonates also accelerate senescence, this seems unlikely. Recently nitrous oxide has received attention as a possible natural plant growth regulator and interestingly, 2,2'-(hydroxynitrosohydrazino)-bisethanamine (DETA/NO) which donates nitrous oxide, was found to be a very effective pulse treatment to increase vase life of several important cut flower species (Badiyan *et al.*, 2004). In leaves nitrous oxide is very effective at reversing ABA- or jasmonate-induced senescence (Hung and Kao, 2003, 2004).

There have been several attempts to investigate the relationship between ethylene and polyamine biosynthesis, mainly because both involve *S*-adenosyl methionine as a precursor thus raising the possibility of competition for this substrate. Studies using exogenous polyamines or measurements of endogenous polyamines suggest that they may delay senescence (Pandey *et al.*, 2000) but when ethylene production was inhibited and flower longevity increased in carnations there was no increase in the levels of endogenous polyamines (Serrano *et al.*, 1999). However, in the same species, others have reported that polyamine application reduced ethylene production and prolonged vase life whilst inhibitors of polyamine biosynthesis increased ethylene production and accelerated senescence (Lee *et al.*, 1997). Thus, the interplay between ethylene and polyamines is probably not simple and further work is required to elucidate the relationship.

In conclusion, at the present time there exist the means to increase floral longevity and quality through specific post-harvest treatments in species where ethylene plays a critical role in determining longevity. However, there are few effective treatments specifically for species in which floral senescence is ethylene-insensitive. In these latter species, other signalling pathways must operate to determine the time of petal collapse. Certainly, in both ethylene-sensitive and ethylene-insensitive species, *de novo* protein synthesis is necessary for the wilting of petals to occur as in carnation where ethylene-induced petal wilting is prevented by application of the protein synthesis inhibitor cycloheximide (CHI) (Wulster *et al.*, 1982; Taverner *et al.*, 2000); a similar situation has been found in Petunia (Jones, Pers. Comm.). In daylily (Lukaszewski and Reid, 1989; Lay-Yee *et al.*, 1992) and iris (Celikel and van Doorn, 1995; Sultan and Farooq, 1997) and a number of other ethylene-insensitive species (Jones *et al.*, 1994), the use of cycloheximide can also considerably delay the collapse of the tepals with fresh weight, protein content and membrane integrity all being maintained in CHI-treated flowers. Despite these studies, however, identification of the proteins, other than certain hydrolytic enzymes involved in the degradation of the tepals, has not yet been achieved. In other species, treatment with CHI causes abnormalities and any benefits to flower longevity are more than eliminated by these 'side' effects. Therefore, it is likely that the synthesis or action of specific proteins will need to be modified to effectively produce longer lived flowers.

11.4 The molecular basis of senescence

The study of petal senescence at the molecular level has changed dramatically over the last 10 years in a way that reflects the technologies that have developed during that time. Early studies were directed by knowledge of chemical, hormonal or physiological treatments that affected the timing of senescence symptoms. The most thoroughly characterised responses are to the plant growth regulator ethylene, primarily because most of the model systems in use (petunia, carnation and *Arabidopsis*) are sensitive to exogenous ethylene and wilt and/or abscise in response to the gas. Extensive physiological experimentation of the effects of ethylene on petal senescence was first supplemented by molecular studies by Woodson (1987) who looked at changing mRNA populations in carnation. With the rise of *Arabidopsis* as the model system of choice for plant genetic and genomic studies, work increasingly concentrated on this plant, particularly in the use of mutant screens to identify plants that were insensitive to ethylene using reverse genetics. The first mutant in the ethylene signalling pathway was *etr1* (Chang, 1993), and this was found to be a dominant negative regulator of ethylene action such that the mutant does not perceive ethylene and the floral organs thus show delayed senescence and abscission. A total of five ethylene receptors have now been isolated in *Arabidopsis* (*ERS1*, *ETR1*, *ETR2*, *ERS2* and *EIN4;* Moussatche and Klee, 2004) and the entire ethylene signalling pathway was reported in this species by Ecker (1995) and updated from a purely molecular perspective by Kieber (1997) and Johnson and Ecker (1998).

The genes identified in *Arabidopsis* have been subsequently used as the basis for gene discovery of homologues in leaf and flower senescence of other species. Carnation has continued to be a widely used model and has been useful in elucidating the role of the gynoecium in triggering petal senescence following pollination (Shibuya *et al.*, 2000). Although other PGRs were investigated in this study (ABA and IAA) it was shown that ethylene is exclusively the inter-organ signal that initiates petal senescence in pollinated and naturally senescing flowers. The cloning of a number of homologues of the ethylene signalling pathway in carnation (Charng *et al.*, 1997; Nagata *et al.*, 2000; Waki *et al.*, 2001) has revealed some interesting differences in gene expression profiles between senescing carnation petals (that wilt as they senescence) and *Arabidopsis*, in which petals abscise. In the latter, the *EIN3* mRNA was not affected by exogenous ethylene application (Chao *et al.*, 1997), but in carnation the levels of *DC-EIL1* fell dramatically during petal senescence. Since regulators such as *DC-EIL1*, *DC-ERS2* and *DC-ETR1* are all negative regulators, a decrease in the mRNA (and by inference, the transcribed protein) causes an increase in sensitivity to ethylene of the petal tissue and hence accelerated senescence (Waki *et al.*, 2001). In this way, the plant is able to remove unwanted petal material rapidly for two purposes – to detract pollinators from visiting the flower and to remobilise components such as phosphorus from nucleic acids.

Modern techniques in biotechnology have allowed the exploitation of senescence regulators such as *ETR1*, and have also provided a route to establish the function of these genes that is not possible by simply following the profiles of selected genes during senescence. Homologues of the *ETR1* gene from *Arabidopsis* have now been cloned and used to produce knockouts in *Petunia* (Wilkinson *et al.*, 1997; Gubrium *et al.*, 2000), and carnation (Bovy *et al.*, 1999). These plants show a delayed floral senescence phenotype and have a potential commercial value, but to date, they have been produced using a constitutive 35S promoter which has caused other, undesirable, pleiotropic effects such as stem breakages (see Section 11.5). The use of tissue specific promoters in future transformations should overcome these difficulties.

Transgenic plants have also been generated to establish the effect of cytokinins on flower senescence. It has long been known that the application of exogenous cytokinin retards the senescence of detached leaves and that sprayed cytokinin prevents natural senescence of attached leaves and extends the shelf life of cut flowers (Van Staden *et al.*, 1988). Linking the cytokinin biosynthesis gene *IPT* to a senescence-enhanced promoter has been shown to significantly delay natural and post-harvest leaf senescence in tobacco (Gan and Amasino, 1995) and lettuce (McCabe *et al.*, 2001; Garratt *et al.*, 2005). A similar *SAG12::IPT* construct was introduced into *Petunia* (Chang *et al.*, 2003; Clark *et al.*, 2004) and was found to also delay flower senescence in this species, suggesting that cytokinin is required to maintain chromoplast as well as chloroplast integrity. More recently, the link between sugar transport, cytokinin and senescence has been examined (Lara *et al.*, 2004) and it was found that extracellular invertase is required in order for cytokinin to delay leaf senescence, presumably in order to maintain export of photoassimilates to sink organs. The same is likely to be true for floral organs since sugar starvation leads to petal senescence (van Doorn, 2004) and the fact that senescing carnation petals have high levels of invertase inhibitor proteins (Halaba and Rudnicki, 1988, 1989).

The PGRs jasmonic acid (JA) and salicylic acid (SA) are also implicated in the control of leaf senescence, but in a way that enhances rather than delays senescence. Early reports (Porat and Halevy, 1993) indicated that flowers as taxonomically diverse as *Petunia* and *Dendrobium* also underwent enhanced senescence following application of methyl jasmonate (MeJA). Rossato *et al.* (2002) postulated that MeJA functions as a 'death hormone' as there was a strong reduction in the uptake of essential nutrients such as nitrogen and potassium post-treatment, which consequently resulted in senescence of the leaves on treated plants. However, senescence that occurs as a result of this regime is reversible, whereas senescence in petals is a tightly programmed series of events that is not reversible once it has started. Therefore, caution should be exercised in directly extrapolating the implications of results from experiments on leaf senescence to senescence in petals. Although JA is strongly implicated in the leaf senescence programme through studies with exogenously applied CK on wild-type and *coi1* (JA insensitive) plants of *Arabidopsis,* it is not clear what the effect is on the endogenous CK response. He *et al.* (2002) reported that *coi1* plants do not show delayed senescence in the absence of exogenous CK, but they do fail to respond to applied CK. The authors attribute the lack of phenotype in *coi1* plants left untreated with JA to the plasticity of senescence, i.e. that many pathways are part of the senescence initiation programme and there is redundancy if one pathway is removed. Unfortunately there are no reports on the timing or appearance of senescence or abscission of floral parts of *coi1 Arabidopsis* mutants, but exogenously applied MeJA was shown to accelerate petal senescence in ethylene sensitive species such as *Dendrobium, Phalaenopsis* and *Petunia* (Porat and Halevy, 1993; Porat *et al.*, 1995). It appears from this work that JA accelerates the production of ethylene and as such JA signalling may be an important part of floral senescence of ethylene sensitive species, but not necessarily of those species insensitive to ethylene.

Studies of SA deficient mutants *nahG*, *pad4* and *npr1* in *Arabidopsis* reveal that a lack of this PGR delays leaf senescence (Morris *et al.*, 2000). However, in *pad4* plants, chlorophyll loss occurs as in wild-type plants but cell death does not proceed in the same manner. Necrosis is associated with cell death events that are not reversible and this result indicates that *PAD4* may function in the terminal cell death pathway. The roles of these genes have not been established in flowers, but since floral senescence is generally not reversible, the prediction would be that *PAD4* is a more likely candidate for this pathway than genes associated with chlorosis.

Both JA and SA are believed to be triggers of senescence in leaves, but the evidence for their downstream targets is patchy. The accumulation of reactive oxygen species (ROS) appears to be a crucial next step after JA/SA induction. However, many of the mutants are under such severe oxidative stress that it is difficult to say if such responses are specific to senescence or part of a general accumulation of ROS that occurs after a multitude of stress conditions (Dat *et al.*, 2000). Studies have shown that both JA and SA induce the expression of genes such as *HvS40* (Krupinska *et al.*, 2002) which is localised to the mesophyll in the presence of ROS. The increased expression of other senescence associated genes (SAGs), e.g. *LSC54*, are also dependent on ROS accumulation (Navabpour *et al.*, 2003). More detailed studies have been made of genes of the *ORE* family such as *ORE9*, which act within a common pathway of senescence initiation following MeJA, SA or ABA treatment (Woo *et al.*, 2001). This may provide the means for the redundancy that is observed between mutants of the hormonal signalling pathways such as *coi1* by sharing downstream elements that can be activated by a variety of hormonal stimuli. *ORE9* is an F-box protein that interacts with part of the ubiquitin E3 ligase complex known as *ASK1* and as such is a gene that ubiquitinates proteins targeted for degradation (Woo *et al.*, 2001). More recently (Woo *et al.*, 2004), individual mutants of the ORE family (*ore1*, *ore3* and *ore9*) were found to have increased tolerance to oxidative stress and much delayed leaf senescence, although the phenotypes of the flowers were not reported.

There are now many large-scale EST and genome screening projects that have produced long lists of SAGs and SEN genes (for review see Smart, 1994; Buchanan-Wollaston, 1997; Gan and Amasino, 1997; Nam, 1997; Weaver *et al.*, 1997; Rubinstein, 2000; Quirino *et al.*, 2000; Buchanan-Wollaston *et al.*, 2003; Gepstein *et al.*, 2003; Lim *et al.*, 2003; Gepstein, 2004; Guo *et al.*, 2004; Jones, 2004). The majority of these studies have focused on leaf senescence, and where putative identities have been assigned for petal-related SAGs, there has been a heavy bias in favour of ethylene-associated genes and proteases. In more recent years techniques such as AFLP and differential display followed by verification through Northern blotting have been superseded by microarray technology followed by real-time PCR.

However, the hypothesis that knowledge gained about gene expression changes in leaves can be applied to petals is, at the very least, questionable. Unfortunately, the microarrays constructed to date for flowers (rose, carnation, *Iris* and *Alstroemeria*) (Channeliere *et al.*, 2002; Breeze *et al.*, 2003; van Doorn *et al.*, 2003) are neither from fully sequenced organisms, nor do they compare leaves and petals and *Arabidopsis* is still the best plant for which all sequence and array information is publicly available. From an analysis of microarray data in the public domain (AtGenExpress) it is possible to compare numbers of genes up- and down-regulated in aging *Arabidopsis* petals and leaves. Stage 12 petals (stage of floral development that begins when the petal height is level with the long stamens) were compared to stage 15 petals (stage of floral development that begins when the stigma extends above the long anthers) and rosette leaf 17 was compared to senescing leaves. For definitions of stages used see ftp://ftp.arabidopsis.org/home/tair/Ontologies/TAIR_Ontology/TAIR_ontology.defs

Data from the microarray slides were imported into Genespring (Silicon Genetics) and experiments to compare young/old petals and young/senescent leaves were set up. In each case the data from three replicate slides per stage were included in each analysis.

From a total of 20 602 targets present on the array, 1447 were up-regulated in senescing leaves and 1314 were up-regulated in old petals. Venn diagram analysis was performed and showed 939, 806 and 508 genes to be uniquely up-regulated in leaves, petals and in both tissues, respectively. Of those genes showing down-regulated expression, 1707 were expressed in leaves but only 902 were expressed in petals. Of these, 1306 were only down-regulated in leaves, 501 only in petals and 401 were down-regulated in both leaves and petals (Figure 11.3). These data indicate that of the up-regulated genes, only around 30% show a similar pattern in leaves and petals whilst for down-regulated genes, the figure is lower – around 25%, with the majority of differentially expressed genes being unique to either petals or leaves. This suggests that predicting genes that might be involved in petal senescence from data derived from leaves is likely to be unreliable.

The lists of genes up- or down-regulated in leaves and/or petals were imported into the TAIR database for gene ontology analysis. The frequency of each biological process was expressed as a percentage of the total and is shown in Figure 11.4. This

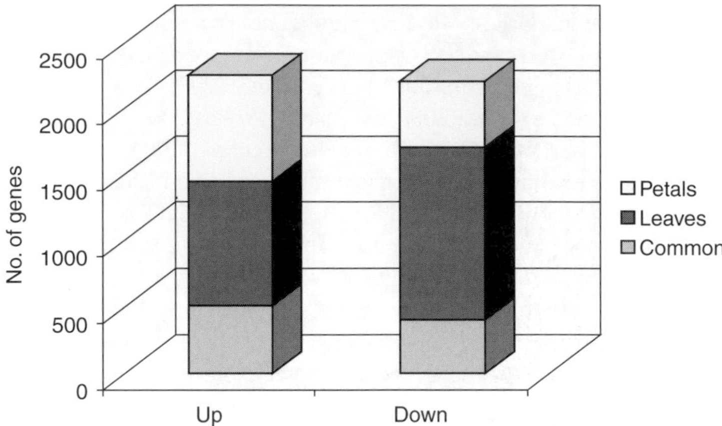

Figure 11.3 Numbers of up- and down-regulated genes in aging leaves and petals. Each experiment was normalised per chip and per gene in Genespring and then the following filters were used: filter on flag (present or marginal in four out of six samples), one way ANOVA with no MTC ($P = 0.05$), fold change of 3.0 or above.

analysis showed that protein metabolism is strongly down-regulated during petal and leaf aging and that transport processes are commonly up-regulated in both tissues. Transcription is more strongly up-regulated in senescing leaves compared to petals, as are genes related to biotic or abiotic stress stimuli. More detailed analysis of molecular functions relating to commonly up- or down-regulated genes shows that of the down-regulated genes, 2% relate to transporter function, but nearly 14% of the up-regulated genes relate to transporter functions. Similarly, transcription factor activity represents 1.4% of the down-regulated genes in leaves and petals but 5.3% of the up-regulated genes in both tissues. Chloroplast located genes are prevalent in both up-and down-regulated lists (18 and 20%, respectively), but mitochondrial genes are more common in the up-regulated (12%) than in the down-regulated (8%) categories. However, genes assigned to the ribosomal functional category are only present in down-regulated lists and comprise 16.7% of the total down-regulated genes in leaves and petals. These data agree with the functional analysis which showed that protein metabolism was strongly down-regulated in both tissues.

Initial papers producing lists of SAG/SEN genes relied on correlations of gene expression with a senescence phenotype, but more recent work has been able to use elegant methods to prove that the genes in question have a functional role to play during senescence (He and Gan, 2002). Many of the workers have been able to distinguish between different triggers of senescence, e.g. darkness (Lee *et al.*, 2001), hormonal (Yoshida *et al.*, 2001) and oxidative stress (Navabpour *et al.*, 2003), to start to separate genes into pathways and identify those that are common to more than one pathway. Post-harvest senescence is also important, particularly for cut flowers and floret vegetables such as broccoli. Studies have revealed that lipid degradation is a key feature of post-harvest degradation (Page *et al.*, 2001; Leverentz *et al.*, 2002)

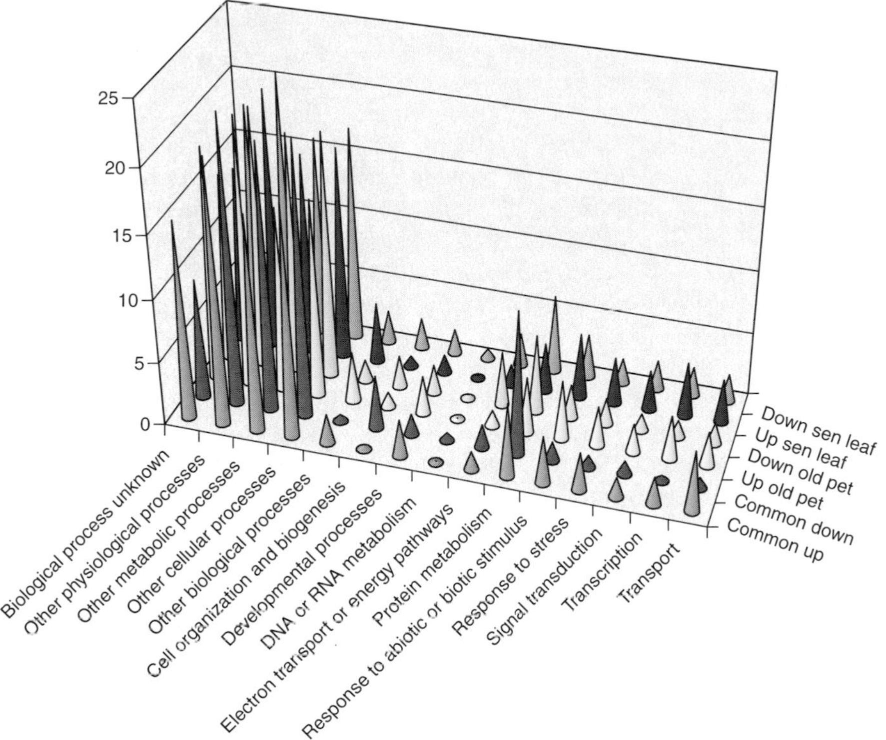

Figure 11.4 The percentage of up- and down-regulated genes represented in the major biological processes in aging leaves and petals. Data derived from Figure 11.3 and analysed using gene ontology definitions from TAIR (www.arabidopsis.org).

as is proteolysis (Valpuesta *et al.*, 1995; Wagstaff *et al.*, 2002; Bhalerao *et al.*, 2003; Pak and van Doorn, 2005) and more particularly genes linked to the terminal phases of senescence which are more reminiscent of PCD (Wagstaff *et al.*, 2003; Coupe *et al.*, 2004).

The large numbers of genes identified as encoding proteases (Jones *et al.*, 1995; Valpuesta *et al.*, 1995; Kim *et al.*, 1999, Eason *et al.*, 2002; Wagstaff *et al.*, 2002; Arora and Singh, 2004) and the fact that protease inhibitors can delay flower senescence (Pak and van Doorn, 2005) has lead to intense speculation that cysteine proteases function as caspases (see Section 11.5.2) in plants. This may be little more than wishful thinking since there is no conclusive evidence for functional caspases in senescing plants, although some caspase inhibitors do have a delaying effect on senescence and some genes, the metacaspases and vacuolar processing enzymes (VPEs), have some homology with animal caspases (Woltering, 2004; Sanmartín *et al.*, 2005). In particular the VPEs have some amino acid sequence homology – sufficient to suggest that they may possess similar active sites to animal caspases (Hara-Nishimura *et al.*, 2005). In *Arabidopsis* VPEs are certainly upregulated during senescence (Kinoshita *et al.,* 1999) and are associated with the death of specific cells during embryogenesis (Nakaune *et al.*, 2005). However, evidence to date shows that metacaspases and VPEs function at a fairly late stage of senescence; in flowers it would seem that this point corresponds to the final collapse of the petals and thus manipulating the activity of these genes would seem to be too late for the prevention of symptoms that result in loss of commercial value and shelf life. Therefore, there is intense interest in genes believed to act earlier in the senescence pathway such as *BFN1* and *Bax Inhibito*r. *BFN1* is a nuclease that is much more highly expressed in flowers of *Arabidopsis* than in leaves (Perez-Amador *et al.*, 2000), and this or other nucleases are up-regulated during senescence of leaves (Perez-Amador *et al.*, 2000) and flowers of petunia (Langston *et al.*, 2005), *Sandersonia* (Eason and Bucknell, 2001) and *Alstroemeria* (Wood and Stead, unpublished). Overall nuclease activity increases during corolla senescence of daylily (Panavas *et al.*, 1998), petunia (Xu and Hanson, 2000) and *Sandersonia* (Eason and Bucknell, 2001), and DNA degradation has also been shown in senescing petals of various other species (van der Kop *et al.*, 2003). Expression of the mammalian *Bax* gene in plants caused increased cell death, the symptoms of which were loss of chloroplast membrane function, increased cell permeability and finally disruption of the vacuole (Yoshinaga *et al.*, 2005). The function of some of these genes is discussed further in Section 5.2.

Many of the papers referred to in the preceding paragraphs and many others, such as Pennel and Lamb (1997), have attempted to compare plant cell death to apoptosis in animal systems. There may indeed be some similarities, particularly as floral senescence is usually non-reversible and less influenced by environmental effects than leaves, but we should keep an open mind to the possibility that plants die in a different way. Plant organs contest a different set of constraints at the end of their useful life compared to animals in that plants uniquely can orchestrate the loss of an enormous percentage of an organism (all the aerial parts in some perennials) and still remain alive. Therefore, the function of senescence in plants has evolved to salvage useful nutrient components (Bleeker and Patterson, 1997), which presumably has required the selection of different processes from that found in animals. Whilst there is so little data available as to which genes or proteins are critical for the induction of senescence, it is hard to speculate as to whether the process is, or is not, analogous to programmed cell death or apoptosis in animals. The most probable areas of commonality are likely to reside in the latter stages of death when senescence ceases and terminal phases of programmed cell death and necrosis take over.

11.5 Molecular events characterising floral senescence

Given that floral longevity is terminated by one of two processes, abscission or wilting, and that in each the processes may be ethylene-sensitive or -insensitive, there must exist at least four pathways leading to the termination of floral life. As we have seen, from the analysis of array data (Section 11.4), these pathways may not have a significant overlap with leaf senescence. It is perhaps not surprising, therefore, that there is no consensus as to whether petals follow a pathway of programmed cell death, apoptosis, autophagy, necrosis or some other form of death (Thomas *et al.*, 2003; van Doorn and Woltering, 2004). Certainly some of the hallmarks of PCD, e.g. TUNEL positive nuclei or DNA laddering, are present in some wilting flowers, both ethylene-sensitive and insensitive, as well as those where abscission appears to terminate floral life. Equally, some reports fail to find some or all of these symptoms despite the fact that the petals are dying, perhaps implicating autophagy as an alternative way to die.

11.6 Autophagy

Autophagy is a major degradation/recycling system and is ubiquitous in eukaryotic cells. It contributes to the turnover of cellular components by delivering portions of the cytoplasm and whole organelles to lysosomes (vacuoles) where they are digested (Klionsky, 2005). This process is mediated by unique double-membrane structures, the autophagosomes. The double membranes grow and engulf part of the cytoplasm, with or without large organelles (see Klionsky and Emr, 2000). The autophagosomes merge with a lysosome (animals) or a vacuole (the plant homologue of a lysosome). Autophagy is dramatically induced by starvation. It then maintains an amino acid pool that enables the continued synthesis of essential proteins. In yeast and *Dictyostelium* e.g. autophagy is induced upon withholding amino acids or sugar from the medium. This would result in early death due to N-starvation or energy starvation, respectively, if it were not for protein recycling by autophagy (Otto *et al.*, 2003, 2004; Yoshimori, 2004).

In yeast, several genes required for autophagy have been identified (for their nomenclature we will follow Klionsky *et al.*, 2003). These yeast genes relate to the induction of autophagy and formation of the autophagosome (a kinase signalling system, and an APG protein conjugation system), to autophagosome size regulation, and to docking and fusion of the autophagosome with the lysosome (Klionsky and Emr, 2000). There is some evidence that these genes may be involved in senescence. In a mammalian cell line, autophagic senescence required the expression of a few mammalian orthologues of some of the yeast autophagy genes (Yu *et al.*, 2004). Autophagy genes were also found to be essential for life span extension – by an insulin-like tyrosine kinase receptor – in the nematode *Ceanorhabditis elegans* (Meléndez *et al.*, 2003). However, other evidence contradicts a destructive role of autophagy genes in plant senescence. In yeast, the autophagy genes are involved in survival during nutrient stress, whereas no evidence has apparently been published that they are involved in inducing cell death. Knock outs in *Arabidopsis,* of orthologues of yeast autophagic genes, resulted in advanced leaf yellowing (rather than the expected delay in yellowing if these genes were important for the induction of the visible senescence symptoms). Knock out of an orthologue of yeast *ATG7* in *Arabidopsis* resulted in hypersensitivity to nutrient limitation, displaying premature leaf senescence. The mRNAs for autophagy genes accumulated as leaves visibly senesced, suggesting that autophagy was up-regulated, but the function seems one of maintenance under conditions of nutrient stress rather than in preventing senescence (Doelling *et al.*, 2002). When starved for nitrogen or carbon, *Arabidopsis* plants carrying a mutation in the orthologue of yeast *ATG9* also showed earlier leaf yellowing than the wild type. Even under nutrient-rich conditions, natural leaf senescence was accelerated in the mutant plants. These phenotypes were complemented by the expression of wild-type gene in the mutant background. Thus, the autophagy gene was required for maintenance of the cellular viability under nutrient-limited conditions and for efficient nutrient use in the whole plant rather than in preventing senescence (Hanaoka *et al.*, 2002). A similar knockout of *ATG4* had the same effect on leaf yellowing (K. Yoshimori, Nagoya, personal communication). These results seem difficult to reconcile with the data (although thus far scant) indicating the importance of yeast *ATG* orthologues in preventing senescence in animal cells (see earlier). The results in plants may be interpreted to indicate that the presence of autophagosomes and their docking to the vacuole is not essential to senescence (at least the senescence of leaves). Rather, plant autophagic senescence may depend on remobilisation in the vacuole, and vacuole rupture, independent of the presence of autophagosomes.

Cell death by autophagy may therefore be considered in a stricter and wider sense. If seen in a strict sense, it would depend on the presence of autophagosomes. Autophagosomes are not numerous and are transient in any cell. In animal cells, they can be made visible by inhibiting their docking on the lysosomal membrane (D. Robinson, Heidelberg, personal communication, 2004). Autophagosomes may for the same reasons not be common in plant cells. Nonetheless, thus far they have been observed in a few experiments (Moriyasu and Ohsomu, 1996, Gunawardena *et al.*, 2001a,b; Surpin *et al.*, 2003) and in an *Arabidopsis* mutant carrying a T-DNA insertion within a yeast *ATG4* orthologue (Yoshimoto, personal communication 2004). Membrane structures have often been observed in vacuoles of senescent cells (Matile and Winkenbach, 1971), which is reminiscent of autophagosomes delivering endoplasmic reticulum or (other) organelles to the lysosome. However, these structures have also been described as artefacts of fixation and termed myelin whorls, whilst in bacteria the existence of similar structures termed mesosomes has been debated (Ebersold, 1981; Hudson, 2003). Nevertheless, the disappearance of the endoplasmic reticulum is one of the first ultrastructural signs of senescence. This is followed by disappearance of large parts of the cytoplasm and most organelles. Only a few mitochondria and the nucleus are seen in relatively late stages. Finally, a small strand of almost empty cytoplasm is usually left, prior to rupture of the tonoplast (Smith *et al.*, 1994; van der Kop *et al.*, 2003). The persistence of the mitochondria is a feature of cell death in animals and leakage of cytochrome c from the mitochondria is often regarded as a sign of PCD, not autophagy. However, the release of cytochrome c from plant mitochondria during senescence has only been reported for tapetal plastids of sunflower (*Helianthus annuus*; Balk and Leaver, 2001).

In a wider sense, autophagy is determined by the activity of a lysosomal compartment involved in degradation of parts of the cell. Senescence in plants, including petal senescence, confirms this. Thus far the data on plant mutants in *ATG* orthologues do not indicate that autophagosomes are essential to visible leaf senescence, but the role of autophagosomes, if any, needs further study. Plant senescence involves cellular degradation while the vacuole is still intact, and involves tonoplast rupture which results in degradation of any cytoplasm and organelles that still remain and in breakdown of the cell wall (Matile and Winkenbach, 1971; Matile, 1997). However, autophagy may operate at several different levels in plants and van Doorn and Woltering (2005) suggest that micro-autophagy may involve small amounts of cytoplasm being sequestrated into the vacuole whilst macro-autophagy might involve the formation of the specialised autophagosomes in which larger amounts of cytoplasm, possible remote from the vacuole, can be directed to the vacuole for degradation. In addition if the tonoplast losses permeability the degradative potential of the enzymes contained therein will be released into the cells causing the cell contents to be completely degraded, this they term mega-autophagy. If indeed there are three distinct forms of autophagy it is likely that each will require the expression of specific genes, especially those that control the behaviour of the tonoplast.

11.6.1 Programmed cell death and apoptosis

Programmed cell death and apoptosis are often used interchangeably, although in animals some authors suggest a distinction. Traditionally caspases have been thought to be crucial in PCD but there now appear to several situations where PCD occurs in a caspase-independent manner. Therefore, a third non-autophagic route to animal cell death has been suggested, namely caspase-independent cell death (CICD; Chipuck and Green, 2005). Whether cell death in plants utilizes similar routes as any of these animal-based pathways remains somewhat controversial.

In plants, homologues to animal caspases have not been identified although other genes with partial homology including the putative functional regions do exist, the so-called metacaspases and the vacuolar processing enzymes (VPEs). Their structural similarity to animal caspases has recently been reviewed and from *Arabidopsis*, VPEγ has the closest structural similarity to caspase-8 (Sanmartín *et al.*, 2005); furthermore VPEγ has been shown to regulate proteolysis in *Arabidopsis* (Rojo *et al.*, 2003). VPE assays indicate that caspase inhibitors can inhibit VPE activity and *vice versa* and the possibility exists that VPEγ binds to inhibitors that block the self-maturation of this enzyme; in doing so activity increases and cells enter a PCD-like process. The difference between VPEs and meta-caspases may be explained by their different sites of activity, for just as in animals, each VPE or caspase appears likely to be located in a different cellular compartment, the metacaspases being either cytosolic (type II) or in mitochondria and chloroplasts (type I) and the VPEs are likely to be located in the endomembrane system (reviewed by Sanmartín *et al.*, 2005).

Whilst the debate continues as to whether metacaspases and VPEs can fulfil the role of the animal caspases, there is little doubt that homologues of other PCD-related genes are to be found in plants e.g. *Bax Inhibitor* (Coupe *et al.*, 2004), Beclin, accession nos CA762920 (rice), BT000508 (*Arabidopsis*) and *Defender Against Apoptotic Death-1* (*DAD-1*) (Orzaez and Granell, 1997; van der Kop *et al.*, 2003; Yamada *et al.*, 2004), but this still means that many of the animal PCD-related genes have no obvious homologues in plants. Curiously, most of the genes that do have homologues in both animals and plants, seem to be of mitochondrial origin and it is the animal cytoplasmic PCD proteins that seem to be absent in plants. As discussed, there are other significant differences between PCD in plants and animals, e.g. blebbing of cells and the formation of apoptotic bodies in animal cells. These structures are then engulfed by other cells and digested by the lysosomal activity of that cell. Such processes seem unlikely to occur in plants due to the cell wall; however, 'self-digestion' by the activity of the vacuolar hydrolytic enzymes might be possible. Such processes might be brought about by either micro- or macro-autophagy as described by van Doom and Woltering (2005). Blebbing of the nucleus is another process rarely seen in plants, although it may not be an essential part of animal cell PCD (Takemura *et al.*, 2001; Chipuck and Green, 2005). Studies using DAPI or other nucleic acid stains often show multiple staining regions within a cell but in many cases this may represent condensation of the chromatin within the nucleus rather than the break up, by whatever process, of the nucleus. In tapetal cells, a unique form of nuclear breakdown appears to occur in which the perinuclear space distends and the two membranes appear to remain connected via numerous highly contorted and granular-filled projections (Stead *et al.*, submitted).

11.7 Progress on manipulating flower senescence using GM

Given the importance of ethylene in controlling senescence, the genes that are involved in ethylene biosynthesis, recognition and signal transduction have been obvious targets for genetic manipulation. The first transgenic plants produced to try and enhance floral longevity were carnations; using antisense technology, the activity of ACC oxidase (ACO) was dramatically reduced and vase life increased (Savin *et al.*, 1995; Kosugi *et al.*, 2002). Similarly, *Torenia* plants expressing antisense ACC synthase (ACS) had increased flower longevity (Aida, 1998). In tobacco, antisense for ACS or ACO reduced stress-induced senescence and increased polyamine content (Wi and Park, 2002) suggesting that reduced ethylene biosynthesis can lead to the accumulation of polyamines. Later the disruption of the *ETR-1* gene, effectively blocking ethylene sensitivity, has been used and shown to increase the vase life of petunia (Wilkinson *et al.*, 1997), carnations (Bovy *et al.*, 1999) and *Nemesia* (Cui *et al.*, 2004) and also to delay corolla abscission in tobacco (Yang, 2005). In petunia, *etr-1* plants show delayed corolla wilting (Wilkinson *et al.*, 1997; Gubrium *et al.*, 2000; Langston *et al.*, 2005) but probably because expression was controlled by the constitutive 35S promotor, other physiological problems are apparent. For example, cuttings of the transgenic plants fail to produce normal adventitious roots and a large proportion of the transgenic plants show sudden but asynchronous wilting. A similar phenotype is encountered in transgenic *etr-1* tobacco plants (Yang, 2005). This response may be related to the increased cell size, or changes in the pattern of lignification observed in *etr-1* petunia plants (Berry, Clark and Stead, unpublished data) as well as in *etr-1 Arabidopsis* plants. Manipulating *EIN3* or *ERS*, both part of the ethylene signal transduction pathway, can also produce petunia plants exhibiting delayed floral senescence (Shaw *et al.*, 2002; Shibuya *et al.*, 2004). Other than manipulating the biosynthesis and/or perception of ethylene, the only other transgenic plants successfully exhibiting increased floral longevity have been those using the *SAG12-IPT* construct which effectively delays petal wilting in *Petunia* (Chang *et al.*, 2003; Clark *et al.*, 2004) and extends floral longevity in tobacco (Schroeder *et al.*, 2001).

Previous studies in which ethylene biosynthesis, sensitivity or signal transduction have been genetically modified by the introduction of antisense or RNAi constructs have, therefore, shown that the phenotype of cut flowers can be modified for the advantage of consumers. However, the genes involved in the collapse, degradation or abscission of petals are often poorly understood and so potential targets for manipulation remain to be elucidated. This is particularly true for ethylene-insensitive species where information as to which gene(s) or protein(s) maybe critical is sadly lacking. Large-scale screening of arrays (Breeze *et al.*, 2003; van Doorn *et al.*, 2003) may identify critical changes in gene expression but, as yet, transformation of these particular species is not easy and so the role of any genes that may appear critical can neither be tested nor exploited by generating transgenic plants with improved floral longevity. Indeed one such gene, prohibitin, has already been identified as being critical in the control of flower senescence in petunia as downregulation of expression in flowers, using VIGS, resulted in flowers with larger corolla cells (fewer cell divisions) and reduced longevity (Chen *et al.*, 2005). In the future it is likely that approaches using transient assays such as VIGS (virus-induced gene silencing) will identify many more critical genes associated with the control of senescence that could provide valuable opportunities to manipulate floral senescence (Chen *et al.*, 2004).

11.8 The future

11.8.1 Improved longevity

If genomic studies fail to provide information on the critically important gene expression stages, it may be that proteomic studies can; several groups have already embarked on the identification of protein changes that occur during the development of ethylene-insensitive flowers such as daylilies or changes that do not occur after treatment with cycloheximide which is known to delay flower senescence significantly in this species (Lay Yee *et al.*, 1992; Bieleski and Reid, 1992). Given that extensive degradation of proteins occur during daylily flower senescence (Valpuesta *et al.*, 1995; Stephenson and Rubinstein, 1998; Mahagamasekera and Leung, 2001), it is perhaps not surprising that one of the first proteins to be identified in this way has peptide fragment sequences that match perfectly with parts of the DNA sequence for a thiol protease (Stead, Kinter and Jones, unpublished data) previously sequenced from daylily (Guerrero *et al.*, 1998).

As more proteins are identified and critical gene expression changes are identified, it should be possible to isolate the gene(s) encoding these and use antisense or RNAi techniques to produce new, longer-lived varieties. Alternatively, knowledge of the proteins that have to be inhibited to extend longevity could lead to the development of specific chemical inhibitors that could be used in floral preservatives, as such proteomic studies could lead to the use of the so-called biorational approach previously favoured for the development of pesticides. Hopefully, such approaches will extend the vase life of existing commonly used cut flower species and increase the range of species that can be used as cut flowers since there are many common garden ornamentals that currently cannot be used as commercial cut flowers, e.g. daylilies, poppies, and so on.

11.8.2 Gene function and control of senescence

Genomic, proteomic and metabolomic studies can, in the future, contribute greatly to our understanding of the control of processes in plants such as senescence; however, to be fully effective such studies require extensive databases to be developed with the nucleotide and amino acid sequences of both genes and proteins available. Without such resources, used in conjunction with transgenic approaches, ascribing function to any gene sequences that might be obtained as a result of, e.g. microarray experiments, will always be slightly fraught, especially when the degree of sequence similarity is low and/or the length over which there is good similarity is only a small fraction of the total gene length. For proteins, the situation is at present far worse than with mRNA with far fewer plant proteins sequenced fully.

11.8.3 Opportunities for 'quality control'

Commercially in the United Kingdom and increasingly so in the United States, it is common to offer the consumer a guarantee as to the potential vase life of the flowers that they purchase (Stead, 2004) although there is little that the grower, wholesaler or retailer can do to determine the potential vase life of a specific crop before offering the flowers for sale. Indeed the only testing possible by the retailer is performed concurrently with that of the consumer! This can lead to disappointments and the need to issue refunds if particular crops do not come up to consumers expectations or retailers guarantees. The development of diagnostic tools to assess potential longevity and/or quality is highly desirable. In medical fields, rapid testing of tissues can be achieved using immunodiagnostics, especially when coupled with some form of biosensor which can be read almost immediately. A similar test that could be indicative of imminent floral senescence could be very useful but as yet it is not clear which protein(s) may provide such indicators although proteomic studies may provide this information in the near future. Alternatively genomic studies can identify changes in the expression of genes that are indicative of imminent senescence and testing for the expression of these particular gene(s) could indicate the potential longevity or quality of a particular batch of flowers. In this way, the guarantee offered to consumers could more accurately reflect the anticipated longevity and allow wholesalers or retailers to reject flowers that have been cold stored for too long, as the appearance of the flowers of some species at the time of sale may not show any detrimental symptoms although their potential quality and/or longevity may have be reduced (Stead *et al.*,

A third way in which a diagnostic might be provided is through the application of hyperspectral imaging; using monochromatic light and recording the reflected spectrum it is possible to detect changes in gene expression and the consequences thereof, long before visible differences can be detected by the naked eye. In this, the nutrient status of pasture grasses can be monitored non-destructively (Gay *et al.*, 2002; Schut *et al.*, 2005). Similarly, the use of near infrared reflectance spectroscopy has been used to predict dry matter yield (Stuth *et al.*, 2003; Halgerson *et al.*, 2004). With cut *Alstroemeria* flowers we have reliably distinguished petals of different ages even though differences were not visible to the naked eye (Gay *et al.*, in preparation); it will be interesting to see if a similar approach can detect stored flowers with a reduced potential vase life.

11.9 Conclusions

Flowers may die, but studying the process is becoming more dynamic and more complex; the future for studies on flower senescence looks very exciting, with genomics and proteomics accelerating our understanding of the mechanisms underlying senescence. The genes and proteins that are implicated can be tested using either transgenic plants or through the use of technologies such as VIGS and then may be used to improve the quality of cut flowers through the generation of transgenic plants. Finally, the use of diagnostics based upon subtle changes in gene expression and/or proteins will allow greatly improved quality control to ensure that the consumer receives a better and a more reliable product.

References

Aida, R., Yoshida, T., Ichimura, K., Goto, R. and Shibata, M. (1998) Extension of flower longevity in transgenic torenia plants incorporating ACC oxidase transgene, *Plant Science*, **138**, 91–101.

Aneja, M., Gianfagna, T. and Ng, E. (1999) The roles of abscisic acid and ethylene in the abscission and senescence of cocoa flowers, *Plant Growth Regulation*, **27**, 149–155.

Arora, A. and Singh, V. P. (2004) Cysteine protease gene expression and proteolytic activity during floral development and senescence in ethylene-insensitive *Gladiolus grandiflora*, *Journal of Plant Biochemistry and Biotechnology*, 13, 123–126.

Badiyan, D., Wills, R. B. H. and Bowyer, M. C. (2004) Use of a nitric oxide donor compound to extend the vase life of cut flowers, *HortScience*, **39**, 1371–1372.

Balk, J. and Leaver, C. J. (2001) The PET1–CMS mitochondrial mutation in sunflower is associated with premature programmed cell death and cytochrome c release, *The Plant Cell*, **13**, 1803–1818.

Bhalerao, R., Keskitalo, J., Sterky, F. *et al.* (2003) Gene expression in autumn leaves, *Plant Physiology*, **131**, 430–442.

Bichara, A. E. and Van Sstaden, J. (1993) The effect of aminooxyacetic acid and cytokinin combinations on carnation flower longevity, *Plant Growth Regulation*, **13**, 161–167.

Bieleski, R. L. and Reid, M. S. (1992) Physiological-changes accompanying senescence in the ephemeral daylily flower, *Plant Physiology*, **98**, 1042–1049.

Bishop, C. (2002) Ethylene: what can the grower do? *Floraculture International*, **12**, 26–28.

Bleeker, A. B. and Patterson, S. E. (1997) Last exit: senescence, abscission, and meristem arrest in *Arabidopsis*, *The Plant Cell*, **9**, 1169–1179.

Bovy, A. G., Angenent, G. C., Dons, H. J. M. and van Altvorst, A. C. (1999) Heterologous expression of the Arabidopsis etr1-1 allele inhibits the senescence of carnation flowers, *Molecular Breeding*, **5**, 301–308.

Breeze, E., Wagstaff, C., Harrison, E. *et al.* (2004) Gene expression patterns to define stages of post harvest senescence in *Alstroemeria* petals, *Plant Biotechnology Journal*, **2**, 155–168.

Buchanan-Wollaston, V. (1997) The molecular biology of leaf senescence, *Journal of Experimental Botany*, **48**, 181–199.

Buchanan-Wollaston, V., Earl, S., Harrison, E. *et al.* (2003) The molecular analysis of leaf senescence – a genomics approach, *Plant Biotechnology Journal*, **1**, 3–22.

Carneiro, T. F., Finger, F. L., dos Santos, V. R., Neves L. L. D. and Barbosa, J. G. (2002) Longevity of *Zinnia elegans* inflorescences affected by sucrose and recuts of the stem, *Pesquisa Agropecuaria Brasileira*, **37**, 1065–1070.

Celikel, F. G. and van Doorn, W. G. (1995) Solute leakage lipid-peroxidation and protein-degradation during the senescence of iris tepals, *Physiologia Plantarum*, **94**, 515–521.

Chanasut, U., Rogers, H. J., Leverentz, M. K. *et al.* (2003) Increasing flower longevity in *Alstroemeria*, *Postharvest Biology and Technology*, **29**, 324–332.

Chang, C., Kwok, S. F., Bleeker, A. B. and Meyerowitz, E. M. (1993) *Arabidopsis* ethylene-response gene *ETR1*: similarity to product to two-component regulators, *Science*, **262**, 539–544.

Chang, H. S., Jones, M. L., Banowetz, G. M. and Clark, D. G. (2003) Overproduction of cytokinins in petunia flowers transformed with P-SAG12-IPT delays corolla senescence and decreases sensitivity to ethylene, *Plant Physiology*, **132**, 2174–2183.

Channeliere, S., Riviere, S., Scalliet, G. *et al.* (2002) Analysis of gene expression in rose petals using expressed sequence tags, *FEBS Letters*, **515**, 35–38.

Chao, Q. M., Rothenberg, M., Solano, R., Roman, G., Terzaghi, W. and Ecker, J. R. (1997) Activation of the ethylene gas response pathway in *Arabidopsis* by the nuclear protein ETHYLENE-INSENSITIVE3 and related proteins, *Cell*, **89**, 1133–1144.

Charng, Y.-E, Sun, C.-W, Chou, S.-J, Chen, Y.-R and Yang, S. F. (1997) cDNA sequence of a putative ethylene receptor from carnation petals (accession no. AF016250) (PGR 97-144), *Plant Physiology*, **115**, 863.

Chen, J. C., Jiang, C. Z., Gookin, T. E., Hunter, D. A., Clark, D. G. and Reid, M. S. (2004) Chalcone synthase as a reporter in virus-induced gene silencing studies of flower senescence. *Plant Molecular Biology*, **55**, 521–530.

Chen, W. S., Liao, L. J., Chen, C. Y. and Huang, K. L. (2001) Kinetin, gibberellic acid and sucrose affect vase life in *Oncidium* spp., *Acta Botanica Gallica*, **148**, 177–181.

Chipuck, J. E. and Green, D. R. (2005) Do inducers of apoptosis trigger caspase-independent cell death? *Nature Molecular Cell Biology Reviews*, **6**, 268–275.

Clark, G. E. and Burge, G. K. (1999) Effects of nitrogen nutrition on *Sandersonia* cut flower and tuber production in a soil-less medium, *New Zealand Journal of Crop Science and Horticulture*, **27**, 145–152.

Clark, D. G., Dervinis, C., Barret, J. E., Klee, H. and Jones, M. (2004) Drought-induced leaf senescence and horticultural performance of transgenic P-SAG12-IPT petunias, *Journal of the American Society for Horticultural Science*, **129**, 93–99.

Comba, L., Corbet, S. A., Hunt, H., Outram, S., Parker, J. S. and Glover, B. J. (2000) The role of genes influencing the corolla in pollination of *Antirrhinum majus*, *Plant Cell and Environment*, **23**, 639–647.

Coupe, S. A., Watson, L. M., Ryan, D. J., Pinkney, T. T. and Eason, J. R. (2004) Molecular analysis of programmed cell death in *Arabidopsis thaliana* and *Brassica oleracea*: cloning LSD1, Bax inhibitor and serine palmitoyltransferase homolgues, *Journal of Experimental Botany*, **55**, 59–68.

Cui, M. L., Takada, K., Ma, B. and Ezura, H. (2004) Overexpression of a mutated melon ethylene receptor gene Cm-ETR1/H69A confers reduced ethylene sensitivity in a heterologous plant, *Nemesia strumosa*, *Plant Science*, **167**, 253–258.

Dat, J., Vandenabeele, E., Vranova, M., Van Montagu, M., Inze, D. and Van Breusegem, F. (2000) Dual action of the active oxygen species during plant stress responses, *Cellular and Molecular Life Sciences*, **57**, 779–795.

Doelling, J. H., Walker, J. M., Friedman, E. M., Thompson, A. R. and Vierstra, R. D. (2002) The APG8/12-activating enzyme APG7 is required for proper nutrient recycling and senescence in *Arabidopsis thaliana*, *Journal of Biological Chemistry*, **277**, 33105–33114.

Doi, M., Miyagawa-Namao, M., Inamoto, K. and Imanishi, H. (1999) Rhythmic changes in water uptake, transpiration and water potential of cut roses as affected by photoperiods, *Journal of the Japanese Society for Horticultural Science*, **68**, 861–867.

Dole, J. A. (1990) Role of corolla abscission in delayed self-pollination of *Mimulas guttatus* (Scrophulariaceae), *American Journal of Botany*, **77**, 1505–1507.

Eason, J. R. (2002) *Sandersonia aurantiaca*: an evaluation of postharvest pulsing solutions to maximise cut flower quality, *New Zealand Journal of Crop and Horticultural Science*, **30**, 273–279.

Eason, J. R. and Bucknell, T. T. (2001) DNA processing during tepal senescence of *Sandersonia aurantica*, *Acta Horticulturae*, **543**, 143–146.

Eason, J. R., Morgan, E. R., Mullan, A. C. and Burge, G. K. (2004) Display life of *Gentiana* flowers is cultivar specific and influenced by sucrose, gibberellin, fluoride, and postharvest storage, *New Zealand Journal of Crop and Horticultural Science*, **32**, 217–226.

Eason, J. R., Ryan, D. J., Pinkney, T. T. and O'Donoghue, E. M. (2002) Programmed cell death during flower senescence: isolation and characterization of cysteine proteinases from *Sandersonia aurantiaca*, *Functional Plant Biology*, **29**, 1055–1064.

Ebersold, H. R., Cordier, J. L. and Luthy, P. (1981) Bacterial mesosomes – method dependent artifacts, *Archives of Microbiology*, **130**, 19–22.

Ecker, J. R. (1995) The ethylene signal-transduction pathway in plants, *Science*, **268**, 667–675.

Finger, F. L., Carneiro, T. F. and Barbosa, J. G. (2004) Postharvest senescence of *Consolida ajacis* inflorescences, *Pesquisa Agropecuaria Brasileira*, **39**, 533–537.

Florack, D. E. A., Stiekema, W. J. and Bosch, D. (1996) Toxicity of peptides to bacteria present in the vase water of cut roses, *Postharvest Biology and Technology*, **8**, 285–291.

Gan, S. and Amasino, R. M. (1997) Making sense of senescence, *Plant Physiology*, **113**, 313–319.

Garratt, L. C., Linforth, R., Taylor, A. J., Lowe, K. C., Power, J. B. and Davey, M. R. (2005) Metabolite fingerprinting in transgenic lettuce, *Plant Biotechnology Journal*, **3**, 165–174.

Gast, K. (1999) Methyl jasmonate and long term storage of fresh cut peony flowers, *Acta Horticulturae*, **543**, 327–330.

Gay, A. P., Donnison, I., Taylor, J., Thomas, H. and Ougham, H. (2002) Hyperspectral imaging as a non-invasive tool to determine crop maturity, nutritional status and stress, Abstracts Society of Chemical Industry, *Conference on Applications of Remote Sensing in Agriculture*, p. 22.

Gepstein, S. (2004) Leaf senescence – not just a 'wear and tear' phenomenon, *Genome Biology*, **5**, art. no. 212.

Gepstein, S., Sabehi, G., Carp, M. J. *et al.* (2003) Large-scale identification of leaf senescence-associated genes, *The Plant Journal*, **36**, 629–642.

Gilissen, L. J. W. (1976) The role of the style as a sense organ in relation to the wilting of the flower, *Planta*, **131**, 201–202.

Gilissen, L. J. W. (1977) Style-controlled wilting of the flower. *Planta*, **133**, 275–280.

Glover, B. J. and Martin, C. (1998) The role of petal cell shape and pigmentation in pollination success in *Antirrhinum majus*, *Heredity*, **80**, 778–784.

Gubrium, E. K., Clevenger, D. J., Clark, D. G., Barrett, J. E. and Nell, T. A. (2000) Reproduction and horticultural performance of transgenic ethylene-insensitive petunias, *Journal of the American Society for Horticultural Science*, **125**, 277–281.

Guerrero, C., de la Calle, M., Reid, M. S. and Valpuesta, V. (1998) Analysis of the expression of two thiolprotease genes from daylily (*Hemerocallis* spp.) during flower senescence, *Plant Molecular Biology*, **36**, 565–571.

Gunawardena, A. H. L. A. N., Pearce, D. M., Jackson, M. B., Hawes, C. R. and Evans, D. E. (2001a) Characterisation of programmed cell death during aerenchyma formation induced by ethylene or hypoxia in roots of maize (*Zea mays* L.), *Planta*, **212**, 205–214.

Gunawardena, A. H. L. A. N., Pearce, D. M. E., Jackson, M. B., Hawes, C. R. and Evans, D. E. (2001b) Rapid changes in cell wall pectic polysaccharides are closely associated with early stages of aerenchyma formation, a spatially localized form of programmed cell death in roots of maize (*Zea mays* L.) promoted by ethylene, *Plant, Cell and Environment*, **24**, 1369–1375.

Guo, Y., Cai, Z. and Gan, S. (2004) Transcriptome of *Arabidopsis* leaf senescence, *Plant, Cell and Environment*, **27**, 521–549.

Halaba, J. and Rudnicki, R. M. (1988) Invertase inhibitor-control of sucrose transport from carnation petals to other flower parts, *Plant Growth Regulation*, **7**, 193–199.

Halaba, J. and Rudnicki, R. M. (1989) Invertase inhibitor in wilting flower petals, *Scientia Horticulturae*, **40**, 83–90.

Halgerson, J. L., Sheaffer, C. C., Martin, N. P., Peterson, P. R. and Weston, S. J. (2004) Near-infrared reflectance spectroscopy prediction of leaf and mineral concentrations in alfalfa, *Agronomy Journal*, **96**, 344–351.

Han, S. S. (1992) Role of sucrose in bud development and vase life of cut *Liatris spicata* (L.) Wild, *HortScience*, **27**, 1198–1200.

Han, S. S. (2001) Benzyladenine and gibberellins improve postharvest quality of cut Asiatic and Oriental lilies. *HortScience*, **36**, 741–745.

Han, S. S. (2003) Role of sugar in the vase solution on postharvest flower and leaf quality of oriental lily 'Stargazer', *HortScience*, **38**, 412–416.

Hanaoka, H., Noda, T., Shirano, Y. *et al.* (2002) Leaf senescence and starvation-induced chlorosis are accelerated by the disruption of an *Arabidopsis* autophagy gene, *Plant Physiology*, **129**, 1181–1193.

Harada, K. and Yasui, K. (2003) Decomposition of ethylene, a flower-senescence hormone, with electrolyzed anode water, *Bioscience Biotechnology and Biochemistry*, **67**, 790–796.

Hara-Nishimura I., Hastsugai N., Nakaune S., Kuroyanagi M. and Nishimura M. (2005) Vacuolar processing enzyme: an executor of plant cell death. *Current Opinion in Plant Biology*, **8**, 404–408.

He, Y. H. and Gan, S. S. (2002) A gene encoding an acyl hydrolase is involved in leaf senescence in Arabidopsis, *The Plant Cell*, **14**, 805–815.

He, Y. H., Fukushige, H., Hildebrand, D. F. and Gan, S. S. (2002) Evidence supporting a role for jasmonic acid in *Arabidopsis* leaf senescence, *Plant Physiology*, **128**, 876–884.

Hicklenton, P. R. (1991) GA3 and benzylaminopurine delay leaf yellowing in cut *Alstroemeria* stems. *HortScience*, **26**, 1198–1199.

Hilioti, Z., Richards, C. and Brown, K. M. (2000) Regulation of pollination-induced ethylene and its role in petal abscission of *Pelargonium x hortorum*, *Physiologia Plantarum*, **109**, 322–332.

Hill, S. E., Stead, A. D. and Nichols, R. (1987) The production of 1-aminocyclopropane-1-carboxylic acid (ACC) by pollen of *Nicotiana tabacum* cv White Burley and the role of pollen-held ACC in pollination-induced ethylene production, *Journal of Plant Growth Regulation*, **6**, 1–13.

Hoeberichts F. A., de Jong A. J. Woltering E. J. (2005) Apoptotic-like cell death marks the early stages of gypsophila (Gypsophila paniculata) petal senescence. *Postharvest Biology and Technology* **35**, 229–236.

Hoekstra, F. A. and Weges, R. (1986) Lack of control by early pistillate ethylene of the accelerated wilting of *Petunia hybrida* flowers, *Plant Physiology*, **80**, 403–408.

Hoogerwerf, A. and van Doorn, W. G. (1992) Numbers of bacteria in aqueous solutions used for postharvest handling of cut flowers, *Postharvest Biology and Technology*, **1**, 295–304.

Hudson, R. G. (2003) Mesosomes and scientific methodology, *History and Philosophy of the Life Sciences*, **25**, 167–191.

Huett, D. O. (1994) Production and quality of sim carnations grown hydroponically in rockwool substrate with nutrient solutions containing different levels of calcium, potassium and ammonium-nitrogen, *Australian Journal of Experimental Agriculture*, **34**, 691–697.

Hung, K. T. and Kao, C. H. (2003) Nitric oxide counteracts the senescence of rice leaves induced by abscisic acid, *Journal of Plant Physiology*, **160**, 871–879.

Hung, K. T. and Kao, C. H. (2004) Nitric oxide acts as an antioxidant and delays methyl jasmonate-induced senescence of rice leaves, *Journal of Plant Physiology*, **161**, 43–52.

Hunter, D. A., Ferrante, A., Vernieri, P. and Reid, M. S. (2004) Role of abscisic acid in perianth senescence of daffodil (*Narcissus pseudonarcissus* 'Dutch Master'), *Physiologia Plantarum*, **121**, 313–321.

Ichimura, K. (1998) Improvement of past-harvest life in several cut flowers by the addition of sucrose, *Japan Agricultural Research Quarterly*, **32**, 275–280.

Ichimura, K. and Goto, R. (2000) Effect of gibberellin A(3) on leaf yellowing and vase life of cut *Narcissus tazetta* var. *chinensis* flowers, *Journal of the Japanese Society for Horticultural Science*, **69**, 423–427.

Ichimura, K., Kawabata, Y., Kishimoto, M., Goto, R. and Yamada, K. (2003) Shortage of soluble carbohydrates is largely responsible for short vase life of cut 'Sonia' rose flowers, *Journal of the Japanese Society for Horticultural Science*, **72**, 292–298.

Islam, N., Patil, G. G. and Gislerod, H. R. (2003) Effects of pre- and postharvest conditions on vase life of *Eustoma grandiflorum* (Raf.) Shinn, *European Journal of Horticultural Science*, **68**, 272–278.

Iwaya-Inoue, M. and Takata, M. (2001) Trehalose plus chloramphenicol prolong the vase life of tulip flowers, *HortScience*, **36**, 946–950.

Izumi, H. (2001) Effects of electrolyzed neutral water on the bacterial populations in a flower vase and in stems of cut roses, *Journal of the Japanese Society for Horticultural Science*, **70**, 599–601.

Johnson, P. R. and Ecker, J. R. (1998) The ethylene gas signal transduction pathway: a molecular perspective, *Annual Review of Genetics*, **32**, 227–254.

Jones, M. L. (2004) Changes in gene expression during senescence, in *Plant Cell Death Processes* (ed. L. D. Nooden), Elsevier, Amsterdam, pp. 51–71.

Jones, M. L., Chaffin G. S., Eason J. R. and Clark D. G. (2005) Ethylene-sensitivity regulates proteolytic activity and cysteine protease gene expression in petunia corollas. *Journal of Experimental Botany*, **56**, 2733–2744.

Jones, M. L., Kim, E. S. and Newman, S. E. (2001) Role of ethylene and 1-MCP in flower development and petal abscission in zonal geraniums, *HortScience*, **36**, 1305–1309.

Jones, M. L., Larsen, P. B. and Woodson, W. R. (1995) Ethylene-regulated expression of a carnation cysteine proteinase during flower petal senescence, *Plant Molecular Biology*, **28**, 505–512.

Jones, R. B., Serek, M., Kuo, C. L. and Reid, M. S. (1994) The effect of protein-synthesis inhibition on petal senescence in cut bulb flowers. *Journal of the American Society for Horticultural Science*, **119**, 1243–1247.

Kebenei, Z., Sisler, E. C., Winkelmann, T. and Serek, M. (2003) Effect of 1-octylcyclopropene and 1-methylcyclopropene on vase life of sweet pea (*Lathyrus odoratus* L.) flowers, *Journal of Horticultural Science and Biotechnology*, **78**, 433–436.

Kebenei, Z., Sisler, E. C., Winkelmann, T. and Serek, M. (2003) Efficacy of new inhibitors of ethylene perception in improvement of display life of kalanchoe (*Kalanchoe blossfeldiana* Poelln.) flowers, *Postharvest Biology and Technology*, **30**, 169–176.

Kerner van Marilaun, A. (1891) Pflanzenleben. Band 2. Verlag des Bibliographisches Instituts, Leipzig.

Kieber, J. J. (1997) The ethylene response pathway in *Arabidopsis*, *Annual Review of Plant Physiology and Plant Molecular Biology*, **48**, 277–296.

Kikuchi, K., Kanahama, K. and Kanayama, Y. (2003) Changes in sugar-related enzymes during wilting of cut delphinium flowers, *Journal of the Japanese Society for Horticultural Science*, **72**, 37–42.

Kim, J. Y., Chung, Y. S., Paek, K. H. *et al.* (1999) Isolation and characterization of a cDNA encoding the cysteine proteinase inhibitor, induced upon flower maturation in carnation using suppression subtractive hybridization, *Molecules and Cells*, **9**, 392–397.

Klionsky, D. J. (2005) The molecular machinery of autophagy: unanswered questions, *Journal of Cell Science*, **118**, 7–18.

Klionsky, D. J. and Emr, S. D. (2000) Cell biology – autophagy as a regulated pathway of cellular degradation, *Science*, **290**, 1717–1721.

Klionsky, D. J., Cregg, J. M., Dunn, W. A. *et al.* (2003) A unified nomenclature for yeast autophagy-related genes, *Developmental Cell*, **5**, 539–545.

Knee, M. (2000) Selection of biocides for use in floral preservatives, *Postharvest Biology and Technology*, **18**, 227–234.

Kosugi, Y., Waki, K., Iwazaki, Y. *et al.* (2002) Senescence and gene expression of transgenic non-ethylene-producing carnation flowers, *Journal of the Japanese Society for Horticultural Science*, **71**, 638–642.

Krupinska, K., Haussuhl, K., Schafer, A. *et al.* (2002) A novel nucleus-targeted protein is expressed in barley leaves during senescence and pathogen infection, *Plant Physiology*, **130**, 1172–1180.

Langston, B. J., Bai, S. and Jones, M. L. (2005) Increases in DNA fragmentation and induction of a senescence-specific nuclease are delayed during corolla senescence in ethylene-insensitive (etr1-1) transgenic petunias, *Journal of Experimental Botany*, **56**, 15–23.

Lara, M. E. B., Garcia, M. C. G., Fatima, T. *et al.* (2004) Extracellular invertase is an essential component of cytokinin-mediated delay of senescence, *The Plant Cell*, **16**, 1276–1287.

LayYee, M., Stead, A. D. and Reid, M. S. (1992) Flower senescence in daylily (*Hemerocallis*), *Physiologia Plantarum*, **86**, 308–314.

Lee, M. M., Lee, S. H. and Park, K. Y. (1997) Effects of spermine on ethylene biosynthesis in cut carnation (*Dianthus caryophyllus* L) flowers during senescence, *Journal of Plant Physiology*, **151**, 68–73.

Lee, R. H., Wang, C. H., Huang, L. T. and Chen S. C. G. (2001) Leaf senescence in rice plants: cloning and characterization of senescence up-regulated genes, *Journal of Experimental Botany*, **52**, 1117–1121.

Leonard, R. T. and Nell, T. A. (2004) Short-term pulsing improves postharvest leaf quality of cut oriental lilies, *HortTechnology*, **14**, 405–411.

Leverentz, M., Wagstaff, C., Rogers, H. *et al.* (2002) Characterisation of a novel lipoxygenase-independent senescence mechanism in *Alstroemeria peruviana* floral tissue, *Plant Physiology*, **130**, 273–283.

Levey, S. and Wingler, A. (2005) Natural variation in the regulation of leaf senescence and relation to other traits in *Arabidopsis*, *Plant, Cell and Environment*, **28**, 223–231.

Lim, P. O., Woo, H. R. and Nam, H. G. (2003) Molecular genetics of leaf senescence in *Arabidopsis*. *Trends in Plant Science*, **8**, 272–278.

Lucas, B. M. (1989) Ethylene and pollination-induced senescence in flowers, Ph.D. thesis, University of London.

Lukaszewska, A. J., Bianco, J., Barthe, P. and Le Page-Degivry, M. T. (1994) Endogenous cytokinins in rose petals and the effect of exogenously applied cytokinins on flower senescence, *Plant Growth Regulation*, **14**, 119–126.

Lukaszewski, T. A. and Reid, M. S. (1989) Bulb-type flower senescence, *Acta Horticulturae*, **261**, 59–62.

Mahagamasekera, M. G. P. and Leung, D. W. M. (2001) Development of leucine aminopeptidase activity during daylily flower growth and senescence, *Acta Physiologiae Plantarum*, **23**, 181–186.

Maier, N. A., Barth, G. E., Bartetzko, M. N., Cecil, J. S. and Chvyl, W. L. (1996) Nitrogen and potassium nutrition of Australian waxflowers grown in siliceous sands. 2. Effect on leaf colour, vase life, and soil pH and conductance, *Australian Journal of Experimental Agriculture*, **36**, 367–371.

Matile, P. (1997) The vacuole and cell senescence, *Advances in Botanical Research Incorporating Advances in Plant Pathology*, **25**, 87–112.

Matile, P. and Winkenbach, F. (1971) Function of lysosomes and lysosomal enzymes in the senescing corolla of the Morning Glory (*Ipomoea purpurea*), *Journal of Experimental Botany*, **22**, 759–771.

McCabe, M. S., Garratt, L. C., Schepers, F. *et al.* (2001) Effects of P-SAG12-IPT gene expression on development and senescence in transgenic lettuce, *Plant Physiology*, **127**, 505–516.

Mckenzie, R. J. and Lovell, P. H. (1992) Flower senescence in monocotyledons – a taxonomic survey, *New Zealand Journal of Crop and Horticultural Science*, **20**, 67–71.

Melendez, A., Talloczy, Z., Seaman, M., Eskelinen, E. L., Hall, D. H. and Levine, B. (2003) Autophagy genes are essential for dauer development and life-span extension in *C-elegans*, *Science*, **301**, 1387–1391.

Mensink, M. G. J. and van Doorn, W. G. (2001) Small hydrostatic pressures overcome the occlusion by air emboli in cut rose stems, *Journal of Plant Physiology*, **158**, 1495–1498.

Mol, R., Filek, M., Machackova, L. and Matthys-Rochon, E. (2004) Ethylene synthesis and auxin augmentation in pistil tissues are important for egg cell differentiation after pollination in maize, *Plant and Cell Physiology*, **45**, 1396–1405.

Molisch, H. (1928) Die Lebensdauer der Pflanzen. Translated by E. H. Fulling (1939) as *The Longevity of Plants*, Science Press, Lancaster, Pennsylvania.

Moriyasu, Y. and Ohsumi, Y. (1996) Autophagy in tobacco suspension-cultured cells in response to sucrose starvation, *Plant Physiology*, **111**, 1233–1241.

Morris, K., A-H-Mackerness, S., Page, T. *et al.* (2000) Salicylic acid has a role in regulating gene expression during leaf senescence, *Plant Journal*, **23**, 677–685.

Mortensen, L. M., Ottosen, C. O. and Gislerod, H. R. (2001) Effects of air humidity and K : Ca ratio on growth, morphology, flowering and keeping quality of pot roses, *Scientia Horticulturae*, **90**, 131–141.

Moussatche, P. and Klee, H. J. (2004) Autophosphorylation activity of the Arabidopsis ethylene receptor multigene family, *The Journal of Biological Chemistry*, **279**, 48734–48741.

Nagata, M., Tanikawa, N, Onazaki, T. and Mori, H. (2000) Ethylene receptor gene (ETR) homolog from carnation, *Journal of the Japanese Society for Horticultural Science*, **69**(Suppl. 1), 407.

Nam, H. G. (1997) The molecular genetic analysis of leaf senescence, *Current Opinions in Biotechnology*, **8**, 200–207.

Navabpour, S., Morris, K., Allen, R., Harrison, E., A-H-Mackerness, S. and Buchanan-Wollaston, V. (2003) Expression of senescence-enhanced genes in response to oxidative stress, *Journal of Experimental Botany*, **54** 2285–2292.

Negre, F., Kish, C. M., Boatright, J. *et al.* (2003) Regulation of methylbenzoate emission after pollination in snapdragon and petunia flowers, *The Plant Cell*, **15**, 2992–3006.

Nuttman, C. and Willmer, P. (2003) How does insect visitation trigger floral colour change? *Ecological Entomology*, **28**, 467–474.

Ohta, K. and Harada, K. (2000) Effect of electrolyzed anode water on the vase life of cut rose flowers, *Journal of the Japanese Society for Horticultural Science*, **69**, 520–522.

Orzaez, D. and Granell, A. (1997) The plant homologue of the defender against apoptotic death gene is down-regulated during senescence of flower petals, *FEBS Letters*, **404**, 275–278.

Otto, G. P., Wu, M. Y., Kazgan, N., Anderson, O. R. and Kessin, R. H. (2003) Macroautophagy is required for multicellular development of the social amoeba *Dictyostelium discoideum*, *Journal of Biological Chemistry*, **278**, 17636–17645.

Otto, G. P., Wu, M. Y., Kazgan, N., Anderson, O. R. and Kessin, R. H. (2004) *Dictyostelium* macroautophagy mutants vary in the severity of their developmental defects, *Journal of Biological Chemistry*, **279**, 15621–15629.

Page, T., Griffiths, G. and Buchanan-Wollaston, V. (2001) Molecular and biochemical characterization of postharvest senescence in broccoli, *Plant Physiology*, **125**, 718–727.

Pak, C. and van Doorn, W. G. (2005) Delay of Iris flower senescence by protease inhibitors, *New Phytologist*, **165**, 473–480.

Pandey, S., Ranade, S. A., Nagar, P. K. and Kumar, N. (2000) Role of polyamines and ethylene as modulators of plant senescence, *Journal of Biosciences*, **25**, 291–299.

Panavas, T., Walker, E. L. and Rubinstein, B. (1998) Possible involvement of abscisic acid in senescence of daylily petals, *Journal of Experimental Botany*, **49**, 1987–1997.

Paull, R. E. and Chantrachit, T. (2001) Benzyladenine and the vase life of tropical ornamentals, *Postharvest Biology and Technology*, **21**, 303–310.

Pennel, R. I. and Lamb, C. (1997) Programmed cell death in plants, *The Plant Cell*, **9**, 1157–1168.

Perez-Amador, M. A., Abler, M. L., Jay De Rocher, E. *et al.* (2000) Identification of BFN1, a bifunctional nuclease induced during leaf and stem senescence in *Arabidopsis*, *The Plant Physiology*, **122**, 169–179.

Philosoph-Hadas, S., Michaeli, R., Reuveni, Y. and Meir, S. (1996) Benzyladenine pulsing retards leaf yellowing and improves quality of goldenrod (*Solidago canadensis*) cut flowers, *Postharvest Biology and Technology*, **9**, 65–73.

Picchioni, G. A., Valenzuela-Vazquez, M. and Murray, L. W. (2002) Calcium and 1-methylcyclopropene delay desiccation of *Lupinus havardii* cut racemes, *HortScience*, **37**, 122–125.

Porat, R. and Halevy, A. H. (1993) Enhancement of petunia and dendrobium flower senescence by jasmonic acid methyl-ester is via the promotion of ethylene production, *Plant Growth Regulation*, **13**, 297–301.

Porat, R., Reiss, N., Atzorn, R., Halevy, A. H. and Borochov, A. (1995) Examination of the possible involvement of lipoxygenase and jasmonates in pollination-induced senescence of Phalaenopsis and Dendrobium orchid flowers,*Physiologia Plantarum*, **94**, 205–210.

Put, H. M. C. and Jansen, L. (1989) The effects on the vase life of cut Rosa cultivar Sonia of bacteria added to the vase water, *Scientia Horticulturae*, **39**, 167–179.

Quirino, B. F., Noh, Y.-S., Himelblau, E. and Amasino, R. M. (2000) Molecular aspects of leaf senescence, *Trends in Plant Science*, **5**, 278–282.

Rattanawisalanon, C., Ketsa, S. and van Doorn, W. G. (2003) Effect of aminooxyacetic acid and sugars on the vase life of *Dendrobium* flowers, *Postharvest Biology and Technology*, **29**, 93–100.

Reid, M., Dodge, L., Celikel, F. and Valle, R. (1999) 1-MCP, a breakthrough in ethylene protection, *Floraculture International*, **9**, 36–40.

Rojo, E., Zouhar, J., Carter, C., Kovaleva, V. and Raikhel, N. V. (2003) A unique mechanism for protein processing and degradation in *Arabidopsis thaliana*, *Proceedings of the National Academy of Sciences of the United States of America*, **100**, 7389–7394.

Rossato, L., MacDuff, J. H., Laine, P., Le Deunff, E. and Ourry, A. (2002) Nitrogen storage and remobilization in *Brassica napus* L. during the growth cycle: effects of methyl jasmonate on nitrate uptake, senescence, growth, and VSP accumulation, *Journal of Experimental Botany*, **53**, 1131–1141.

Rubenstein, B. (2000) Regulation of cell death in flower petals, *Plant Molecular Biology*, **44**, 303–318.

Sanmartín M, Jaroszewski, L., Raikhel, N. V. and Rojo, E. (2005) Caspases, regulating death since the origin of life? Plant Physiology, **137**, 841–847.

Savin, K. W., Baudinette, S. C., Graham, M. W. *et al.* (1995) Antisense ACC oxidase RNA delays carnation petal senescence, *HortScience*, **30**, 970–972.

Schroeder, K. R., Stimart, D. P. and Nordheim, E. V. (2001) Response of *Nicotiana alata* to insertion of an autoregulated senescence-inhibition gene, *Journal of the American Society for Horticultural Science*, **126**, 523–530.

Schut, A. G. T., Lokhorst, C., Hendriks, M. M. W. B., Kornet, J. G. and Kasper, G. (2005) Potential of imaging spectroscopy as tool for pasture management, *Grass and Forage Science*, **60**, 34–45.

Serek, M. and Andersen, A. S. (1993) AOA and BA influence on floral development and longevity of potted victory parade miniature rose, *HortScience*, **28**, 1039–1040.

Serek, M. and Sisler, E. C. (2001) Efficacy of inhibitors of ethylene binding in improvement of the postharvest characteristics of potted flowering plants, *Postharvest Biology and Technology*, **23**, 161–166.

Serrano, M., Amoros, A., Pretel, M. T., Martinez-Madrid, M. C. and Romojaro, F. (2001) Preservative solutions containing boric acid delay senescence of carnation flowers, *Postharvest Biology and Technology*, **23**, 133–142.

Serrano, M., Martinez-Madrid, M. C. and Romojaro, F. (1999) Ethylene biosynthesis and polyamine and ABA levels in cut carnations treated with aminotriazole, *Journal of the American Society for Horticultural Science*, **124**, 81–85.

Setyadit, Joyce, D. C., Irving, D. E. and Simons, D.H. (2004) Effects of 6-benzylaminopurine treatments on the longevity of harvested *Grevillea* 'Sylvia' influorescences, *Plant Growth Regulation*, **43**, 9–14.

Sexton, R., Laird, G. and van Doorn, W. G. (2000) Lack of ethylene involvement in tulip tepal abscission, *Physiologia Plantarum*, **108**, 321–329.

Shaw, J. F., Chen, H. H., Tsai, M. F., Kuo, C. I. and Huang, L. C. (2002) Extended flower longevity of *Petunia hybrida* plants transformed with boers, a mutated ERS gene of *Brassica oleracea*, *Molecular Breeding*, **9**, 211–216.

Shibuya, K., Barry, K. G., Ciardi, J. A. *et al.* (2004) The central role of PhEIN2 in ethylene responses throughout plant development in petunia, *Plant Physiology*, **136**, 2900–2912.

Shibuya, K., Yoshioka, T., Hashiba, T. and Satoh, H. (2000) Role of the gynoecium in natural senescence of carnation (*Dianthus caryophyllus* L.) flowers, *Journal of Experimental Botany*, **51**, 2067–2073.

Shimamura, M., Ito, A., Suto, K., Okabayashi, H. and Ichimura, K. (1997) Effects of alpha-aminoisobutyric acid and sucrose on the vase life of hybrid Limonium, *Postharvest Biology and Technology*, **12**, 247–253.

Smart, C. M. (1994) Gene expression during leaf senescence, *New Phytologist*, **126**, 419–448.

Smith, M. T., Saks, Y. and van Staden, J. (1992) Ultrastructural-changes in the petals of senescing flowers of *Dianthus caryophyllus* L, *Annals of Botany*, **69**, 277–285.

Spanjers, A. W. (1978) Voltage variation in *Lilium longiflorum* pistils induced by pollination, *Experientia*, **34**, 36–37.

Stead, A. D. (1985) The relationship between pollination, ethylene production and flower senescence, in *Ethylene and Plant Development* (eds J. Tucker and J. Roberts), Butterworths, pp. 71–82.

Stead, A. D. (2004) Cut flower sales in the US – can Europe teach us anything? *OFA Bulletin*, July/August 2004, No. 885.

Stead, A. D. and Duckett J. G. (2004) Plastid ontogeny in the corolla cells of *Digitalis purpurea* L. Foxy. *Annals of Botany*, **46**, 549–555.

Stead, A. D. and van Doorn, W. (1994) Strategies of flower senescence, in *Molecular and Cellular Aspects of Plant Reproduction* (eds R. Scott and A. D. Stead), SEB Symposia 55, Cambridge University Press, Cambridge, pp. 215–237.

Stead, A. D. and Moore, K. G. (1977) Flower development and senescence in *Digitalis purpurea* cv Foxy, *Annals of Botany*, **41**, 283–292.

Stead, A. D. and Moore, K. G. (1979) Studies on flower longevity in *Digitalis*: I. Pollination induced corolla abscission in Digitalis flowers, *Planta*, **146**, 409–414.

Stead, A. D. and Moore, K. G. (1983) Studies on flower longevity in *Digitalis*: II. The role of ethylene in corolla abscission. *Planta*, **157**, 15–21.

Stead, A. D. and Reid, M.S. (1990) The effect of pollination and ethylene on the color-change of the banner spot of *Lupinus albifrons* (Bentham) flowers, *Annals of Botany*, **66**, 655–663.

Stead, A. D., Gritsch, C. S. and Wagstaff, C. Programmed cell death in tapetal cells – unique ultrastructural changes associated with nuclear breakdown, submitted to Protoplasma.

Stephenson, P. and Rubinstein, B. (1998) Characterization of proteolytic activity during senescence in daylilies, *Physiologia Plantarum*, **104**, 463–473.

Stuth, J., Jama, A. and Tolleson, D. (2003) Direct and indirect means of predicting forage quality through near infrared reflectance spectroscopy, *Field Crops Research*, **84**, 45–56.

Sultan, S. M. and Farooq, S. (1997) Effect of cycloheximide on some physiological changes associated with senescence of detached flowers of *Iris germanica* L, *Acta Physiologiae Plantarum*, **19**, 41–45.

Surpin, M., Zheng, H. J., Morita, M. T. *et al.* (2003) The VTI family of SNARE proteins is necessary for plant viabilityand mediates different protein transport pathways, *The Plant Cell*, **15**, 2885–2899.

Takemura, G., Kato, S., Aoyama, T. *et al.* (2001) Characterization of ultrastructure and its relation with DNA fragmentation in Fas-induced apoptosis of cultured cardiac myocytes, *Journal of Pathology*, **193**, 546–556.

Taverner, E. A., Letham, D. S., Wang, J. and Cornish, E. (2000) Inhibition of carnation petal inrolling by growth retardants and cytokinins, *Australian Journal of Plant Physiology*, **27**, 357–362.

Thomas, H., Ougham, H. J., Wagstaff, C. and Stead, A. D. (2003) Defining senescence and death, *Journal of Experimental Botany*, **54**, 1127–1132.

Valpuesta, V., Lange, N. E., Guerrero, C. and Reid, M. S. (1995) Up-regulation of a cysteine protease accompanies the ethylene-insensitive senescence of daylilies (*Hemerocallis*) flowers, *Plant Molecular Biology*, **28**, 575–582.

van der Kop DAM, Ruys, G., Dees, D., van der Schoot, C., de Boer, A. D. and van Doorn, W. G. (2003) Expression of defender against apoptotic death (DAD-1) in Iris and Dianthus petals, *Physiologia Plantarum*, **117**, 256–263.

van der Meulen-Muisers, J. J. M., van Oeveren, J. C., van der Plas, L. H. W. and van Tuyl, J. M. (2001) Postharvest flower development in Asiatic hybrid lilies as related to tepal carbohydrate status, *Postharvest Biology and Technology*, **21**, 201–211.

van Doorn, W. G. (1997) Effects of pollination on floral attraction and longevity, *Journal of Experimental Botany*, **48**, 1615–1622.

van Doorn, W. G. (2001) Categories of petal senescence and abscission: a re-evaluation, *Annals of Botany*, **87**, 447–456.

van Doorn, W. (2002a) Does ethylene treatment mimic the effects of pollination on floral lifespan and attractiveness? *Annals of Botany*, **89**, 375–383.

van Doorn, W. G. (2002b) Effect of ethylene on flower abscission: a survey, *Annals of Botany*, **89**, 689–693.

van Doorn, W. G. (2004) Is petal senescence due to sugar starvation? *Plant Physiology*, **134**, 35–42.

van Doorn, W. G. and Cruz, P. (2000) Evidence for a wounding-induced xylem occlusion in stems of cut chrysanthemum flowers, *Postharvest Biology and Technology*, **19**, 73–83.

van Doorn, W. G. and Vaslier N. (2002) Wound-induced xylem occlusion in stems of cut chrysanthemum flowers: roles of peroxidase and catechol oxidase. *Postharvest Biology and Technology*, **26**, 275–284.

van Doorn, W. G. and Woltering, E. J. (2004) Senescence and programmed cell death: substance or semantics? *Journal of Experimental Botany*, **55**, 2147–2153.

van Doorn, W. G. and Woltering, E. J. (2005) Many ways to exit? Cell death categories in plants. *Trends in Plant Science*, **10**, 117–122.

van Doorn, W. G., Balk, P. A., van Houwelingen, A. M. *et al.* (2003) Gene expression during anthesis and senescence in *Iris* flowers, *Plant Molecular Biology*, **53**, 845–863.

van Doorn, W. G., de Witte, Y. and Harkema, H. (1995) Effect of high numbers of exogenous bacteria on the water relations and longevity of cut carnation flowers, *Postharvest Biology and Technology*, **6**, 111–119.

van Doorn, W. G., de Witte, Y. and Perik, R. J. J. (1990) Effect of anti-microbial compounds on the number of bacteria in stems of cut rose flowers, *Journal of Applied Bacteriology*, **68**, 117–122.

van Ieperen, W., van Meeteren, U. and Nijsse, J. (2002) Embolism repair in cut flower stems: a physical approach, *Postharvest Biology and Technology*, **25**, 1–14.

Van Staden, J., Cook, E. L. and Nooden, L. D. (1988) Cytokinins and senescence, in *Senescence and Aging in Plants* (ed. L. D. Nooden), Academic Press, London, pp. 281–328.

Veen, H. and Van der Geijn, S. C. (1978) Mobility and ionic form of silver as related to longevity of cut carnations, *Planta*, **140**, 93–96.

Verlinden, S. and Garcia, J. J. V. (2004) Sucrose loading decreases ethylene responsiveness in carnation (*Dianthus caryophyllus* cv. White Sim) petals, *Postharvest Biology and Technology*, **31**, 305–312.

Wada, H., Iwaya-Inoue, M., Akita M. and Nonami H. (2005) Hydraulic conductance in tepal growth and extension of vase life with trehalose in cut tulip flowers. *Journal of the American Society for Horticultural Science*, **130**, 275–286.

Wagstaff, C., Chanasut, U., Harren, F. J. M. *et al.* (2005) Ethylene and flower longevity in *Alstroemeria*: relationship between tepal senescence, abscission and ethylene biosynthesis, *Journal of Experimental Botany*, **56**, 1007–1016.

Wagstaff, C., Leverentz, M., Griffiths, G. *et al.* (2002) Cysteine protease gene expression and proteolytic activity during senescence of *Alstroemeria* petals, *Journal of Experimental Botany*, **53**, 233–240.

Wagstaff, C., Malcolm, P., Rafiq, A. *et al.* (2003) Structural and molecular evidence for programmed cell death (PCD) in *Alstroemeria* petals, *New Phytolologist*, **160**, 49–59.

Wainwright, C. M. (1978) The floral biology and pollination ecology of two desert lupines, *Bulletin of the Torrey Botanical Society*, **105**, 24–38.

Waithaka, K., Dodge, L. L. and Reid, M. S. (2001) Carbohydrate traffic during the opening of *Gladiolus* flowers, *Journal of Horticultural Science and Biotechnology*, **76**, 120–124.

Waki, K., Shibuya, K., Yoshioka, T., Hashiba, T. and Satoh, S. (2001) Cloning of a cDNA encoding EIN3-like protein (DC-EIL1) and decrease in its mRNA level during senescence in carnation flower tissue, *Journal of Experimental Botany*, **52**, 377–379.

Weaver, L. M., Himelblau, E. and Amasino, R. M. (1997) Leaf senescence: gene expression and regulation, in *Genetic Engineering: Principles and Methods*, Vol. 19 (ed. J. K. Setlow), Plenum Press, New York, pp. 215–234.

Weiss, M. R. (1991) Floral color changes as cues for pollinators, *Nature*, **354**, 227–229.

Whitehead, C. S., Halevy, A. H. and Reid, M. S. (1984) Roles of ethylene and 1-aminocyclopropane-1-carboxylic acid in pollination and wound induced senescence of *Petunia hybrida*, *Physiologia Plantarum*, **61**, 643–648.

Wi, S. J. and Park, K. Y. (2002) Antisense expression of carnation cDNA encoding ACC synthase or ACC oxidase enhances polyamine content and abiotic stress tolerance in transgenic tobacco plants, *Molecules and Cells*, **13**, 209–220.

Wilkinson, J. Q., Lanahan, M. B., Clark, D. G. *et al.* (1997) A dominant mutant receptor from Arabidopsis confers ethylene insensitivity in heterologous plants, *Nature Biotechnology*, **15**, 444–447.

Williamson, V. G., Faragher, J. D., Parsons, S. and Franz P. (2002) Inhibiting the postharvest wound response in wildflowers. Rural Industries Research and Development Corporation (RIRDC) Publication No. 02/114.

Woltering, E. J. (2004) Death proteases come alive, *Trends in Plant Science*, **9**, 469–472.

Woltering, E. J. and van Doorn, W. G. (1988) Role of ethylene in senescence of petals – morphological and taxonomical relationships, *Journal of Experimental Botany*, **39**, 1605–1616.

Woltering, E. J., deVrije, T., Harren, F. and Hoekstra, F. A. (1997) Pollination and stigma wounding: same response, different signal? *Journal of Experimental Botany*, **48**, 1027–1033.

Woo, H. R., Chung, K. M., Park, J. H. *et al.* (2001) ORE9, an F-box protein that regulates leaf senescence in *Arabidopsis*, *The Plant Cell*, **13**, 1779–1790.

Woo, H. R., Kim, J. H., Nam, H. G. and Lim, P. O. (2004) The delayed leaf senescence mutants of *Arabidopsis*, ore1, ore3, and ore9 are tolerant to oxidative stress, *Plant and Cell Physiology*, **45**, 923–932.

Woodson, W. R. (1987) Changes in protein and mRNA populations during carnation petal senescence, *Physiologia Plantarum*, **71**, 495–502.

Wulster, G., Sacalis, J. and Janes, H. (1982) The effect of inhibitors of protein synthesis on ethylene induced senescence in isolated carnation petals, *Journal of the American Society for Horticultural Science*, **107**, 112–115.

Xu, Y. and Hanson, M.R. (2000) Programmed cell death during pollination-induced petal senescence in petunia, *Plant Physiology*, **122**, 1323–1333.

Yamada, T., Takatsu, Y., Kasumi, M., Marubashi, W. and Ichimura, K. (2004) A homolog of the defender against apoptotic death gene (DAD1) in senescing gladiolus petals is down-regulated prior to the onset of programmed cell death, *Journal of Plant Physiology*, **161**, 1281–1283.

Yang, T. (2005) The role of ethylene in the regulation of senescence in *Arabidopsis* and *Nicotiana*, Ph.D. thesis, University of Nottingham.

Yoshida, S., Ito, M., Nishida, I. and Watanabe, A. (2001) Isolation and RNA gel blot analysis of genes that could serve as potential molecular markers for leaf senescence in *Arabidopsis thaliana*, *Plant and Cell Physiology*, **42**, 170–178.

Yoshimori, T. (2004) Autophagy: a regulated bulk degradation process inside cells, *Biochemical and Biophysical Research Communications*, **313**, 453–458.

Yoshinaga, K., Arimura, S., Hirata, A., Niwa, Y., Yun, D. J., Tsutsumi, N., Uchimiya, H., Kawai-Yamada, M. (2005). Mammalian Bax initiates plant cell death through organelle destruction. *Plant Cell Reports*, **24**, 408–417.

Yu, L., Alva, A., Su, H.*et al.* (2004) Regulation of an ATG7-beclin 1 program of autophagic cell death by caspase-8, *Science*, **304**, 1500–1502.

Zhang, W., Zhang, H., Gu, Z. P. and Zhang, J. J. (1991) Cause of senescence of nine sorts of flowers, *Acta Botanica Sinica*, **33**, 429–436.

Zhang, Z. M. and Leung, D. W. M. (2001) Elevation of soluble sugar levels by silver thiosulfate is associated with vase life improvement of cut gentian flowers, *Journal of Applied Botany-Angewandte Botanik*, **75**, 85–90.

Zubko, E., Adams, C. J., Machaekova, I., Malbeck, J., Scollan, C. and Meyer, P. (2002) Activation tagging identifies a gene from *Petunia hybrida* responsible for the production of active cytokinins in plants, *The Plant Journal*, **29**, 797–808.

Zhou, Y., Wang, C. Y., Ge, H., Hoeberichts, F. A. and Visser, P. B. (2005) Programmed cell death in relation to petal senescence in ornamental plants. *Journal of Integrative Plant Biology*, **47**, 641–650.

Index